Food Safety, Plastics and Sustainability

Scrivener Publishing
100 Cummings Center, Suite 541J
Beverly, MA 01915-6106

Publishers at Scrivener
Martin Scrivener (martin@scrivenerpublishing.com)
Phillip Carmical (pcarmical@scrivenerpublishing.com)

Food Safety, Plastics and Sustainability

Materials, Chemicals, Recycling and the Circular Economy

Johannes Karl Fink
Montanuniversität Leoben, Austria

Scrivener Publishing

WILEY

This edition first published 2023 by John Wiley & Sons, Inc., 111 River Street, Hoboken, NJ 07030, USA
and Scrivener Publishing LLC, 100 Cummings Center, Suite 541J, Beverly, MA 01915, USA

For more information about Scrivener publications please visit www.scrivenerpublishing.com.

Wiley Global Headquarters
111 River Street, Hoboken, NJ 07030, USA

For details of our global editorial offices, customer services, and more information about Wiley products visit us at www.wiley.com.

Library of Congress Cataloging-in-Publication Data

ISBN 978-1-394-17456-0

Cover image: Pixabay.com
Cover design by Russell Richardson

Set in size of 11pt and Minion Pro by Manila Typesetting Company, Makati, Philippines

Printed in the USA

10 9 8 7 6 5 4 3 2 1

Contents

Preface **xi**

1 Food Safety **1**

1.1 Food Security 2
1.2 Migration of Substances from Packaging 3
1.2.1 Modeling of Migration 4
1.2.2 Sample Pretreatment Methods 6
1.2.3 Special Chemicals 7
1.2.4 Safety of Recycled HDPE and PP 11
1.3 Food Safety and Hygiene 12
1.3.1 Sensors for Amine Detection 13
1.4 Impact of Microplastics on Humans and the Environment 14
1.4.1 Polychlorinated Biphenyl Compounds 17
1.4.2 Polycyclic Aromatic Hydrocarbon Compounds 17
1.4.3 Organochlorine Pesticides 21
1.4.4 Carcinogens 30
1.5 Methods of Food Packaging 42
1.6 Recycling Safety 58
References 58

2 Regulations **67**

2.1 European Strategy for Plastics 67
2.2 Basic Principles of Packaging 68
2.3 Colorants and Optical Brighteners 68
2.4 Non-intentionally Added Substances in Food Contact Material 69
2.4.1 Analysis Methods 71
2.5 Recycling 74
2.5.1 Recycled Plastics in Food Packaging 74
2.5.2 Planning and Decision-Making Process 76
2.5.3 Mechanical Recycled Plastics 76

2.6 Standards 77
2.6.1 Gloves 78
2.6.2 Specifications for Compostable Plastics 78
2.6.3 Containers for Packaging of Liquid Foodstuffs 79
2.6.4 Williams-Landel-Ferry Equation 79
References 80

3 Testing Methods 83
3.1 Risk Assessment 84
3.2 Antimicrobial Packaging 84
3.3 Food Colorants 96
3.3.1 Milk Freshness Testing 101
3.3.2 Meat Freshness Testing 102
3.3.3 Fish Freshness Testing 103
3.4 Standards 103
3.4.1 Odor and Taste Transfer from Packaging Film 106
3.4.2 Annealed Aluminum and Aluminum-Alloy Foil for Food Contact 106
3.4.3 Detection of Leaks in Packages 107
3.4.4 Changes in Sensory Attributes 111
3.4.5 Irradiated Food Contact Materials 112
3.4.6 Oxygen Headspace Analysis 114
References 114

4 Food Packaging 121
4.1 Automated Method 121
4.2 Multilayer Packaging 126
4.3 Flexible Packaging 127
4.4 Multicomponent Package 129
4.4.1 Barrier Films 134
4.5 Chlorine Dioxide Gas-Releasing Package 140
4.6 Sustainable Food Packaging 142
4.7 Ventable Food Packages 142
4.7.1 Ventable Microwave-Safe Food Package 142
4.7.2 Self-Venting Microwave-Safe Food Package 143
4.7.3 Smart and Active Food Packaging 143
4.8 Packages for Pourable Food 146
4.9 Food Storage 149
4.9.1 Trays 149
4.9.2 Storage Container 151
4.9.3 Storage System 154

4.10 Special Materials for Food Packaging 156
4.10.1 Film-Forming Biodegradable Classes 156
4.10.2 Cytotoxicity 158
4.10.3 Microwavable Food Packaging Materials 160
4.10.4 Film-Forming Materials 161
4.10.5 Antimicrobial Biodegradable Food Packaging 164
4.10.6 Multilayer Films for Food Packaging 170
4.10.7 Bilayer Film with Dihydromyricetin 170
References 174

5 Materials 191
5.1 Nanocomposites 191
5.1.1 PVC/Ag Nanocomposites 192
5.1.2 Montmorillonite Clays 193
5.1.3 Graphene Nanoplates 197
5.1.4 Carbon Nanotubes 197
5.1.5 Micro- and Nanoplastics 197
5.2 Biopolymers 198
5.2.1 Proteins 199
5.2.2 Gelatin 203
5.2.3 Storage Polymers in Bacterial Cells 207
5.2.4 Bio-Based Nanocomposites 208
5.2.5 Starch 209
5.2.6 Dry Heating Treatment 210
5.2.7 Thermoplastic Starch 210
5.2.8 Modified Starch 211
5.2.9 Starch-Based Foams 212
5.2.10 Residual Starch from Potato 213
5.2.11 Cellulose 213
5.2.12 Biodegradable Polymer Reinforced with Cellulose Nanocrystals 214
5.2.13 Cellulose Nanofibrils 221
5.2.14 Chia Seeds 224
5.2.15 Poly(lactic acid) 226
5.2.16 Poly(hydroxyalkanoate)s 227
5.2.17 Poly(hydroxybutyrate) 229
5.2.18 Poly(caprolactone) 230
5.3 Microplastics 231
5.3.1 Microplastics in the Environment 232
5.3.2 Microplastics in Food 233

5.3.3 Microplastics in Agroecosystems 234
5.3.4 Release of Microplastics 235
5.4 Single-Use Materials 235
5.5 Edible Films 236
5.5.1 Poly(saccharide)s 236
5.5.2 Cellulose 238
References 239

6 Additives 255
6.1 Chemical Additives 255
6.1.1 Hazardous Metal Additives 259
6.2 Mathematical Models 262
6.3 Evaluation of Models 264
6.4 Plastic Additives by Swelling 264
6.5 Fatty Acid Amide 277
6.5.1 Plasticizers 277
6.5.2 Antioxidants 277
6.5.3 Monomers and Oligomers 278
6.5.4 Light Stabilizers 278
6.5.5 Slip Additives 278
6.6 Perfluorinated Compounds 279
6.7 Toxic Materials in Plastics 284
6.8 Aflatoxins 284
6.8.1 Sorbents for Aflatoxin 285
6.9 Migration of Additives from Plastics into Food 286
6.10 Marine Environments 287
6.11 Surfactants 288
6.12 Salting-Out Agents 289
6.12.1 Determination of Food Stimulants 289
6.12.2 Determination of Biogenic Amines 289
6.12.3 Liquid-Liquid Microextraction 290
6.12.4 Determination of Aromatic Amines 292
References 296

7 Applications 303
7.1 Glaze-Like Coatings 303
7.2 Functional Food Applications 306
7.2.1 Fructan 306
7.2.2 Zogale 306
7.2.3 *Garcinia indica* 307
7.2.4 Mulberry Plants 309

7.2.5 Citrus Seeds 310
7.2.6 Vine Tea 310
7.2.7 Anti-Stress Drugs 316
7.2.8 Microalgae 318
7.2.9 Phenolics 318
7.2.10 Liposomes 319
7.2.11 Diacylglycerol Functional Oil 319
7.3 Prevention of Oxidation 320
7.3.1 Photo-oxidation 320
7.3.2 Lipid Oxidation 322
7.3.3 Natural Antioxidants 322
7.3.4 Extraction Methods for Natural Antioxidants 330
7.4 Heterocyclic Aromatic Amines in Prepared Food 339
7.5 Radiation-Curable Compositions 341
7.6 Antimycotics 344
References 347

8 Recycling 357
8.1 Identification of the Materials 359
8.1.1 Spectroscopic Methods 359
8.1.2 Real Time Mass Spectrometry 359
8.1.3 Optical Identification 360
8.1.4 Chemical Safety Aspects 361
8.1.5 Migration of Contaminants 364
8.1.6 Migration of Plasticizers 366
8.1.7 Migration of Aluminum and Silicon 366
8.2 Recycling Methods 367
8.2.1 Delamination 367
8.2.2 Separation 369
8.2.3 Mechanical Recycling 369
8.2.4 Selective Dissolution 372
8.2.5 Dissolution-Reprecipitation Technique 372
8.2.6 Compatibilization 374
8.2.7 Feedstock Recycling 374
8.2.8 Closed-Loop Recycling 375
8.2.9 Supercritical Ethanol 377
8.3 Post-Consumer Polyolefins 377
8.3.1 Predictive Models 377
8.3.2 Safety Concerns 378
8.3.3 Multilayer Composites 380
8.3.4 Recyclability Enhancement of Food Containers 384

8.4 Plastic Waste Generation 384
8.5 Recycled Plastics in Food Contact 385
8.5.1 Improving Safety and Quality 385
References 385

Index **391**
Acronyms 391
Chemicals 394
General Index 404

Preface

This book focuses on plastics for food safety and materials and chemicals, methods and applications of these polymers.

The book begins with a chapter about food safety. Here, food security and the issues of migration of substances from packaging into the corresponding food are discussed.

Then, in the next chapter, regulations, standards, and specifications are detailed. Furthermore, in another chapter, testing methods, such as risk assessment, freshness testing of food, and food colorants are discussed.

In the chapter entitled "Food Packaging," the methods that can be used for these issues are given. Also, materials, special uses, and finally, methods for recycling are given.

The subsequent chapters and their subject matter are:

Chapter 1: Food Safety
Chapter 2: Regulations
Chapter 3: Testing Methods
Chapter 4: Food Packaging
Chapter 5: Materials
Chapter 6: Additives
Chapter 7: Applications
Chapter 8: Recycling

The text focuses on the literature of the past decade. Beyond education, this book will serve the needs of industry engineers and specialists who have only a passing knowledge of the plastics and composites industries but need to know more.

How to Use This Book

Utmost care has been taken to present reliable data. Because of the vast variety of material presented here, however, the text cannot be complete in all aspects, and it is recommended that the reader study the original literature for more complete information.

The reader should be aware that mostly US patents have been cited where available, but not the corresponding equivalent patents in other countries. For this reason, the author cannot assume responsibility for the completeness, validity or consequences of the use of the material presented herein. Every attempt has been made to identify trademarks; however, there were some that the author was unable to locate.

Index

There are three indices: an index of acronyms, an index of chemicals, and a general index.

In the index of chemicals, compounds that occur extensively, e.g., 'acetone,' are not included at every occurrence, but rather when they appear in an important context.

Acknowledgements

I am indebted to our university librarians, Dr. Christian Hasenhüttl, Margit Keshmiri, Friedrich Scheer, Christian Slamenik, Renate Tschabuschnig, and Elisabeth Groß for their support in literature acquisition. I also want to express my gratitude to all the scientists who have carefully published their results concerning the topics dealt with herein. This book could not have been otherwise compiled.

Last, but not least, I want to thank the publisher, Martin Scrivener, for his abiding interest and help in the preparation of the text. In addition, my thanks go to Jean Markovic, who made the final copyedit with utmost care.

Johannes Fink
Wiener Neustadt, January 6, 2023

1

Food Safety

With the world's growing population, the provision of a safe, nutritious and wholesome food supply for all has become a major challenge (1). To achieve this, effective risk management based on sound science and unbiased information is required by all stakeholders, including the food industry, governments and consumers themselves. In addition, the globalization of the food supply requires the harmonization of policies and standards based on a common understanding of food safety among authorities in countries around the world. With some 280 chapters, the *Encyclopedia of Food Safety* provides unbiased and concise overviews which form in total a comprehensive coverage of a broad range of food safety topics, which may be grouped under the following general categories (1):

1. History and basic sciences that support food safety,
2. Foodborne diseases, including surveillance and investigation,
3. Foodborne hazards, including microbiological and chemical agents,
4. Substances added to food, both directly and indirectly,
5. Food technologies, including the latest developments,
6. Food commodities, including their potential hazards and controls,
7. Food safety management systems, including their elements and the roles of stakeholders.

The Encyclopedia provides a platform for experts from the field of food safety and related fields, such as nutrition, food science and technology and environment, to share and learn from state-of-the art expertise with the rest of the food safety community (1).

Plastic plays a significant and growing role in modern day society, delivering many benefits, particularly in food safety and preservation, and can help to reduce food waste (2).

Advances in packaging can not only reduce losses but also improve food quality and safety. To offer the best food protection the FAO suggests that a packaging solution could include more, but better packaging, rather than less packaging.

A novel approach for food safety, which integrates a statistical study and subjective discussion, was adopted to review the recent advances in the environment and food safety (3).

Here, a scientometric-based statistical study was conducted based on 4904 publications collected from the Web of Science Core Collection database. It was found that the research on the environment and food safety grew steadily from 2001 to 2020. The statistical analysis of most-cited papers, titles, abstracts, keywords, and research areas revealed that the research on the environment and food safety was diverse and multidisciplinary.

In addition to the scientometric study, strategies to protect the environment and ensure the safety of food were critically discussed, followed by a discussion on the emerging research topics, including emerging contaminates, e.g., microplastics, rapid detection of contaminants, e.g., biosensors, and environmentally friendly food packaging materials, e.g., biodegradable polymers (3).

1.1 Food Security

Plastic pollution arising from food systems is driving policies for reduction, removal, reuse and recycling (4).

A systematic scoping review was used to describe the extent, range and nature of published evidence since 2000 on seven major plastic types used at any point within food systems.

The majority of the publications focus on the agricultural production; relatively fewer publications relate to retail, household and food waste disposal plastics (4).

Some plastic types were more frequently linked to certain outcomes than others: Poly(ethylene) and poly(propylene) (PP) were frequently explored in relation to effects on the nutrient or bacterial content of crops and food (n = 445 and 274, respectively), but only one study considered the effects of PP (commonly used in infant feeding bottles) on the presence of plastics or associated chemicals in the human body.

Among the studies that looked at plastics and chemicals in the human body, as well as human health states and/or physiological changes, the most commonly explored types were unspecified plastics, miscellaneous plastics and poly(ethylene terephthalate) (PET). Despite finding evidence of each major plastic type in our sample, a lack of chemical specification of the plastic exposures was very common (4).

1.2 Migration of Substances from Packaging

The usage of new food packaging materials has increased the number of occurring hazards due to the migration from packaging material to the packaged food (5). Although polymers have mainly monopolized the interest of migration testing and experimentation, recent studies have revealed that migration also occurs from traditional materials generally considered to be safe, such as paper, carton, wood, ceramic, and metal. The regulations and the directives of the EU tend to become stricter in this respect.

The emphasis is on reaching a consensus in terms of food simulants and testing conditions for migration studies. Furthermore, the list of hazardous monomers, oligomers, and additives continues to be augmented in order to ensure that the consumer safety is in current agreement with the hazard analysis critical control points (HACCP), which is continuously gaining ground (5).

Food and beverages can be very aggressive chemical milieu and may interact strongly with materials that they touch (6). Whenever food is placed in contact with another substance, there is a risk that chemicals from the contact material may migrate into the food. These chemicals may be harmful if ingested in large quantities, or impart a taint or odor to the food, thereby negatively affecting the food quality. Food packaging is the most obvious example of a food contact material. As the demand for prepackaged foods increases, so might the potential risk to consumers from the release of chemicals into the food product.

Chemical Migration and Food Contact Materials reviews the latest controls and research in this field and how they can be used to ensure that food is safe to eat.

In a monograph, the regulation and quality control of chemical migration into food are discussed (6). Then, the latest developments in areas such as exposure estimation and analysis of food contact materials are reviewed. Finally, specific chapters on major food contact materials and packaging types are presented, such as recycled plastics, metals, paper and board, multilayer packaging and intelligent packaging.

The large number of synthetic materials that are used for the manufacture of packages makes the evaluation of the food package interactions complicated (7).

Different parameters, such as the nature of the food of interest, the type of food package contact, the time and temperature of contact, the packaging materials used, the properties of the migrating substances, as well as the amount of potential migrants contained in the packaging materials, can drastically affect the migration rate and the extent of migration.

Due to the extreme variety of foods used, several food simulants have been suggested and applied for testing the migration phenomenon under various laboratory conditions (7). The use of many of those simulants is defined by national and international legislation. The main migration phenomena, which are related to the most commonly used packaging materials, have been detailed (7). In the study, it was clearly demonstrated that the complexity of the migration phenomena requires more research to establish internationally accepted risk management procedures and standardized testing methods.

1.2.1 Modeling of Migration

The potential for the use of migration modeling for studying polyolefin packaging materials, such as low-density and high-density poly(ethylene) and poly(propylene), was summarized and demonstrated with practical examples (8).

For these polymers, an upper limit of migration into foodstuffs can be predicted with a high degree of statistical confidence. The only analytical information needed for modeling in such cases is the initial concentration of the migrant in the polymer matrix.

For polyolefins of unknown origin or newly developed materials with new properties, a quick experimental method has been described for obtaining the characteristic matrix parameter needed for the process of migration modeling.

For easy handling of both the experimental results and the diffusion model, a user-friendly software has been developed. An additional aim of the described method is the determination of the migrant partition between polymer and food or food simulant and the specific contribution of the migrant molecular structure on the diffusion coefficient. For migration modeling of packaging materials with multilayer structures, a numerical solution of the diffusion equation is described. This procedure has also been applied for modeling the migration into solid or high-viscosity foodstuffs (8).

The permeation through and diffusion/migration from high-density poly(ethylene), poly(butylene), poly(propylene) and crosslinked poly(ethylene) films was investigated experimentally with three different methods to determine diffusion coefficients in these polyolefins for a series of additives, their degradation products and other organic substances in the 20°C to 60°C temperature range (9).

The experimental methods used were dynamic permeation through additive-free polymer films, kinetic desorption from additivated films into water and kinetic migration from additivated into additive-free polymer films. It was found that in general the temperature dependence of the obtained diffusion coefficients was well represented by the Arrhenius law.

Some of these results also suggested that the contact of the polyolefins with water had an influence on the magnitude of the diffusion coefficients and on their apparent activation energy of diffusion (9).

The migration of phthalate from PET bottles containing non-alcoholic beer was done by performing an adaptive neuro-fuzzy inference system analysis (10).

The data showed that the storage temperature, contact surface and storage period correlate with the rate of migration. The migration of phthalate increases with storage duration gradually and reduces under different temperatures and contact surface. Moreover, increased temperature and storage duration resulted in an increase in migration level ranging from 0.6 $\mu g\, l^{-1}$ to 2.9 $\mu g\, l^{-1}$ (10).

The study used an adaptive neuro-fuzzy inference system analysis architecture, which consists of three inputs (temperature, surface and storage period), Gaussian-bell membership functions for each input variable and one output layer, which represent the migration level. The validation and training models showed an excellent match between the experimental and predicted values of adaptive neuro-fuzzy inference system analysis.

The analysis of the model showed that adaptive neuro-fuzzy inference system analysis is a powerful tool for predicting phthalate migration from bottles containing non-alcoholic beer (10).

1.2.2 *Sample Pretreatment Methods*

An excessive absorption of migrating substances from food contact materials can affect human health. Thus, it is essential to analyze the migration of contaminants from food contact materials (11).

However, comprehensive analysis has been challenged by low concentration of migrating substances and manifold and complex matrix interference of food contact materials. Therefore, appropriate sample pretreatment methods should be applied before instrumental detection, which is essential to improve the analytical efficiency, sensitivity, and reliability.

The development of sample pretreatment methods for the analysis of migrating substances from food contact materials has been reviewed in the past decade (11). To extract volatile and semi-volatile substances, headspace extraction, headspace solid phase microextraction, and a purge and trap technique are discussed here.

For non-volatile substances, solid-liquid extraction and field-assisted extraction are usually used to extract them from food contact materials; while liquid-liquid extraction, solid phase extraction, and their corresponding microextraction techniques play important roles in the enrichment process (11).

Also, new progress in the development of sample pretreatment methods of food contact materials was summarized, covering new devices, specific adsorbents, and sample preparation methods for rapid detection (11).

1.2.3 *Special Chemicals*

1.2.3.1 *Styrene*

Poly(styrene) (PS) is extensively used in diverse forms for packaging of many food products such as meat, dairy and bakery products. There is a potential migration of styrene monomer from PS packages into the foods which are in contact with them (12).

The representative styrene migration from PS packaging material into the corresponding foods was detailed. Also, the addressed parameters affecting the styrene migration were discussed. The analytical methods for detecting the styrene monomer in food products and PS packaging materials was also covered in the study. The possible safety and health issues related to the styrene monomer migration were also covered.

The quality of PS packaging materials in terms of their styrene monomer residue level and the storage conditions of foods can greatly affect styrene migration. Also, the food characteristics, such as fat content and pH, can significantly affect the migration of styrene. Although styrene is considered a nontoxic compound, its migration into foods can downgrade sensorial properties as well as result in health problems. In some cases, the presence of styrene in foods can cause carcinogenic, hematological, cytogenetic, and neurotoxic issues (12).

A sensitive, accurate and fast headspace-solid phase microextraction-gas chromatography-tandem mass spectrometry (HS-SPME-GC-MS/MS) method was developed and validated for the determination of styrene in various food matrices. The mean recovery ranged from 90% to 116% with a relative standard deviation of =11% (13).

This method was used for the determination of the concentration of styrene in 23 foodstuffs packed in PS containers, as well as the levels of styrene migrating into various foods (water, milk, cheese or cream) from 14 tableware or kitchenware articles made of styrene plastics. All samples were collected from the Greek market in 2020.

The concentrations of styrene in the packaged foods ranged from 0.4 to 160 $ng\, g^{-1}$, with the highest concentration found in a meat product packed in a foamed PS tray. It is worth noting that 56% of PS packaged dairy products and desserts had a styrene concentration of higher than 10 $ng\, g^{-1}$. Particularly high levels of styrene that have not previously been reported, of up to 46 $ng\,g^{-1}$, were found in dairy products for children. The highest level of styrene migration from tableware or kitchenware articles, 89 $ng\,g^{-1}$, was observed when disposable cups from foamed PS were filled with milk at 70°C for 2 *h* (13).

1.2.3.2 Bisphenol A

Bisphenol A is mainly used in the production of poly(carbonate), a material with high durability and strength. This chemical is also used in the production of epoxy resins, applied in the coating of metal surfaces in contact with food. Moreover, it is employed in thermal paper (14).

Bisphenol A has been widely reported by the media, being the aim of diverse scientific studies, since it is a xenoestrogen with a chemical structure identical to that of *β*-estradiol, which allows it to interact with human estrogen receptors. These compounds are shown in Figure 1.1.

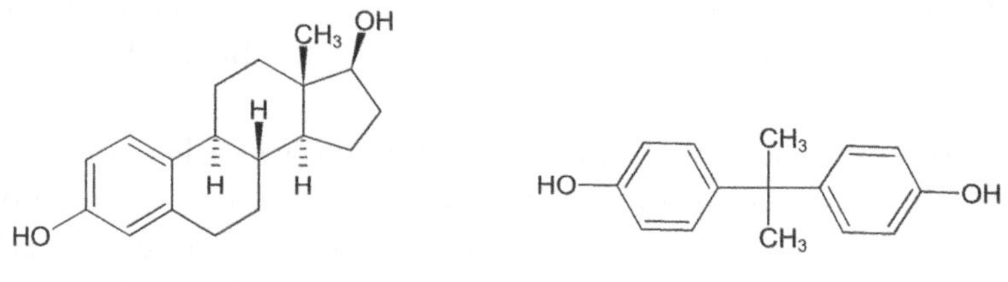

β-Estradiol   Bisphenol A

Figure 1.1 Xenoestrogens.

Bisphenol A monitoring is essential to avoid a potential health risk of the population. The European Union has recently updated the specific migration limit for bisphenol A at 0.05 $mg\,kg^{-1}$ of food.

Regarding the levels of migration determined in different types of food contact materials, it was found that, despite the great majority of studies showing positive samples, none of them exceeded the value of specific migration limit established by the European Union at the time of the studies (14).

1.2.3.3 *Benzophenone*

Benzophenone may be present in cartonboard food-packaging materials as a residue from UV-cured inks and lacquers used to print on the packaging (15). It may also be present if the cartonboard is made from recycled fibers recovered from printed materials.

A method has been developed to test for benzophenone in cartonboard packaging materials and to test the migration levels in foodstuffs.

Here, the packaging material is extracted with a solvent containing d10-benzophenone as the internal standard. Foods are extracted with solvent containing d10-benzophenone and the extract is defatted using hexane. The extracts are analyzed by gas chromatography (GC)-mass spectroscopy (MS). d10-Benzophenone is the deuterium labeled benzophenone. It is shown in Figure 1.2.

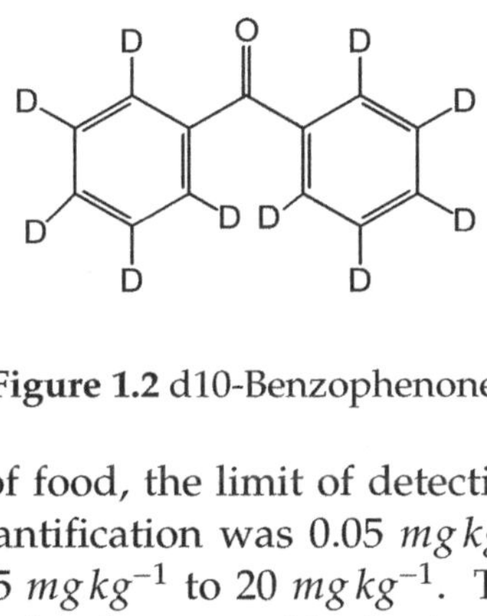

Figure 1.2 d10-Benzophenone.

For the analysis of food, the limit of detection was 0.01 $mg\,kg^{-1}$ and the limit of quantification was 0.05 $mg\,kg^{-1}$. The calibration was linear from 0.05 $mg\,kg^{-1}$ to 20 $mg\,kg^{-1}$. The method for food analysis was validated in-house and it also returned satisfactory results in a blind check-sample exercise organized by an independent laboratory.

The methods were applied to the analysis of 350 retail samples that used printed cartonboard packaging. A total of 207 (59%) packaging samples had no significant benzophenone (<0.05 $mg\,dm^{-2}$). Seven (2%) were in the range 0.05 $mg\,dm^{-2}$ to 0.2 $mg\,dm^{-2}$, 60 (17%) were from 0.2 $mg\,dm^{-2}$ to 0.8 $mg\,dm^{-2}$ and 76 (22%) were from 0.8 $mg\,dm^{-2}$ to 3.3 $mg\,dm^{-2}$.

A total of 71 samples were then selected at random from the 143 packaging samples that contained benzophenone, and the food itself was analyzed. Benzophenone was detected in 51 (72%) of the foods. Two food samples (3%) were in the range 0.01 $mg\,kg^{-1}$ to 0.05 $mg\,kg^{-1}$. A total of 29 (41%) were from 0.05 $mg\,kg^{-1}$ to 0.5 $mg\,kg^{-1}$, 17 (24%) were from 0.5 $mg\,kg^{-1}$ to 5 $mg\,kg^{-1}$ and three (4%) food samples exceeded 5 $mg\,kg^{-1}$. The highest level of benzophenone in food was 7.3 $mg\,kg^{-1}$ for a high-fat chocolate confectionery product packaged in direct contact with cartonboard, with room temperature storage conditions and with a high contact area:food mass ratio. When the mass fraction of the migration of benzophenone was calculated for the different contact and storage regimes involved, the attenuation effects of indirect contact and of low temperature storage were cumulative. Thus, there was a sixfold reduction in migration for indirect contact compared with direct contact, a sixfold reduction for chilled/frozen storage compared with ambient storage, and 40-fold reduction for the two contact conditions combined (15).

1.2.3.4 *Perfluorochemicals*

Perfluorochemicals are widely used in the manufacturing and processing of a vast array of consumer goods, including electrical wiring, clothing, household and automotive products. Furthermore, relatively small quantities of perfluorochemicals are also used in the manufacturing of food contact substances that represent potential sources of oral exposure to these chemicals (16).

The most recognizable products to consumers are the uses of perfluorochemicals in non-stick coatings (poly(tetrafluoroethylene) (PTFE)) for cookware and also their use in paper coatings for oil and moisture resistance. Recent epidemiology studies have demonstrated the presence of two particular perfluorochemicals, perfluorooctane sulfonate and perfluorooctanoic acid, in human serum at very low part per billion levels. These perfluorochemicals are biopersistent and are the subject of numerous studies investigating the many possible sources of human exposure. Among the various uses of these two chemicals, perfluorooctane sulfonate is a residual impurity in some paper coatings used for food contact and perfluorooctanoic acid is a processing aid in the manufacture of PTFE used for many purposes, including non-stick cookware.

Very little information is available on the types of perfluorochemicals that have the potential to migrate from perfluoro coatings into food. One obstacle to studying the migration is the difficulty in measuring perfluorochemicals by routine conventional analytical techniques such as GC/MS or LC-UV. Many perfluorochemicals used in food contact substances are not detectable by these conventional methods. As liquid chromatography-mass spectrometry (LC/MS) develops into a routine analytical technique, potential migrants from perfluoro coatings can be more easily characterized. In a study, data is presented on the types of perfluorochemicals that are used in food packaging and cookware. Additionally, research is presented on the migration or potential for migration of these chemicals into foods or food-simulating liquids.

The results from migration tests show $mg\,kg^{-1}$ amounts of perfluoro paper additives/coatings transfer to food oil. Analysis of PTFE cookware shows residual amounts of perfluorooctanoic acid in the low $\mu g\,kg^{-1}$ range. Perfluorooctanoic acid is present in microwave popcorn bag paper in amounts as high as 300 $\mu g\,kg^{-1}$ (16).

1.2.4 Safety of Recycled HDPE and PP

An analytical protocol was set up and successfully applied to study the food safety of recycled high-density poly(ethylene) (HDPE) and PP crates (17). A worst-case scenario was applied that focused not only on overall migration and specific migration of accepted starting materials but also on migratable degradation products of polymers and additives that may be formed during mechanical recycling.

The analytical protocol was set up to cover a wide variety of possible migrants. Identification and semi-quantification were possible for almost all migrants that increased significantly with increasing mechanical recycling steps for both the HDPE and PP crates.

It was concluded that the analytical protocol was suitable to study the influence of (multiple) recycling on the food safety of plastic materials. The protocol can be applied to other plastic foodcontact materials and provides valuable information on the food safety of the recycling process and the resulting recycled foodcontact materials in addition to challenge testing (17).

1.3 Food Safety and Hygiene

Food hygiene is the conditions and measures necessary to certify the safety of food from production to consumption. Food can become contaminated at any point during slaughtering or harvesting, processing, storage, distribution, transportation and preparation.

All conditions and measures that are required during production, processing, storage, distribution and preparation of food to ensure that it is safe, wholesome and fit for human consumption were defined in 1984 by the World Health Organization (WHO).

A lack of requisite food hygiene can lead to foodborne diseases and death of the consumer. Foodborne illness has been associated with improper storage or reheating (50%), food stored inappropriately (45%) and cross contamination (39%). Proper food preparation can prevent most foodborne diseases.

More than 200 known diseases can be transmitted through food (18). Some of these diseases are collected in Table 1.1.

Table 1.1 Known diseases (19).

Disease or Agent	
Bacterial	Bacterial
Bacillus cereus	Botulism, foodborne
Brucella spp.	*Campylobacter* spp.
Clostridium perfringens	*Escherichia coli*
Listeria monocytogenes	*Salmonella typhi*
Salmonella, nontyphoidal	*Shigella* spp.
Staphylococcus	*Vibrio cholerae*
Vibrio vulnificus	*Yersinia enterocolitica*
Parasitic	Parasitic
Cryptosporidium parvum	*Cyclospora cayetanensis*
Giardia lamblia	*Toxoplasma gondii*
Trichinella spiralis	
Viral	Viral
Norwalk-like viruses	Rotavirus
Astrovirus	Hepatitis A

Hazard analysis and critical control points is a systematic preventive approach to food safety from biological, chemical, and physical hazards in production processes that can cause the finished product to be unsafe and designs measures to reduce these risks to a safe level. Food hygiene and safety usually refer to contamination with microorganisms or microbes.

1.3.1 Sensors for Amine Detection

Biogenic amines are good indicators of food freshness because they are products of microbial fermentation (20). In the process of food spoilage, microbes break down amino acids via deaminization to generate ammonia, and via decarboxylation to generate biogenic amines such as cadaverine, putrescine, spermidine, spermine, and others. These amines are shown in Figure 1.3.

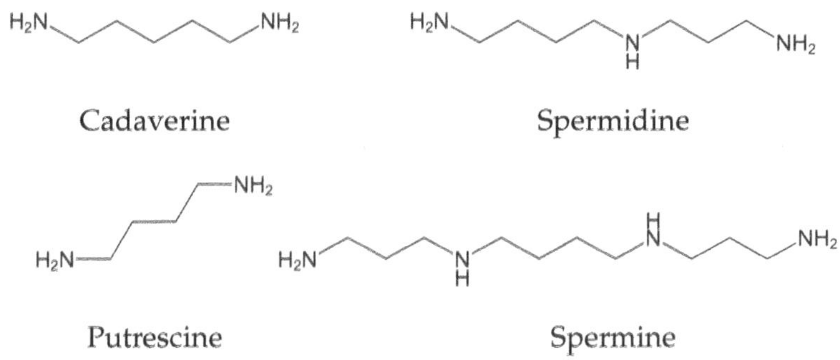

Figure 1.3 Biogenic amines.

These biogenic amines not only signal food spoilage, but also have an adverse impact on human health and physiological functions. Thus, monitoring biogenic amines in food is important both because the chemical species can have toxic effects, and because they signify food spoilage by microbes. When compared to time-temperature indicators, which only respond to temperature changes, a system detecting the presence of biogenic amines offers a more direct method of monitoring food safety and hygiene.

Aggregation-induced emission (AIE) active chemosensors exhibit a change for UV-vis absorption and become non-luminescent upon protonation (20). Upon deprotonation, the chemosensors revert to their original absorption and emission. This deprotonation process can be triggered in the presence of amines, and specifically, biogenic amines. So, the chemosensors can detect amine species, e.g., biogenic amines produced during food fermentation, quickly and with a high sensitivity.

As chemicals, 1,2-dihydroquinoxaline derivates can be used (20, 21).

Some derivates are shown in Table 1.2 and Figure 1.4.

Table 1.2 1,2-Dihydroquinoxaline derivates.

Compound
3-Methyl-2-prop-2-enyl-1,2-dihydroquinoxaline
3-Methoxy-6-(trifluoromethyl)-1,2-dihydroquinoxaline
2-Phenyl-1,2-dihydroquinoxaline
3-Phenyl-1,2-dihydroquinoxaline
3-Phenyl-1,2-dihydroquinoxaline-2-carboxylic acid
3-Methyl-2-propan-2-yl-1,2-dihydroquinoxaline
5-[6-[4-(Trifluoromethyl)phenyl]pyrimidin-4-yl]oxy-1,2-dihydroquinoxaline
5-Bromo-3-[4-(trifluoromethyl)phenyl]-1,2-dihydroquinoxaline
3-[Chloro(phenyl)methyl]-1,2-dihydroquinoxaline
3-(1H-Indol-3-yl)-1,2-dihydroquinoxaline
6-Chloro-1,2-dihydroquinoxaline
3-(3-Nitrophenyl)-1,2-dihydroquinoxaline
Ethyl 7-amino-3-ethoxy-6-(trifluoromethyl)-1,2-dihydroquinoxaline-2-carboxylate
8-Amino-2-ethyl-5-(hydroxymethyl)-1,2-dihydroquinoxaline-6-carboxylic acid
2,3-Dimethyl-1,2-dihydroquinoxaline
2-Chloro-3-(5-chloro-1H-pyrazol-3-yl)-1,2-dihydroquinoxaline
2-Bromo-5-[6-[4-(trifluoromethyl)phenyl]pyrimidin-4-yl]oxy-1,2-dihydroquinoxaline

1.4 Impact of Microplastics on Humans and the Environment

Micro- and nanoplastics have the potential to be transferred between trophic levels and, therefore, the risk characterization and the assessment of dietary exposure to them constitutes a current challenge for food safety alongside the study of the role of plastics as vectors of other contaminants and pathogenic microorganisms (22).

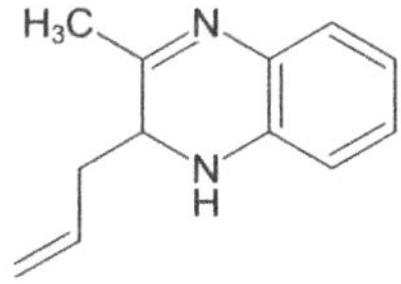

3-Methyl-2-prop-2-enyl-
1,2-dihydroquinoxaline

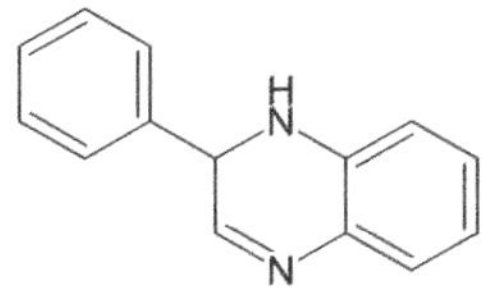

2-Phenyl-
1,2-dihydroquinoxaline

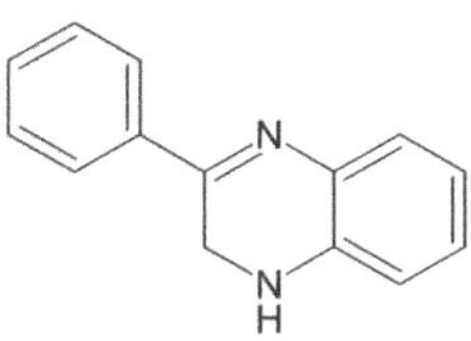

3-Phenyl-
3-Phenyl-1,2-dihydroquinoxaline

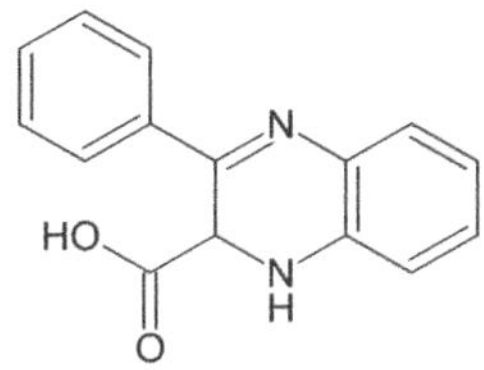

3-Phenyl-1,2-dihydroquinoxaline-
2-carboxylic acid

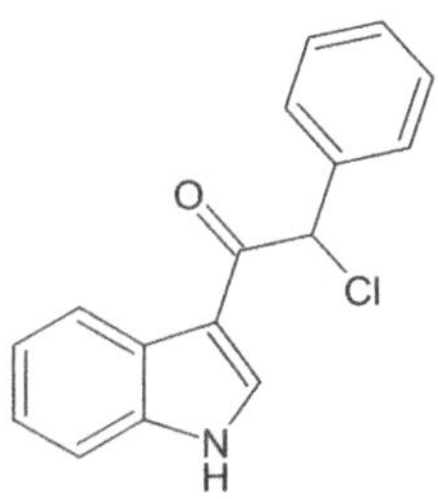

3-(1H-Indol-3-yl)-
1,2-dihydroquinoxaline

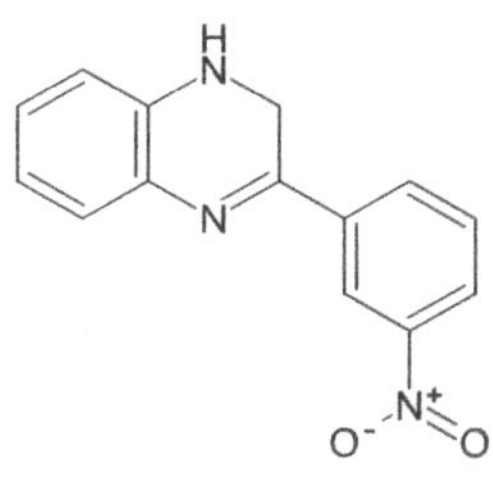

3-(3-Nitrophenyl)-
1,2-dihydroquinoxaline

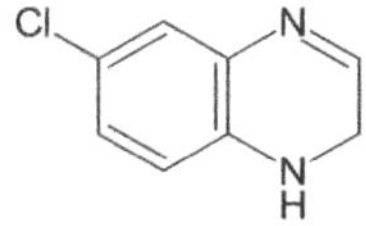

6-Chloro-1,2-dihydroquinoxaline

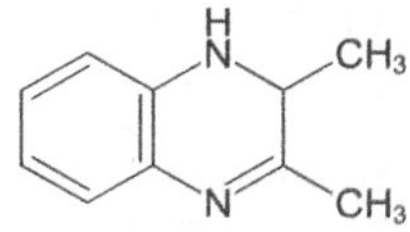

2,3-Dimethyl-
1,2-dihydroquinoxaline

Figure 1.4 1,2-Dihydroquinoxaline derivates.

The risk posed by microplastics to humans and the environment, has become a hot topic (23). The concern is focused not only on the effect of microplastics as such but also on additives and chemical contaminants absorbed by microplastics that may be released and negatively affect animals and environmental health. Despite several works having been written on this topic, a number of knowledge gaps still should be filled to enable a correct risk assessment of this important issue. For example, the relevance of microplastics for food safety has not yet been fully established and scientific results aimed at establishing a possible health risk for contaminants associated with microplastics are rather controversial. The risk assessment of microplastics in foodstuff is still at a very early stage and very few studies on the monitoring of microplastics in foodstuff and their effects on human health are available. Additionally, it is difficult to compare results from different studies as methodologies and study designs are not uniform. For this reason, it is not always possible to reach some definitive conclusion (23).

Studies have shown high concentrations of chemical contaminants that adsorb microplastics from the surrounding environment (24).

The compounds are listed in Table 1.3.

Table 1.3 Contaminants in microplastics (24).

Compound	Amount in pallets/[$ng\ g^{-1}$]
Polycyclic aromatic hydrocarbons	52.1 – 17,023.6
Polychlorinated biphenyl	0.9 – 2285.8
Organochlorine pesticides	0.4 – 13,488.7
UV filters	0 – 37,740.3
Brominated diphenyl ethers	0 – 180.58
Organophosphorus flame retardants	20.0 – 378.0
Compound	**Amount in fragments/[$ng\ g^{-1}$]**
Polycyclic aromatic hydrocarbons	35.1 – 8725.8
Polychlorinated biphenyl	1.6 – 772.5
Organochlorine pesticides	0.4 – 3778.8
UV filters	3.7 – 2169.3
Brominated diphenyl ethers	0.06 – 3923.9
Organophosphorus flame retardants	22.6 – 7013.9

1.4.1 *Polychlorinated Biphenyl Compounds*

Details of polychlorinated biphenyl compounds are shown in Table 1.4 and in Figure 1.5

Table 1.4 Polychlorinated biphenyl compounds.

Abbreviation	Name
PCB 1	2-Chlorobiphenyl
PCB 2	3-Chlorobiphenyl
PCB 3	4-Chlorobiphenyl
PCB 5	2,3-Dichlorobiphenyl
PCB 7	2,4-Dichlorobiphenyl
PCB 11	3,3′-Dichlorobiphenyl
PCB 12	3,4-Dichlorobiphenyl
PCB 14	3,5-Dichlorobiphenyl
PCB 15	4,4′-Dichlorobiphenyl
PCB 18	2,2′,5-Trichlorobiphenyl
PCB 28	2,4,4′-Trichlorobiphenyl
PCB 29	2,4,5-Trichlorobiphenyl
PCB 31	2,4′,5-Trichlorobiphenyl
PCB 52	2,2′,5,5′-Tetrachlorobiphenyl
PCB 77	3,3′,4,4′-Tetrachlorobiphenyl
PCB 101	2,2′,4,5,5′-Pentachlorobiphenyl
PCB 105	2,3,3′,4,4′-Pentachlorobiphenyl
PCB 138	2,2′,3,4,4′,5′-Hexachlorobiphenyl
PCB 149	2,2′,3,4′,5′,6-Hexachlorobiphenyl
PCB 153	2,2′,4,4′,5,5′-Hexachlorobiphenyl
PCB 170	2,2′,3,3′,4,4′,5-Heptachlorobiphenyl
PCB 180	2,2′,3,4,4′,5,5′-Heptachlorobiphenyl
PCB 209	Decachlorobiphenyl

1.4.2 *Polycyclic Aromatic Hydrocarbon Compounds*

Fossil fuels and other organic materials formed during incomplete burning can produce and release a complex cluster of emerging pollutants into the environment known as polycyclic aromatic hydrocarbons (25, 26).

Also, human activities have been multiplying the waste and effluents generated in the environment containing different polycyclic aromatic hydrocarbons. All sorts of contaminants including poly-

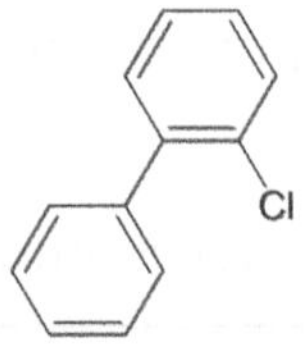

2-Chlorobiphenyl

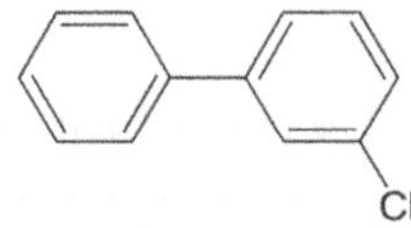

3-Chlorobiphenyl

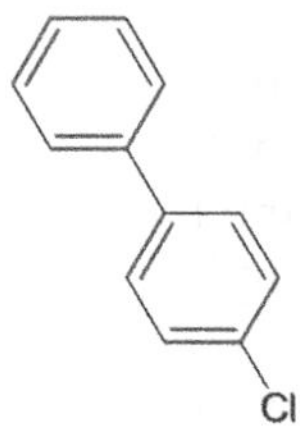

4-Chlorobiphenyl

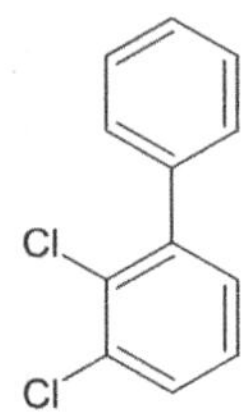

2,3-Dichlorobiphenyl

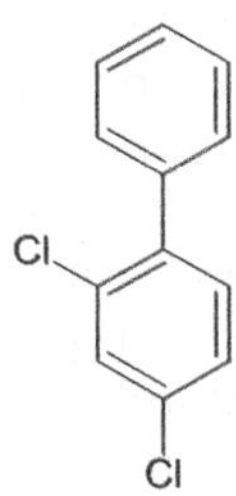

2,4-Dichlorobiphenyl

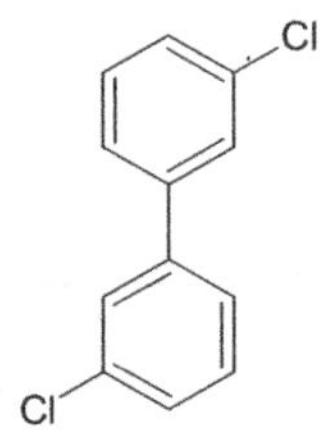

3,3′-Dichlorobiphenyl

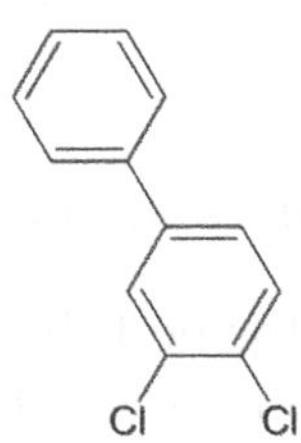

3,4-Dichlorobiphenyl

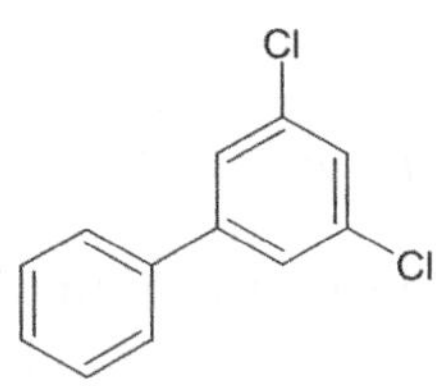

3,5-Dichlorobiphenyl

Figure 1.5 Polychlorinated biphenyl compounds.

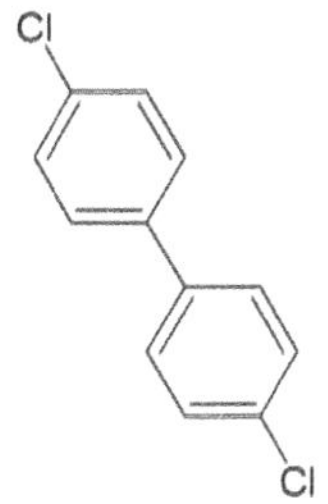

4,4′-Dichlorobiphenyl

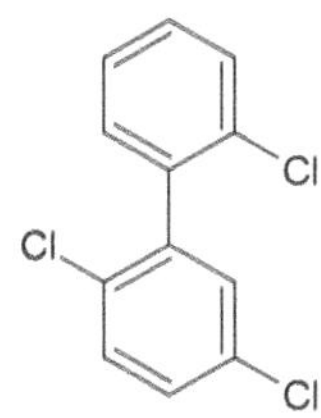

2,2′,5-Trichlorobiphenyl

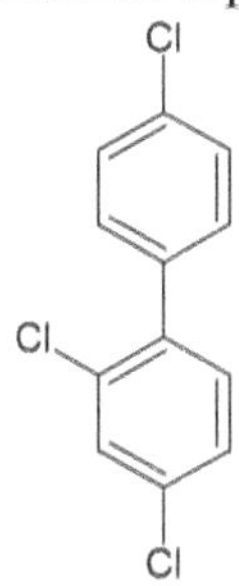

2,4,4′-Trichlorobiphenyl

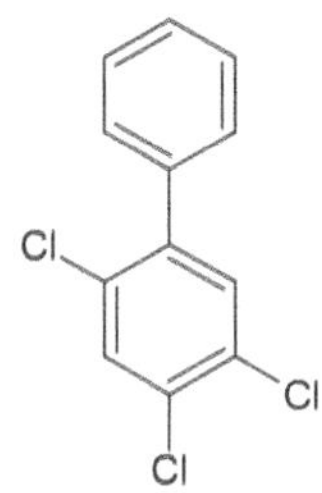

2,4,5-Trichlorobiphenyl

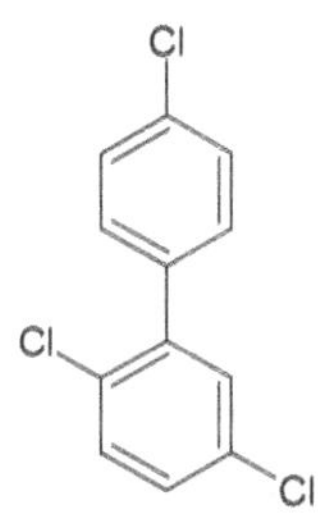

2,4′,5-Trichlorobiphenyl

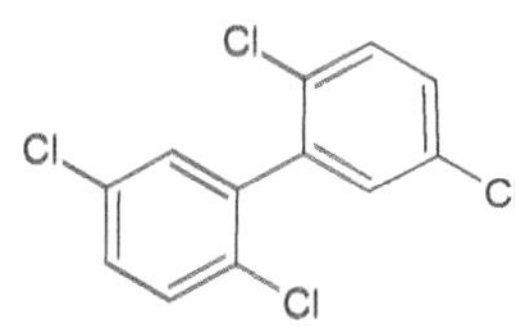

2,2′,5,5′-Tetrachlorobiphenyl

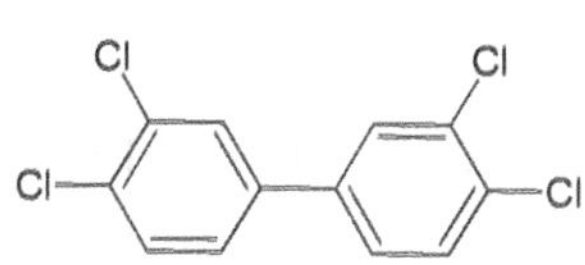

3,3′,4,4′-Tetrachlorobiphenyl

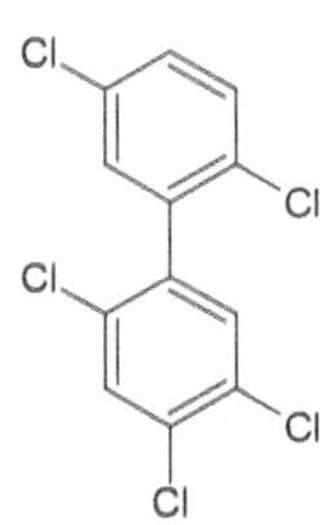

2,2′,4,5,5′-Pentachlorobiphenyl

Figure 1.5 (cont) Polychlorinated biphenyl compounds.

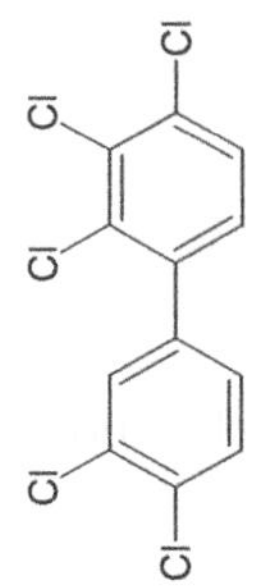

2,3,3′,4,4′-Pentachlorobiphenyl

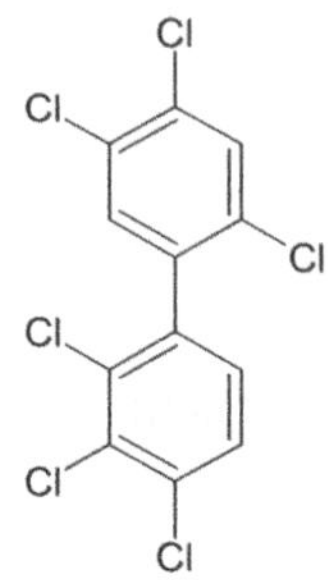

2,2′,3,4,4′,5′-Hexachlorobiphenyl

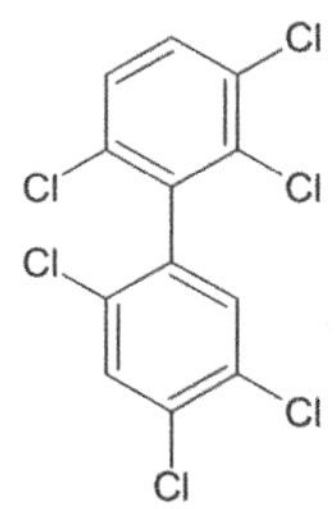

2,2′,3,4′,5′,6-Hexachlorobiphenyl

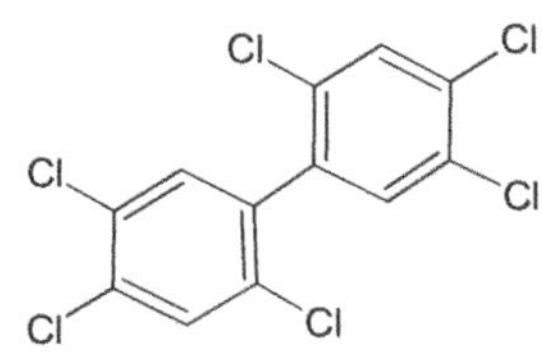

2,2′,4,4′,5,5′-Hexachlorobiphenyl

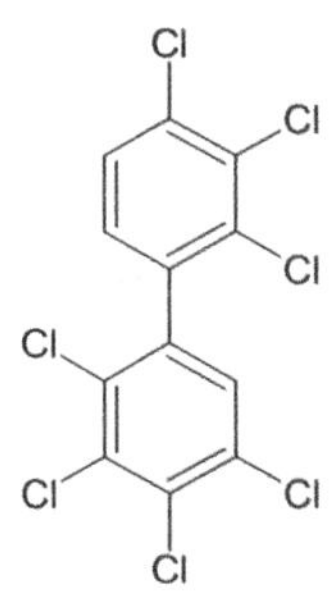

2,2′,3,3′,4,4′,5-Heptachlorobiphenyl

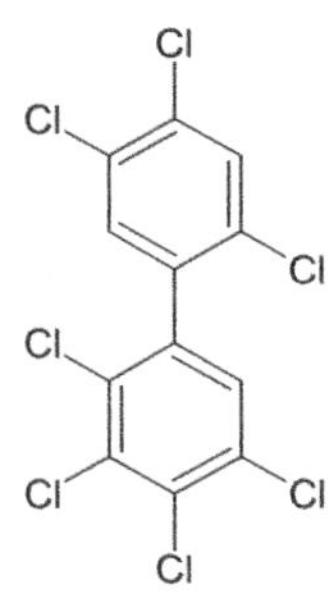

2,2′,3,4,4′,5,5′-Heptachlorobiphenyl

Figure 1.5 (cont) Polychlorinated biphenyl compounds.

cyclic aromatic hydrocarbons ultimately find their way into the aquatic ecosystem and thereby deplete the quality and standards of the aquatic habitats directly or indirectly (26).

Details of polycyclic aromatic hydrocarbon compounds are shown in Table 1.5 and in Figure 1.6

Table 1.5 Polycyclic aromatic hydrocarbons.

Name	Name
Benz[a]anthracene	Benzo[b]fluoranthene
Benzo[j]fluoranthene	Benzo[k]fluoranthene
Benzo[a]pyrene	Chrysene
Dibenz[a,h]acridine	Dibenz[a,j]acridine
Dibenz[a,h]anthracene	7H-Dibenz[c,g]carbazole
Dibenzo[a,e]pyrene	Dibenzo[a,h]pyrene
Dibenzo[a,i]pyrene	Indeno[1,2,3-cd]pyrene

Benz[a]anthracene is produced during incomplete combustion of organic matter. It is one of the carcinogenic constituents of tobacco smoke (27).

Also, benzo[j]fluoranthene is present in fossil fuels and is released during the incomplete combustion of organic matter. It has been traced in the smoke of cigarettes, exhaust from gasoline engines, emissions from the combustion of various types of coal and emissions from oil heating, as well as an impurity in some oils such as soybean oil (28).

The primary sources of dibenzopyrenes in the environment are combustion of wood and coal (29), gasoline and diesel exhaust (30), and tires (31). Dibenzo[a,l]pyrene is a constituent of tobacco smoke.

1.4.3 Organochlorine Pesticides

Details of organochlorine pesticides are shown in Table 1.6 and in Figure 1.7

Fungal contamination of animal feed is often unavoidable and is a serious concern given that many of these contaminants include toxic metabolites known as mycotoxins (32). Mycotoxin contamination can occur in a crop growing in the field, or contamination may be introduced during harvesting, storage and/or processing of the animal feed for use in raising monogastric and ruminant animals.

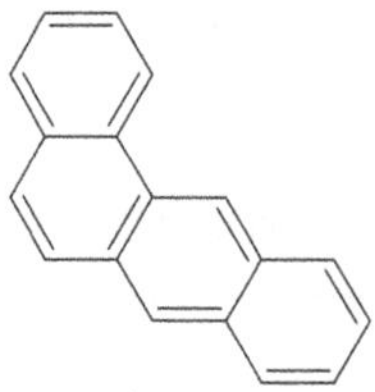

Benz[a]anthracene

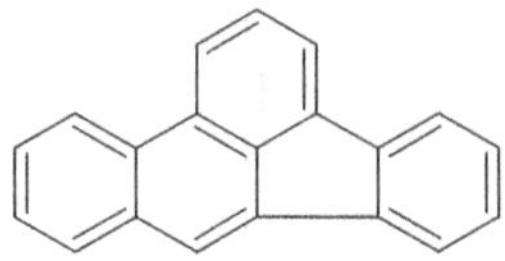

Benzo[b]fluoranthene

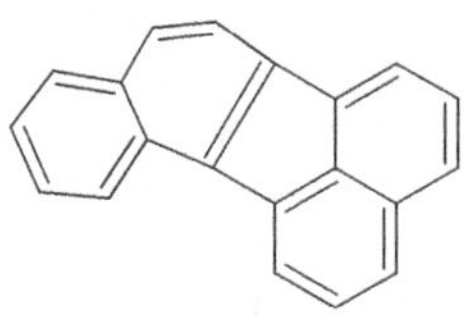

Benzo[j]fluoranthene

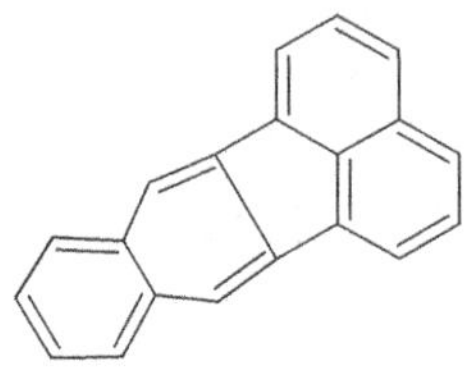

Benzo[k]fluoranthene

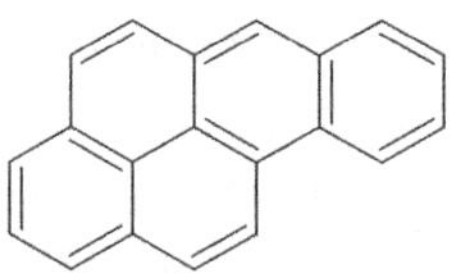

Benzo[a]pyrene

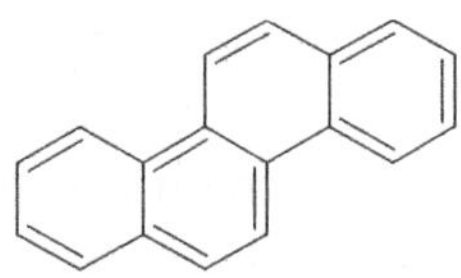

Chrysene

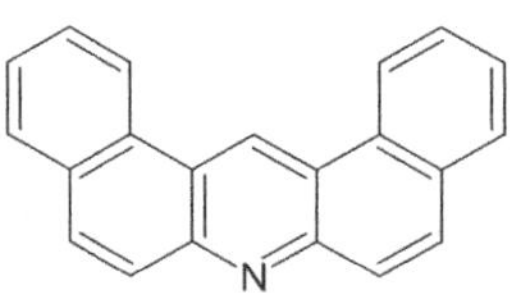

Dibenz[a,j]acridine

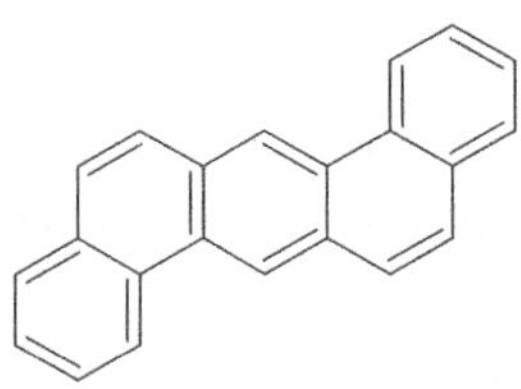

Dibenz[a,h]anthracene

Figure 1.6 Polycyclic aromatic hydrocarbons.

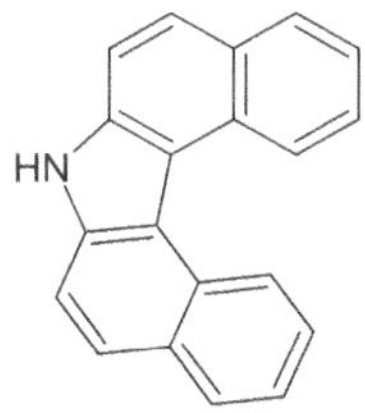

7H-Dibenz[c,g]carbazole

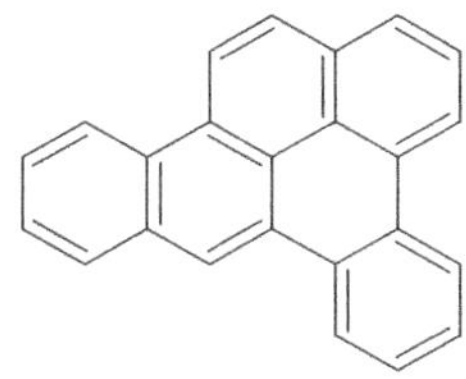

Dibenzo[a,e]pyrene

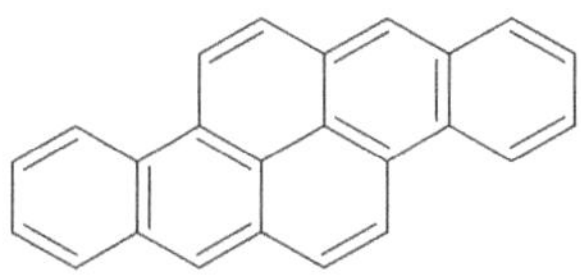

Dibenzo[a,h]pyrene

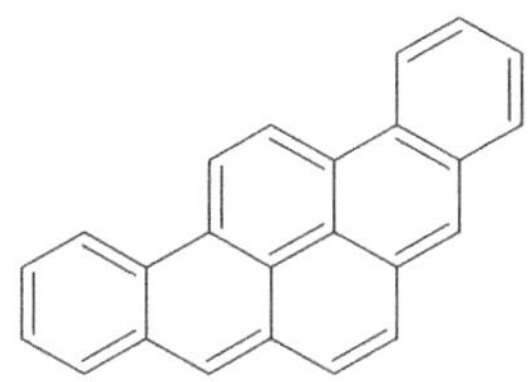

Dibenzo[a,i]pyrene

Figure 1.6 (cont) Polycyclic aromatic hydrocarbons.

Table 1.6 Organochlorine pesticides.

Name	Name
Aflatoxin	Fumonisin
Ochratoxin	Zearalenone

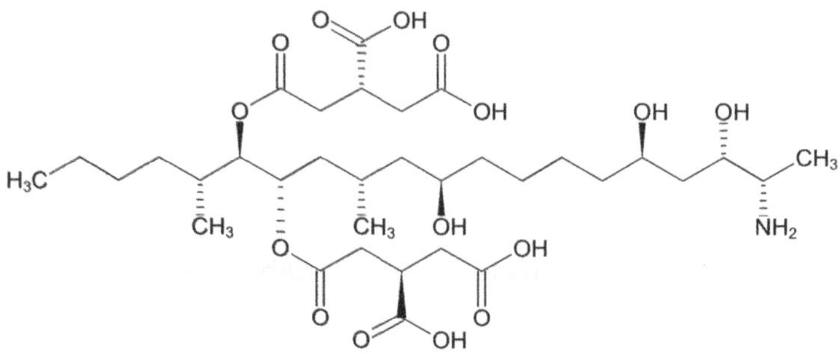

Fumonisin

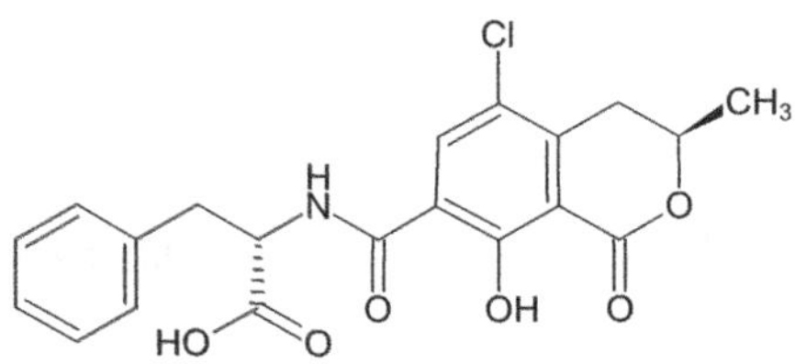

Ochratoxin A

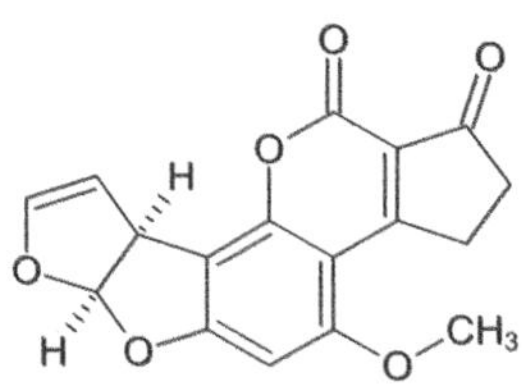

Aflatoxin

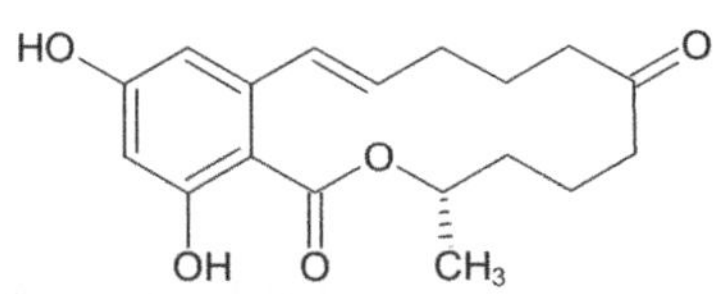

Zearalenone

Figure 1.7 Organochlorine pesticides.

Mycotoxins are fairly stable compounds often found in animal feed for monogastric and ruminant animals, and they are a known cause of a wide variety of deleterious effects in these animals. Pesticides are other common contaminants of animal feed. Endotoxins are not natural contaminants of feedstuffs. Feedstuffs can be contaminated with endotoxins when mixed with products of animal origin.

Endotoxins are another type of toxin of bacterial origin, which are commonly found in the gastrointestinal tract of livestock; and pesticides are commonly found in the foodstuffs fed to various types of livestock, and as a result, have been known to have toxic effects on livestock (32).

Mycotoxins are known to cause toxic, teratogenic, mutagenic, and carcinogenic effects, and have been linked to a depression of the animal's immune system. Furthermore, mycotoxins can affect different organs in an animal, including urinary, digestive, nervous, reproductive, and immune systems, and as such, it makes it more difficult to establish a precise diagnosis once an animal is affected. The effects of mycotoxins depend on the level of contamination, the presence of one or more toxins, the type of animal, its age, the time of exposure, genetic makeup, and its nutritional and health status at the time of exposure to contaminated feed.

The most dangerous mycotoxins affecting poultry are aflatoxin, ochratoxin, T-2 toxin, fumonisin, and deoxynivalenol, also known as DON. These mycotoxins, along with other trichothecene mycotoxins, can also affect monogastric and ruminaNZ amimals, to greater or lesser degrees (32).

1.4.3.1 Aflatoxin

Aflatoxins are various poisonous carcinogens and mutagens that are produced by certain molds, particularly *Aspergillus* species (33). The fungi grow in soil, decaying vegetation and various staple foodstuffs and commodities such as hay, sweetcorn, wheat, millet, sorghum, cassava, rice, chili peppers, cottonseed, peanuts, tree nuts, sesame seeds, sunflower seeds, and various spices. In short, the relevant fungi grow on almost any crop or food. When such contaminated food is processed or consumed, the aflatoxins enter the general food supply. They have been found in both pet and human foods, as well as in feedstocks for agricultural animals. Animals fed contaminated food can pass aflatoxin transformation products into eggs, milk products, and meat (34).

For example, contaminated poultry feed is the suspected source of aflatoxin-contaminated chicken meat and eggs in Pakistan (35). Children are particularly affected by aflatoxin exposure, which is associated with stunted growth (36), delayed development (37), liver damage, and liver cancer.

An association between childhood stunting and aflatoxin exposure has been reported in some studies but could not be detected in all studies (38).

Adults have a higher tolerance to exposure, but are also at risk. No animal species is immune. Aflatoxins are among the most carcinogenic substances known. After entering the body, aflatoxins may be metabolized by the liver to a reactive epoxide intermediate or hydroxylated to become the less harmful aflatoxin M1.

Aflatoxin poisoning most commonly results from ingestion, but the most toxic aflatoxin compound, B1, can permeate the skin.

The United States Food and Drug Administration (FDA) action levels for aflatoxin present in food or feed is 20 to 300 *ppb* (39). The FDA has had occasion to declare both human and pet food recalls as a precautionary measure to prevent exposure.

The term *aflatoxin* is derived from the name of the species *Aspergillus flavus*, in which some of the compounds first were discovered. The word was coined around 1960 after its discovery as the source of Turkey X disease (40). Aflatoxins form one of the major groupings of mycotoxins, and apart from *Aspergillus flavus*, various members of the group of compounds occur in species such as *Aspergillus parasiticus*, *Aspergillus pseudocaelatus*, *Aspergillus pseudonomius*, and *Aspergillus nomius* (41).

1.4.3.2 Fumonisin

The fumonisins are a group of mycotoxins derived from *Fusarium* of the *Liseola* section (42). They have strong structural similarity to sphinganine, the backbone precursor of sphingolipids. Sphinganine is shown in Figure 1.8.

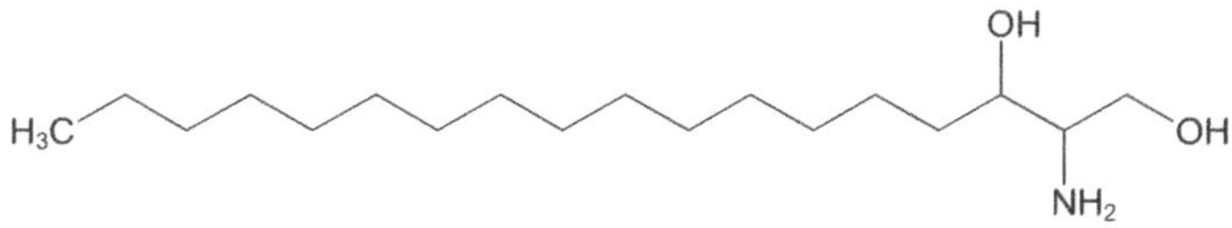

Figure 1.8 Sphinganine.

1.4.3.3 Ochratoxin

Ochratoxins are a group of mycotoxins produced by some *Aspergillus* species (mainly *A. ochraceus* and *A. carbonarius*, but also by 33% of *A. niger* industrial strains) and some *Penicillium* species, especially *P. verrucosum* (43). Ochratoxin A is the most prevalent and relevant fungal toxin of this group, while ochratoxins B and C are of lesser importance.

Ochratoxin A is known to occur in commodities such as cereals, coffee, dried fruit, and red wine. It is possibly a human carcinogen and is of special interest as it can be accumulated in the meat of animals. Exposure to ochratoxins through diet can cause acute toxicity in mammalian kidneys. Exposure to ochratoxin A has been associated with Balkan endemic nephropathy, a kidney disease with high mortality in people living near tributaries of the Danube River in Eastern Europe (44).

It has been suggested that carriers of alleles associated with phenylketonuria may have been protected from spontaneous abortion caused by ochratoxin exposure, providing a heterozygous advantage for the alleles despite the possibility of severe intellectual disability in the more rare instance of inheritance from both parents (43).

1.4.3.4 Zearalenone

Zearalenone (ZEN), also known as RAL and F-2 mycotoxin, is a potent estrogenic metabolite produced by some *Fusarium* and *Gibberella* species (45). The chemical name of zearalenone is 6-[10-hydroxy-6-oxo-trans-1-undecenyl]-β-resorcyclic acid lactone. Specifically, the *Gibberella zeae*, the fungal species where zearalenone was initially detected, in its asexual/anamorph stage is known as *Fusarium graminearum* (46). Several *Fusarium* species produce toxic substances of considerable concern to livestock and poultry producers, namely deoxynivalenol, T-2 toxin, HT-2 toxin, diacetoxyscirpenol and zearalenone.

In particular, zearalenone is produced by *Fusarium graminearum, Fusarium culmorum, Fusarium cerealis, Fusarium equiseti* (47), *Fusarium verticillioides* (48), and *Fusarium incarnatum*. Zearalenone is the primary toxin that binds to estrogen receptors, causing infertility, abortion or other breeding problems, especially in swine (48). Often, zearalenone is detected together with deoxynivalenol in contaminated samples and its toxicity needs to be considered in combination with the presence of other toxins (49).

Studies in animal models suggest that zearalenone is metabolized primarily to α-zearalenol and β-zearalenol. α-Zearalenolis metabolized predominantly into β-zearalenol and, to a lesser extent, into zearalanone. Some compounds are shown in Figure 1.9.

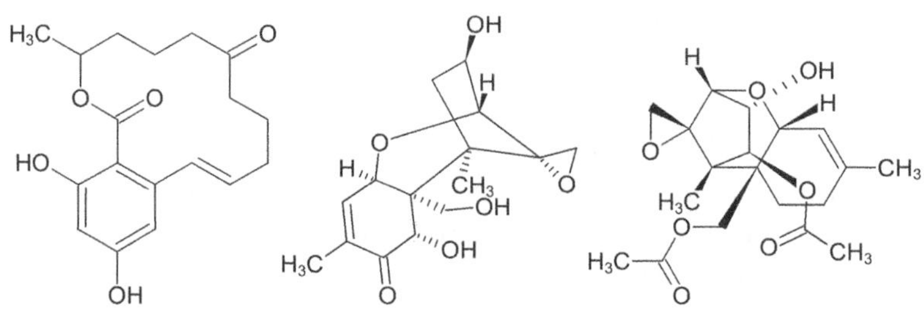

Zearalenone Deoxynivalenol Diacetoxyscirpenol

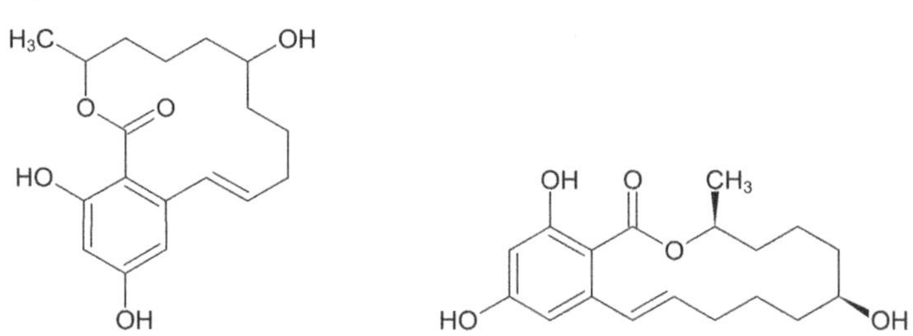

α-Zearalenol β-Zearalenol

Figure 1.9 Estrogenic metabolites.

Zearalenone is heat-stable and is found worldwide in a number of cereal crops such as maize, barley, oats, wheat, rice, and sorghum (50–52).

Its production increases when the climate is warm with air humidity at or above twenty percent (48). The environmental pH plays also a role in the toxin's production. When temperatures fall to 15°C, alkaline soils still support ZEN production. At the preferred *Fusarium* temperature, which ranges between 25°C and 30°C, neutral pH results in the greatest toxin production (53).

1.4.3.5 Brominated Diphenyl Ethers

Details of brominated diphenyl ethers are shown in Table 1.7 and in Figure 1.10

Table 1.7 Brominated diphenyl ethers.

Abbreviation	Name
BDE 15	4,4′-Dibromodiphenyl ether
BDE 28	2,4,4′-Tribromodiphenyl ether
BDE 36	3,3′,5-Tribromodiphenyl ether
BDE 37	3,4,4′-Tribromodiphenyl ether
BDE 47	2,2′,4,4′-Tetrabromodiphenyl ether
BDE 49	2,2′,4,5′-Tetrabromodiphenyl ether
BDE 66	2,3′,4,4′-Tetrabromodiphenyl ether
BDE 71	2,3′,4′,6-Tetrabromodiphenyl ether
BDE 75	2,4,4′,6-Tetrabromodiphenyl ether
BDE 77	3,3′,4,4′-Tetrabromodiphenyl ether
BDE 85	2,3′,4,4′,6-Pentabromodiphenyl ether
BDE 99	2,2′,4,4′,5-Pentabromodiphenyl ether
BDE 100	2,2′,4,4′,6-Pentabromodiphenyl ether
BDE 119	2,3′,4,4′,6-Pentabromodiphenyl ether
BDE 126	3,3′,4,4′,5-Pentabromodiphenyl ether
BDE 138	2,2′,3,4,4′,5′-Hexabromodiphenyl ether
BDE 153	2,2′,4,4′,5,5-Hexabromodiphenyl ether
BDE 154	2,2′,4,4′,5,6′-Hexabromodiphenyl ether
BDE 181	2,2′,3,4,4′,5,6-Heptabromodiphenyl ether
BDE 183	2,2′,3,4,4′,5′,6-Heptabromodiphenyl ether
BDE 203	2,2′,3,4,4′,5,5′,6-Octabromodiphenyl ether
BDE 205	2,3,3′,4,4′,5,5′,6-Octabromodiphenyl ether
BDE 206	2,2′,3,3′,4,4′,5,5′,6-Nonabromodiphenyl ether
BDE 207	2,2′,3,3′,4,4′,5,6,6′-Nonabromodiphenyl ether
BDE 209	Decabromodiphenyl ether

On the other hand, plastics have additives that are intentionally added during their manufacture to give them color, or other properties, such as flexibility, termoresistance, UV resistance, and fire resistance.

Antimicrobial food packaging incorporating triclosan (5-chloro-2-(2,4-dichlorphenoxy)-phenol) was banned in the European Union in 2010 (54). In other countries, regulations on the use of triclosan in food contact materials are unclear.

In this context, compliance with the EU ban on triclosan was investigated in commercial antibacterial PP food containers sold through online sales platforms. Triclosan was present in all food containers tested. Migration studies to food simulants under different conditions (e.g., conventional and microwave heating, single and repeated use) have been performed according to Regulation 10/2011.

The highest triclosan migration levels corresponded to 95% ethanol at 40°C for 10 *d*, although migration rates varied between different containers. Furthermore, microplastics were found to leach from the surface of antibacterial food containers when exposed to microwave heating or oven heating. Therefore, microplastics containing triclosan might enter food in contact with the containers during use, posing a potential health risk for consumers (54).

1.4.4 Carcinogens

1.4.4.1 Carcinogen Contamination

Carcinogen contamination in the food chain, for example, heavy metal ions, pesticides, acrylamide, and mycotoxins, have caused serious health problems (55). A major objective of food safety research is the identification and prevention of exposure to these carcinogens, because of their impossible-to-reverse tumorigenic effects.

However, the detection of carcinogens is difficult because of their trace-level presence in food. Thus, reliable and accurate separation and determination methods are essential to protect food safety and human health.

The state of the art was summarized for separation and determination methods for analyzing carcinogen contamination, especially the advances in biosensing methods. Furthermore, the application

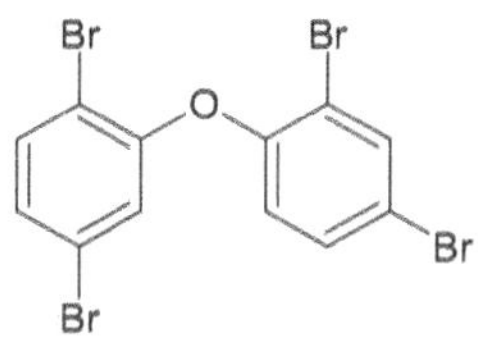

2,2′,4,5′-Tetrabromo-diphenyl ether

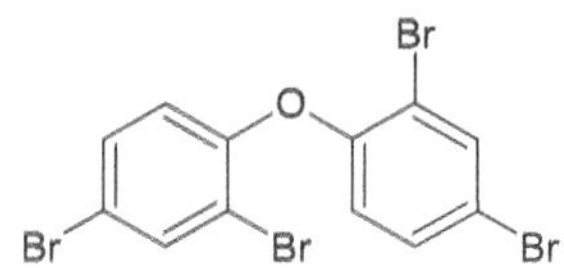

2,2′,4,4′-Tetrabromo-diphenyl ether

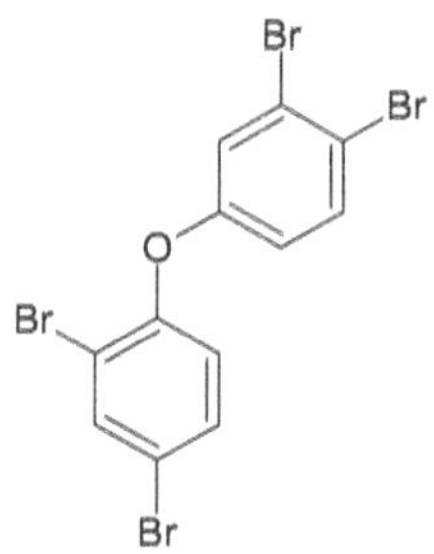

2,3′,4,4′-Tetrabromo-diphenyl ether

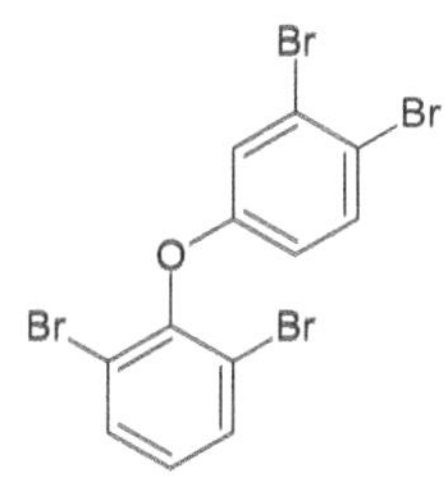

2,3′,4′,6-Tetrabromo-diphenyl ether

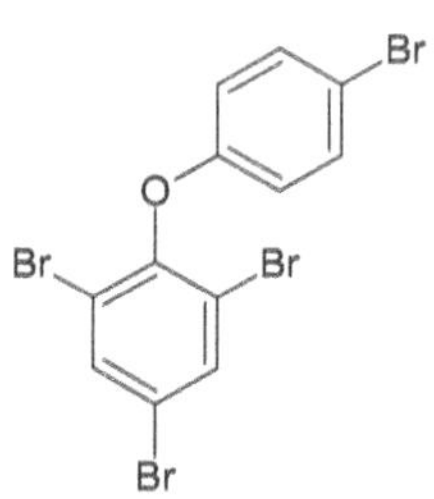

2,4,4′,6-Tetrabromo-diphenyl ether

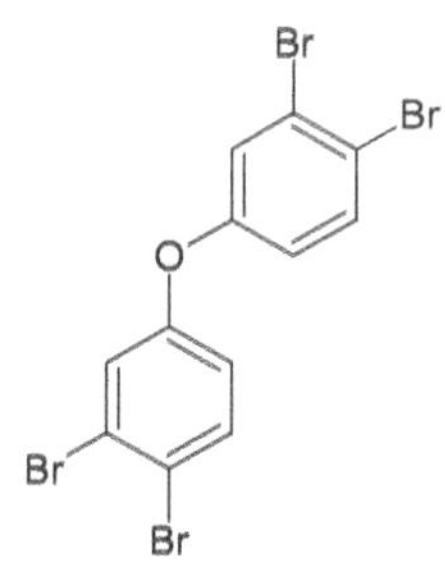

3,3′,4,4′-Tetrabromo-diphenyl ether

Figure 1.10 Brominated diphenyl ethers.

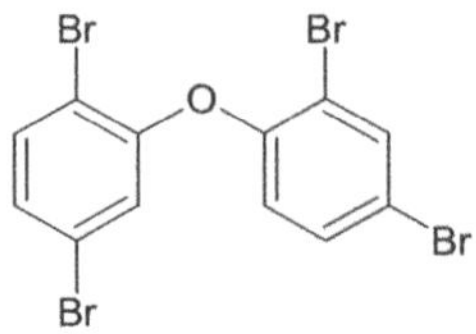

2,2′,4,5′-Tetrabromo-
diphenyl ether

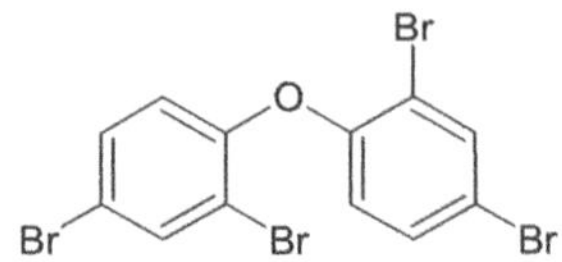

2,2′,4,4′-Tetrabromo-
diphenyl ether

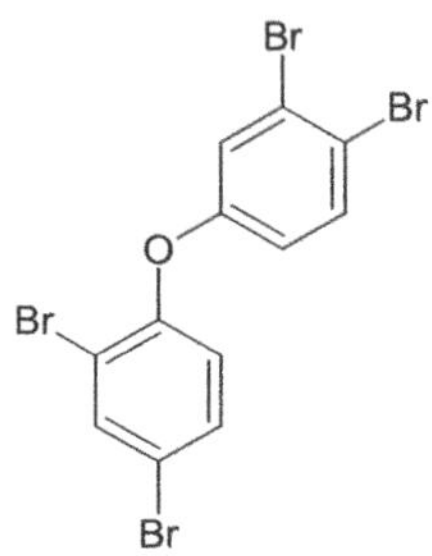

2,3′,4,4′-Tetrabromo-
diphenyl ether

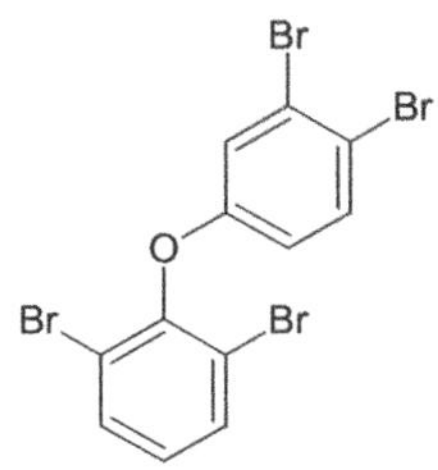

2,3′,4′,6-Tetrabromo-
diphenyl ether

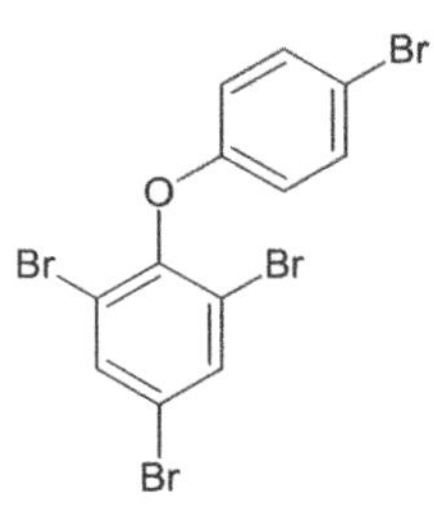

2,4,4′,6-Tetrabromo-
diphenyl ether

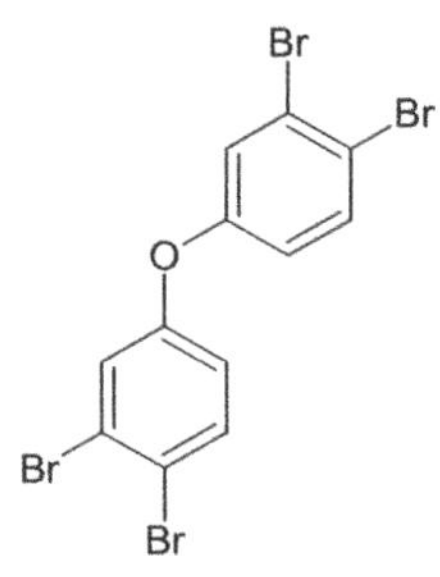

3,3′,4,4′-Tetrabromo-
diphenyl ether

Figure 1.10 (cont) Brominated diphenyl ethers.

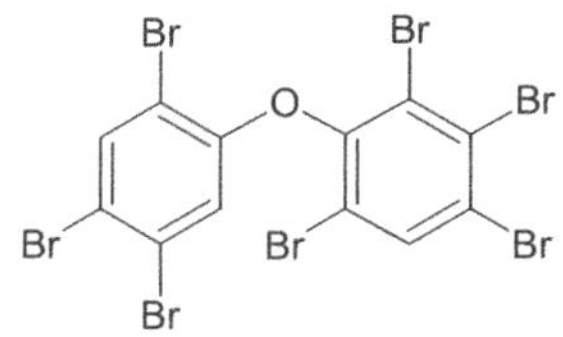

2,2′,3,4,4′,5′,6-Heptabromo-
diphenyl ether

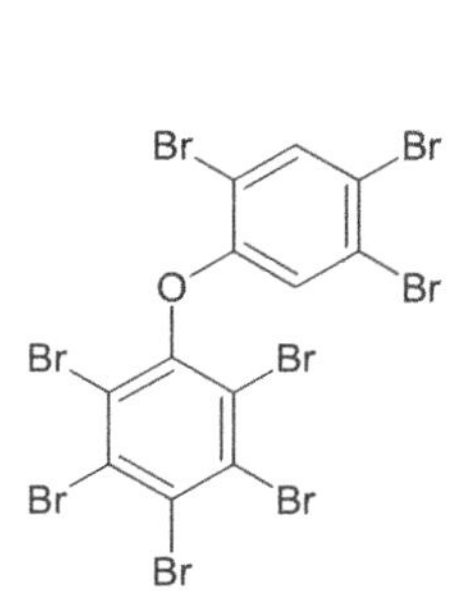

2,2′,3,4,4′,5,5′,6-Octabromo-
diphenyl ether

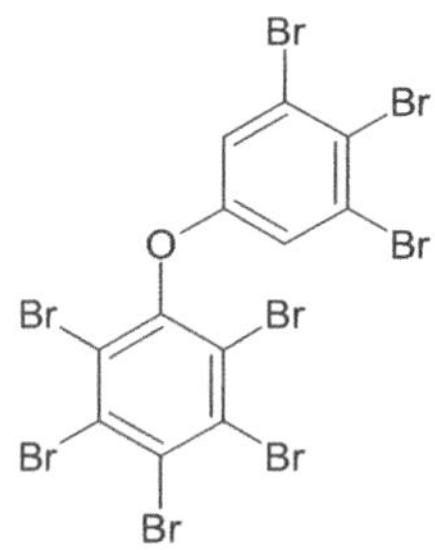

2,3,3′,4,4′,5,5′,6-Octabromo-
diphenyl ether

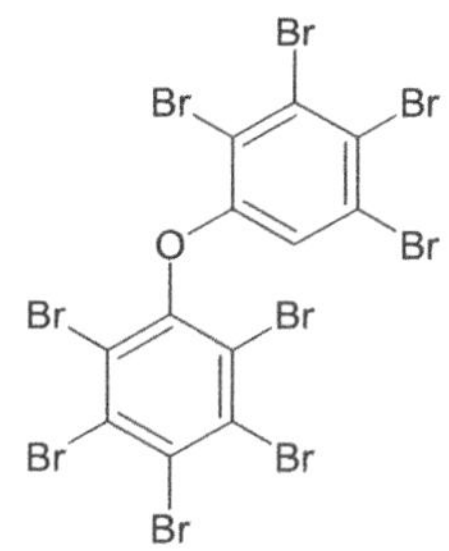

2,2′,3,3′,4,4′,5,5′,6-Nonabromo-
diphenyl ether

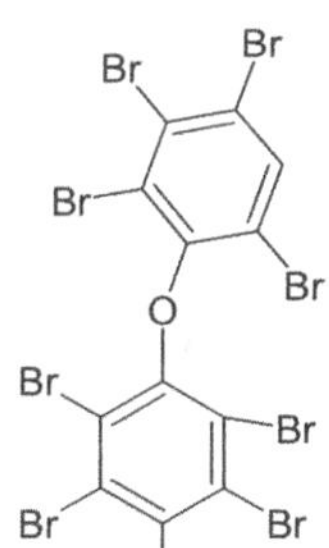

2,2′,3,3′,4,4′,5,6,6′-Nonabromo-
diphenyl ether

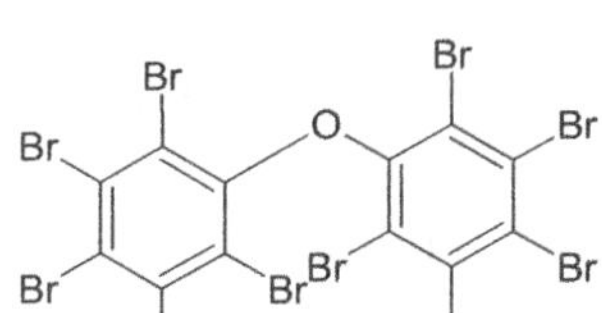

Decabromo-
diphenyl ether

Figure 1.10 (cont) Brominated diphenyl ethers.

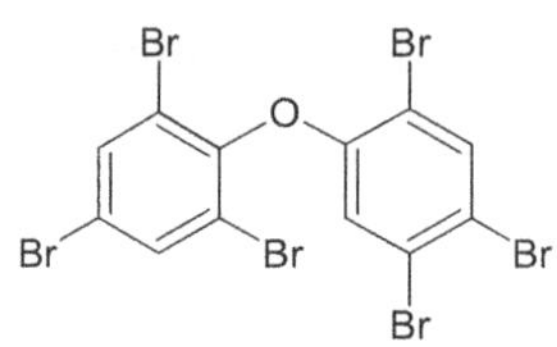

2,2′,4,4′,5,6′-Hexabromo-
diphenyl ether

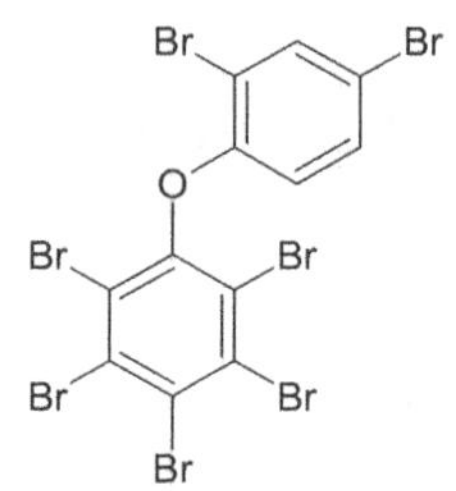

2,2′,3,4,4′,5,6-Heptabromo-
diphenyl ether

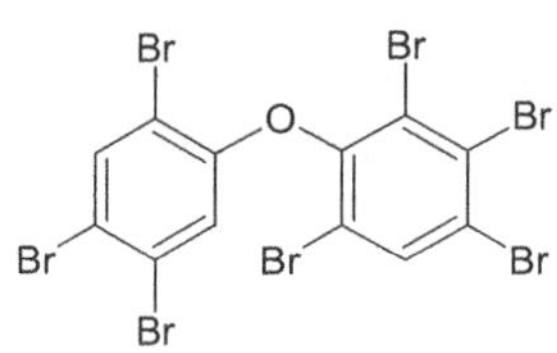

2,2′,3,4,4′,5′,6-Heptabromo-
diphenyl ether

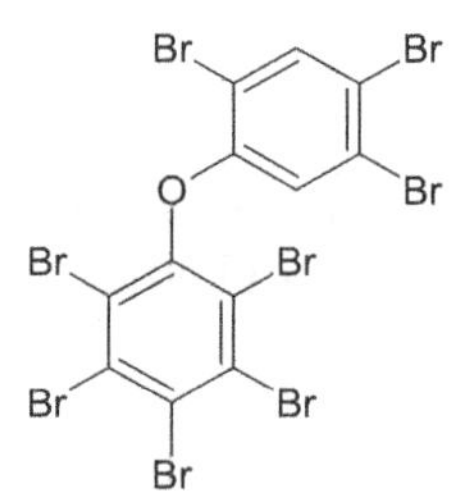

2,2′,3,4,4′,5,5′,6-Octabromo-
diphenyl ether

Figure 1.10 (cont) Brominated diphenyl ethers.

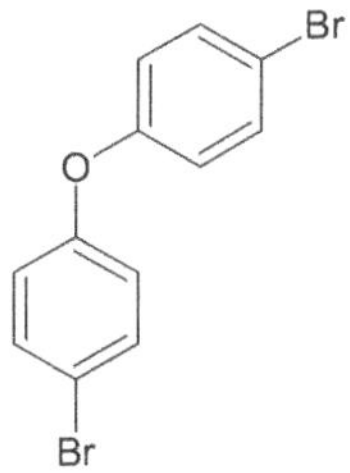

4,4′-Dibromo-
diphenyl ether

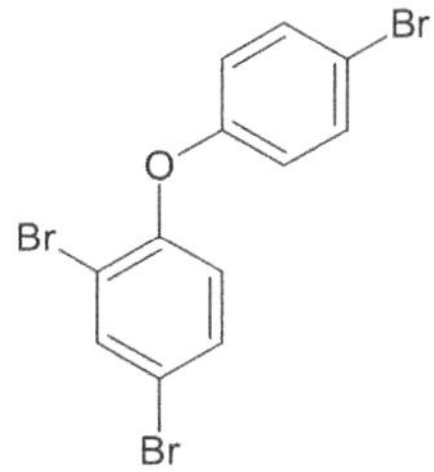

2,4,4′-Tribromo-
diphenyl ether

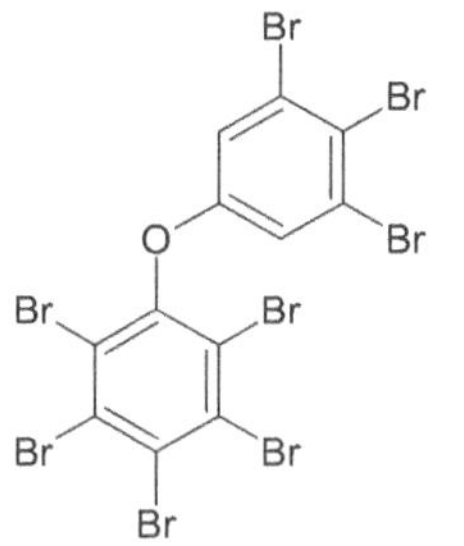

2,3,3′,4,4′,5,5′,6-Octabromo-
diphenyl ether

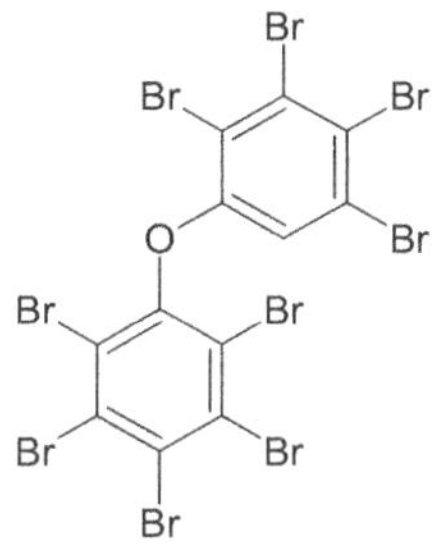

2,2′,3,3′,4,4′,5,5′,6-Nonabromo-
diphenyl ether

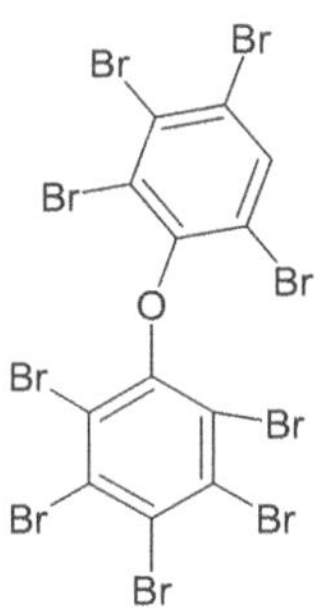

2,2′,3,3′,4,4′,5,6,6′-Nonabromo-
diphenyl ether

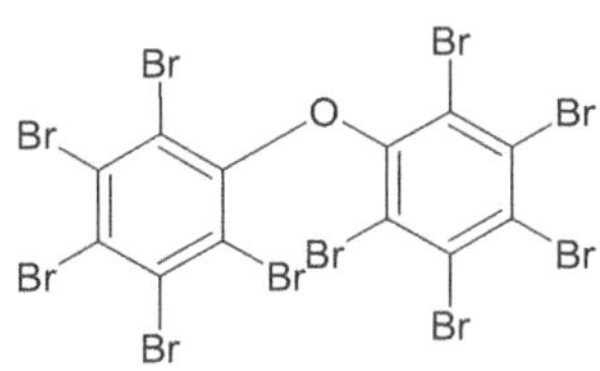

Decabromo-
diphenyl ether

Figure 1.10 (cont) Brominated diphenyl ethers.

5

of promising technology including nanomaterials, imprinted polymers, and microdevices was detailed (55).

1.4.4.2 *Reduction of Carcinogens*

A wide variety of foods have tested positive for the presence of acrylamide (56). Acrylamide has especially been found primarily in carbohydrate food products that have been heated or processed at high temperatures. Examples of foods that have tested positive for acrylamide include coffee, cereals, cookies, potato chips, crackers, french-fried potatoes, breads and rolls, and fried breaded meats.

In general, relatively low contents of acrylamide have been found in heated protein-rich foods, while relatively high contents of acrylamide have been found in carbohydrate-rich foods, compared to non-detectable levels in unheated and boiled foods. Reported levels of acrylamide found in various similarly processed foods include a range of 330-2,300 $\mu g\, kg^{1}$ in potato chips, a range of 300-1100 $\mu g\, kg^{1}$ in french fries, a range of 120-180 $\mu g\, kg^{1}$ in corn chips, and levels ranging from not detectable up to 1400 $\mu g\, kg^{1}$ in various breakfast cereals.

It is believed that acrylamide is formed from the presence of amino acids and reducing sugars. For example, it is believed that a reaction between free asparagine, an amino acid commonly found in raw vegetables, and free reducing sugars accounts for the majority of acrylamide found in fried food products. Asparagine is shown in Figure 1.11. Asparagine accounts for approximately 40% of the total

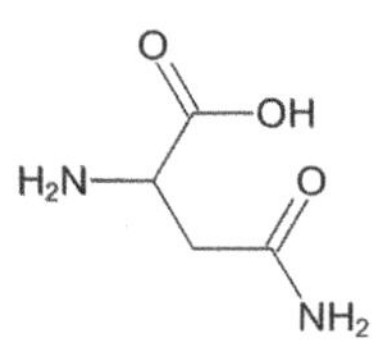

Figure 1.11 Asparagine.

free amino acids found in raw potatoes, approximately 18% of the total free amino acids found in high protein rye, and approximately 14% of the total free amino acids found in wheat.

The most likely route for the formation of acrylamide involves a Maillard reaction (56). The Maillard reaction is a chemical reaction between amino acids and reducing sugars that gives browned food its distinctive flavor. It is named after the French chemist Louis Camille Maillard, who first described it in 1912 while attempting to reproduce biological protein synthesis (57).

The Maillard reaction has long been recognized in food chemistry as one of the most important chemical reactions in food processing and can affect flavor, color, and the nutritional value of the food (56). The Maillard reaction requires heat, moisture, reducing sugars, and amino acids.

The first step of the Maillard reaction involves the combination of a free amino group from free amino acids or proteins, with a reducing sugar, e.g., glucose, to form Amadori or Heyns rearrangement products.

The second step involves the degradation of the Amadori or Heyns rearrangement products via different alternative routes involving deoxyosones, fission, or Strecker degradation. A complex series of reactions, including dehydration, elimination, cyclization, fission, and fragmentation, results in a pool of flavor intermediates and flavor compounds.

The third step of the Maillard reaction is characterized by the formation of brown nitrogenous polymers and copolymers.

Figure 1.12 shows the pathways for the formation of acrylamide starting with asparagine and glucose.

A process and a method for reducing the amount of acrylamide in thermally processed foods has been developed (56). This allows the production of foods with significantly reduced levels of acrylamide.

The method provides a continuous feed of peeled and sliced raw potatoes and contacting the continuous feed of raw potato slices with an aqueous solution at about 60°C for about 5 *min* to reduce the amount of the acrylamide precursors in the raw potato slices (56).

Using another method, a continuous feed of peeled and sliced raw potatoes, par-frying the raw potato slices at about 171°C to about 182°C until the moisture content is reduced to 3% to 10%, then oven-drying the par-fried slices at less than about 120°C until the moisture content is further reduced to about 1% to 2%. Also, some other methods have been detailed (56).

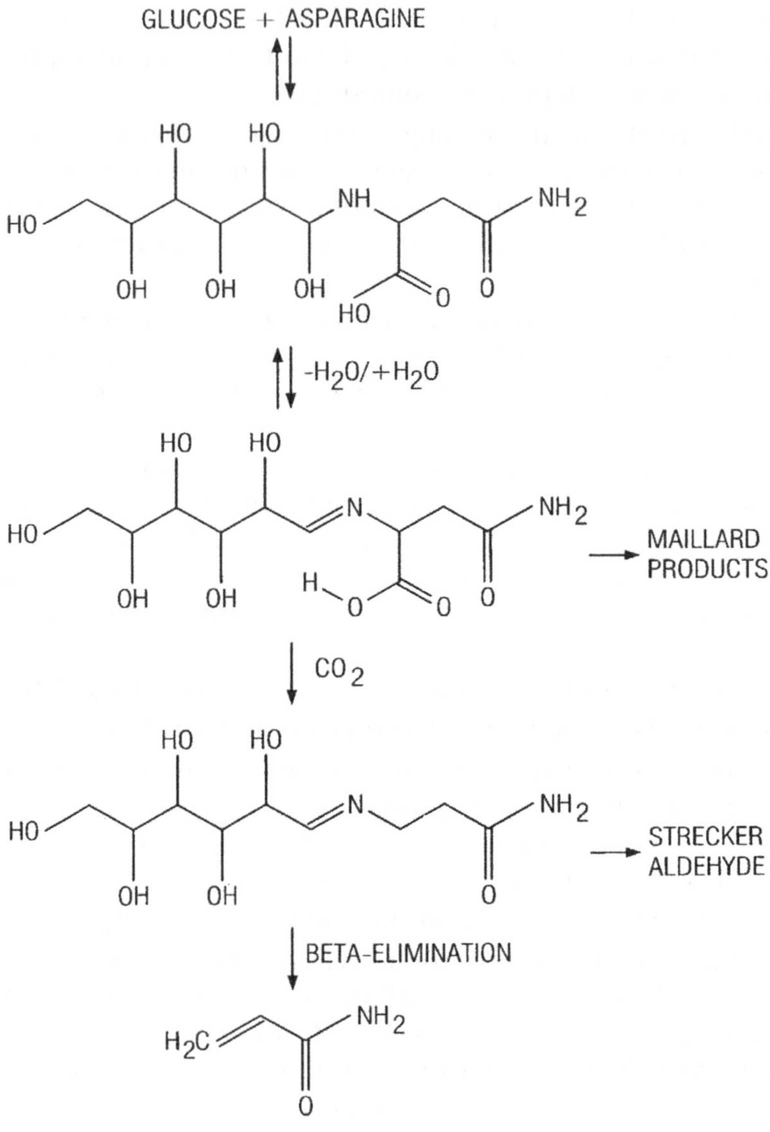

Figure 1.12 Formation of acrylamide (56).

L-Tryptophan can be applied to a foodstuff to prevent the development of mutagens and carcinogens (58). L-Tryptophan is shown in Figure 1.13. Before cooking of a foodstuff, such as hamburger,

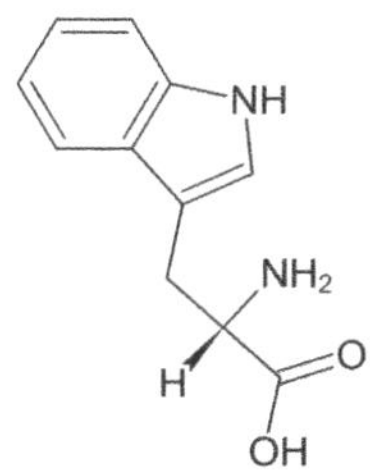

Figure 1.13 L-Tryptophan.

L-tryptophan is applied to the surfaces thereof to inhibit the generation of IQ-type carcinogens (58).

The L-tryptophan can be sprinkled on the surface of the foodstuff or incorporated into a sauce which is applied to the foodstuff or put into solution in water (58). The L-tryptophan is preferably applied in an amount greater than the order of magnitude of 1.0 $mg\,cm^{-2}$.

For testing, patties with a thickness of 3 *mm* and 9.6 *cm* were prepared from 50 *g* of store-bought 85% lean ground beef using a glass Petri dish cover with 1.5×9.6 *cm* as a mold. Inhibitor-treated and untreated patties were then either fried or broiled for 5 *min*.

When the concentration of added L-tryptophan was varied from 1.75 to 140 *mM* in the complete liquid reflux mode, IQ-type mutagenicity was inhibited, in a dose-dependent fashion, over a range of 15 to 100% (58).

1.4.4.3 Biosensors

Common substances in food can pose a great threat to human health, including foodborne pathogens, heavy metals, mycotoxins, pesticides, herbicides, veterinary drugs, allergens and illegal additives (59). In order to develop rapid, low-cost, portable and on-site detection methods of those contaminants and allergens to ensure food safety, gold nanoparticles of versatile shapes and morphologies, such as nanorods, nanoclusters, nanoflowers, nanostars, nanocages, nanobipyramids and nanowires, have been employed as probes because they possess extraordinary properties that can be used to design biosensors that enable the detection of various contaminants and allergens.

By means of a surface modification, gold nanoparticles can directly or indirectly sense specific targets based on different mechanisms, such as hydrogen bonds, nucleic acid hybridization, aptamer-target binding, antigen antibody recognition, enzyme inhibition, and enzyme mimicking activity. Gold nanoparticles can induce a distinct color change from red to blue when they transform from a monodispersed state to an aggregated state in liquid solution, which can be observed by the naked eye. If Raman molecules are functionalized on gold nanoparticles, their aggregation will alter the interparticle distance and induce the surface-enhanced Raman scattering that can be employed for highly sensitive detection.

Ultra-small gold nanoparticles, such as gold nanoclusters, also feature in fluorescence that enable a fluorescent readout. The formats of gold nanoparticles for food safety detection in real-world range broadly include but are not limited to films, fibers, liquid solutions, tapes, chips and lateral flow strips (59).

1.4.4.4 Anti-Carcinogens

The proliferation of cancer cells by deoxyribonucleic acid synthesis is considered as a key biological index that is observed at the progression stage of carcinogenesis (60).

Therefore, substances that can inhibit such a process are expected to inhibit the progression stage of carcinogenesis, thereby preventing as well as inhibiting the development of cancer (61–63).

A variety of foods, such as coffee, tea, coke and chocolate, contain caffeine, c.f. Figure 5.4, a xanthine derivative.

Theobromine is also a xanthine derivative. Unlike caffeine, it is abundant in cacao bean or cacao bean husk. In general, cacao bean contains about 1.5%-3% of theobromine and cacao bean husk contains about 2% of theobromine. Xanthine derivates are shown in Figure 1.14.

Processed food such as chocolate, which is made from cacao bean, contains about 0.2% to 0.5% of theobromine and this content is about 7–10 times larger than that of caffeine (64).

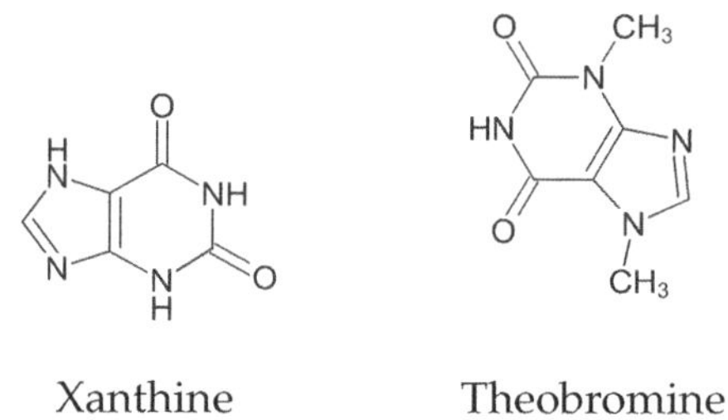

Figure 1.14 Xanthine derivates.

The general pharmacological functions of xanthine derivatives include stimulation of the central nervous system, skeletal muscle and cardiac muscle as well as relaxation of smooth muscle and coronary artery, accentuation of secretion of gastric juice, and diuresis.

By contrast, although caffeine is present in a variety of foods and has a wide scope of pharmacological activities, theobromine is only abundantly present in cacao bean and cacao bean husk, and thus the studies carried out on theobromine have not been extensive. Caffeine has been reported to have anti-carcinogenic activities; for example, a synergistic effect on therapeutic agent for gastric cancer (65), an inhibitory effect on lung cancer (66), and an inhibitory effect on 12-*o*-tetradecanoyphorbol-13-acetate-induced carcinogenesis (67, 68). 12-*o*-Tetradecanoyphorbol-13-acetate is shown in Figure 1.15.

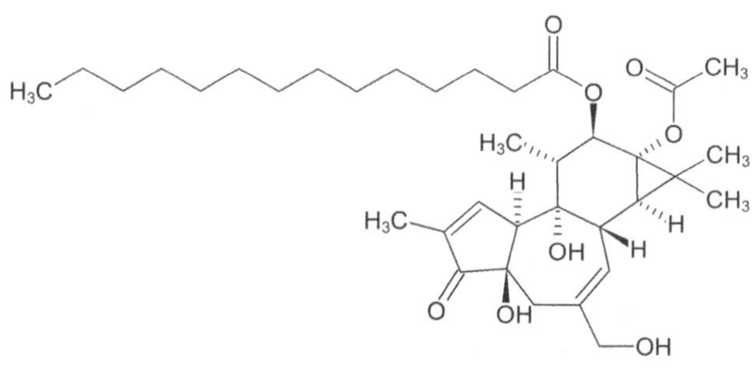

Figure 1.15 12-*o*-Tetradecanoyphorbol-13-acetate.

The isolation of theobromine from cacao bean can be done as

follows (60):

Preparation 1–1: Cocoa butter was removed from cacao bean containing cacao bean husk. Then 6 parts by weight of 50% acetone solution was added to 1 part by weight of the remaining fraction (cacao mass) and agitation extraction was performed for 5 *h* at 60°C while refluxing. The extract was centrifuged for 30 *min* at 4°C at a rate of 8,000 *rpm* and the resulting supernatant was collected. The extraction was repeated once for the remnant and all the supernatant was combined. Theobromine was isolated from the extract by high-performance liquid chromatography according to the method of Kreiser.

1.5 Methods of Food Packaging

Food packaging is an essential part of modern food processing. The issues of food packaging have been described in a chapter of a monograph (69).

During recent years, there has been a continuing modernization of agricultural activity, focused on the increase of productivity and quality, which has allowed a continuous development and the capacity to adjust the offer to the evolution of consumer habits and preferences (70, 71). Within this context, there has been special interest in the incorporation of quality, food health and safety standards, the strengthening of local zoo and phytosanitary heritage, environmental concern, corporate social responsibility, technological innovation and market management. All of which are instruments that favor production differentiation. Fresh exportable products present high perishability, which dramatically affects their quality when arriving at distant markets, thus reducing their value. Therefore, to extend their lifespan and maintain food safety, the quality of fresh products to be exported is a constant challenge to the industry.

Basically, two problems must be faced when seeking to keep fruits and vegetables fresh. First, they are alive and many chemical and biochemical reactions are occurring. Some reactions, if not controlled, may rapidly lead to the aging of vegetables and the loss of freshness. Secondly, the risk of microbiological contamination must be delayed as long as possible, since it compromises food safety, as well as the product final quality. Microbiological growth may cause serious alterations such as smell, funny taste and/or alteration in color and texture of vegetables (72).

Berries are particularly highly perishable and present important post-harvest physiological activity. As a consequence, they have short ripening and aging periods, which makes their commercialization a real challenge. They're susceptible to developing post-harvest diseases, to the point that some symptoms can be observed only 12 hours after being kept at room temperature, especially if they are wet. In general, the lifespan of berries is limited due to diseases caused by fungi, particularly *Botrytis cinerea* (70).

Studies in the USA showed that annual loss of fresh berries through the commercialization chain can rise to 40% of national production, which causes many millions of dollars in loss, even though that country is known for its high technology and for presenting different means for refrigerated transport. *Botrytis cinerea* is the agent that causes the gray mold rot, which infects more than 200 different vegetal species, producing serious economic loss before and after picking. This pathogen may attack the crop in any of its developing states and can infect any part of the plant.

Due to the considerable impact of this pathogen and to the economic consequences on important crops such as grapes, tomatoes, strawberries, raspberries, cranberries, among others, there have been an important number of studies on the biology of this fungus, the interactions in which it participates and on possible methods to control it (73).

Although berries may be of a high quality, they have a short lifespan, mainly because of fungal growth (*Botrytis*), which causes an important loss due to product rejection and the poor quality image in the target markets. On the other hand, berries are exported in packages not in accordance with international market trends. Considering international trends, it is essential to develop recyclable and/or active biodegradable packaging with antimicrobial capacity, allowing for an increase in the lifespan of berries, as well as their quality and safety, at a competitive price, while taking into consideration the packaging traditionally used by the industry nowadays. All this is done without significantly modifying the present packaging production line or their common use form.

Controlling physiological processes and the growth of microorganisms is the key to the preservation and long lifespan of fruits and vegetables. That way, packaging systems and low temperatures are technologies that contribute to the extension of the lifespan, quality and safety of these products.

Fresh fruits are mostly exported using modified atmosphere packaging (MAP) technology in order to reduce metabolic processes and bioburden. Usually, entire coverage for pallets is used to contain the modified atmosphere. These conditions, however, are complex in practice, expensive, and may be affected by external factors, such as blackouts or leaks, often making them a nonviable process to maintain for periods over 30-35 days, as in the case of blueberries. The main difficulty of this technology informed by some berry exporting companies is the variation in the gaseous mixtures in different pallets, making the homogenization of the exported load a difficult task (70).

At present, one of the main requirements in the food industry is the development of noninvasive procedures that may cause minimum alteration in food physicochemical and sensory characteristics. In this context, active packages have emerged as an excellent alternative to complement these requirements, since they extend the lifespan of food, minimize additive incorporation and improve food safety with minimum alteration of its characteristics, providing the chance to develop new products.

Besides which, active packages can be combined with other technologies, adding a synergic effect, increasing the quality and lifespan of the product. Not long ago, packaging had a merely passive function, which consisted of separating the product from the environment, acting as a shield against physical and/or mechanical effects, chemical reactions and the invasion of microorganisms, as well as preventing migration of their own compounds into food, with minimal interaction with the food, and therefore becoming an inert shield. In the last few years, a dramatic step forward has occurred in the area of food packaging concerning new packaging techniques.

From a traditional system focused on the search for the maximum inertia between package and food, there has been a turn to the development of a kind of packaging called active packaging; that is, packages which interact with the food in order to improve its properties. Thus, active packaging can be defined as a system in which food, packaging and environment act positively to increase the lifespan or improve some property of the food. This has eventually led to the concept of packaging as an emerging technology in food preservation (70).

However, in recent years, with the demand for fresh and better quality products, the concept of packaging has changed to offer active protection to the product. Thus, active packaging can be defined as a system in which food, packaging and the environment act positively to increase the lifespan or improve the sole property of the food (74).

The packaging can be called active when it performs a desired role, as well as providing an inert shield against external conditions. The word *desired* is important for this definition, because it makes a clear difference between desired and non-desired effects. The most important active packaging concepts are: antimicrobial packaging; edible covering; oxygen, ethylene, odor and taste absorbers; moisture regulators; carbon dioxide, antimicrobial materials, antioxidant and flavor releasers. Their applications are many and their use is clearly growing. The active packaging system has been used for years in Japan and in recent years its use has been growing in Europe and the United States. Consequently, active packaging technology is an emerging promising area in food technology, which may contribute several preservation benefits in a broad variety of products.

The current concern as to how to improve and guarantee safety in processed food has raised great interest in the use of active antimicrobial packaging. Lately, prominent food packaging research teams worldwide along with leading companies in the development of packaging are expending significant effort to develop active packages with antimicrobial capacity. Nevertheless, many patented developments are targeted to the use of compounds whose use in food packaging is conditioned, since the legislation in many countries forbids such compounds in direct contact with the food, as is the case of silver ions, for instance.

Within the concept of active packaging, antimicrobial packaging is defined as packaging which incorporates active antimicrobial agents instead of the same substances added directly on the food. The concept of antimicrobial packaging assumes the gradual release of the antimicrobial agent from the packaging to the food, inhibiting or delaying microorganism growth on its surface (75).

As the microbiological contamination of most food occurs mainly on the surface due to the processing and subsequent manipulation, efforts have been made to improve safety, delaying food deterioration through the use of antimicrobial materials applied with aerosols or by immersion. However, direct superficial application of antibacterial substances on food has limited benefits, since active substances are neutralized upon contact, or are rapidly diffused from the surface to the interior of food. On the other hand, incorporation of a bactericidal or bacteriostatic substance in food formulations may result in partial inactivation of the active substances by constituents of the product itself, and a limited effect on the surface microflora can be expected (76).

Therefore, the use of active packaging which contains antimicrobial agents could be more efficient, due to the slow migration of agents from the packaging material to the surface of the product, helping to keep high concentrations where they are more necessary. If an antimicrobial agent can migrate from the packaging for a long period of time, its activity can also be extended to the stages of transportation and storage during product distribution. From the point of view of food security, antimicrobial substances used in developing active packaging must be approved to be in contact with the food, and must, therefore, be considered *generally considered as safe* by the FDA. In this regard, it is important to highlight that antimicrobial materials must be incorporated into food packaging in such concentrations so that the released concentration is allowed by the regulation.

The most commonly used food packaging antimicrobial agents are: sorbic acid, propionic acid, potassium sorbate, benzoic acid, sodium benzoate and citric acid (76). Bacteriocins, such as nisin and pediocin, have also been used (77), besides other compounds, among which are enzymes, such as peroxidase and lysozyme, the latter used to inhibit microbial growth of Gram-positive bacteria (78); natural poly(saccharide)s with antimicrobial properties such as chitosan (79); phenolic compounds extracted from plants; chelating agents such as ethylenediamine tetraacetic acid (EDTA), and metals such as silver ions. Ethylenediamine tetraacetic acid is shown in Figure 1.16.

The triclosan compound (2,4,4′-trichloro-2′-hydroxydiphenyl ether), c.f. Figure 6.1, was approved for food contact use in the European Union by the Scientific Committee on Food (SCF, 2000) with a limit of 5 *mg/kg* of food. Antimicrobial agents in packaging

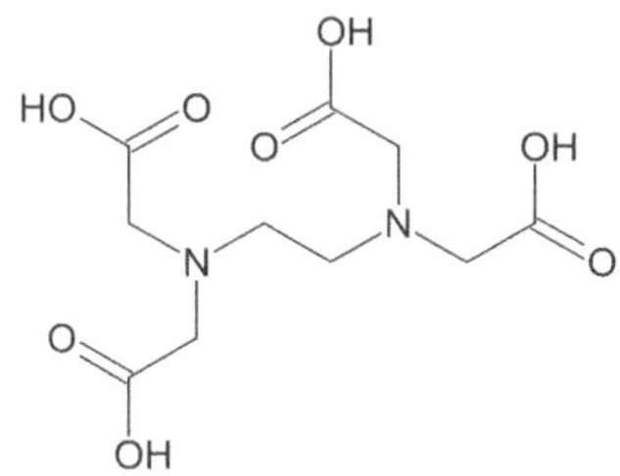

Figure 1.16 Ethylenediaminetetraacetic acid.

are able to control microbiological contamination by use of three mechanisms: reducing microorganism growth rate, increasing the lag phase, or inactivation by contact (76). Antimicrobial materials can act by inhibiting essential metabolic routes, or by intervening in the microorganism's reproductive genetics, while other compounds modify the functioning of the cell membrane or the cell wall structure, as is the case in lysozyme, which destroys the cell wall without modifying metabolic routes, while lactoferrin and EDTA act as essential cation chelating agents (80). Considering that most antimicrobial agents have different action mechanisms, it is thought that mixing different kinds of antimicrobial agents could increase antimicrobial activity through synergistic reactions, provided that they don't interfere with each other. Therefore, it is desirable to experiment with combinations of several antimicrobial agents to extend their activity in a wider range of microorganisms, maximizing the effectiveness and safety of packaging systems with antimicrobial agents. The method employed to incorporate the active agent with antimicrobial capacity into the packaging can be achieved in different ways.

In some cases, the antimicrobial agent is incorporated as a coating in the internal part of the film which will be in direct contact with the food. Also, the antimicrobial agent can be incorporated into the polymer matrix itself, in which case, the agent diffuses through all the amorphous zones of the polymer, or through the holes and empty spaces between polymeric chains. Undoubtedly, incorporation into the polymeric matrix requires studies on the subject that define the effective concentration levels of antimicrobial agents, kinetics of the agent diffusion, distribution of the agent in the matrix and the effective action kinetics on microorganisms.

On the other hand, the methodology to incorporate the active agent into the plastic film limits the possibilities to use antimicrobial agents. Thus, when the agent is incorporated into the polymeric matrix by using extrusion, its resistance to high temperatures and high pressure becomes essential. Active packaging technology is an emerging promising area in food technology that can contribute several benefits to the preservation of a broad range of products. Within the concept of active packaging, antimicrobial packaging is defined as packaging that incorporates active antimicrobial agents in its structure, avoiding the addition of these substances directly on the food. The concept of antimicrobial packaging consists of the gradual release of the antimicrobial agent from the package to the food, inhibiting or delaying microorganism growth on the food surface. Active packaging application for preserving food has been used for extending the lifespan and increasing the quality of various products such as carrots, broccoli, strawberries, cheese, garlic, among others.

At present, there is great international interest in the use of active packaging related to the food industry, among which is packaging with antimicrobial capacity. Nonetheless, most research studies are focused on the incorporation of antimicrobial agents in non-recyclable or non-biodegradable plastic packaging, or on edible coatings applied directly on the food. The globalization of international trade opens clear possibilities for the strengthening of production development and commercialization of fruit production, particularly those species recently incorporated or grown, that may become relevant in world markets, not as much due to their volume, but due to some special attributes, whether it is quality, opportunity, environmental issues or others.

In this way, quality and sustainability issues represent relatively recent concepts, which have, however, gained prominence in the list of consumer's concerns, and in the minds of authorities, producers and academicians around the world, along with other actors in society.

This new reality, which has gradually started taking over the markets and commercialization of production has involved developing two basic regulatory categories:

1. Those related to product quality, where regulations seek to control harmful effects, both in consumer health and the environment, of the importing country, and
2. Those related to processes, where the regulations include requirements that seek to develop production processes that are harmless to the environment and the safety of workers in the importing countries.

Added to the above is the requirement of processing techniques that can guarantee the harmlessness of the products.

Although this phenomenon has opened new niche markets, such as green markets, ecological markets, environmental seals, quality certifications, among others, it is also true that it is generating new demands for industrial, agricultural, forestry and fishing production, which forces it to promote a strategy based on quality in order to broaden the access to markets and to improve competitive conditions through technological innovation and new investments. Today, one of the most serious environmental concerns is the amount of waste produced and stored as a consequence of packaging and boxing used. Most parts of the materials used by packaging industries come from fossil fuels, which are known to have low degradability. On account of this, food packaging materials, as any other short-term packaging material, represent a serious problem worldwide (81). Thus, it is thought that 25% of world waste comes from packaging. The current global consumption of plastic is more than 200 million tons, with an annual growth of around 5%, which represents the biggest application field of crude oil (82).

In the food packaging industry, to remain competitive and to offer better plastic solutions means being able to deal with some key elements, such as the approval of food contact material by the FDA and waste management, within most of the legislations. Growing environmental awareness demands environmentally friendly use attributes from packaging and processes. As a consequence, recyclability and/or biodegradability of packaging materials are not only functional requirements, but also important environmental attributes. The recycling of solid waste is a common practice in countries committed to sustainable development.

The European Parliament and Council Directive 94/62/EC, of 20 Dec. 1994, requires that a minimum of 15% plastic from packaging and packaging waste be recycled by the end of 2005, while Directive 2004/12/EC of the European Parliament and Council, of 11 Feb. 2004, establishes that a minimum of 22.5% must be recycled by the end of 2011 (83). At present there is great interest in recyclable materials, which are those materials reused by the industry in order to minimize environmental waste or even biodegradable materials, which are defined as those materials that degrade due to the action of microorganisms naturally present in the environment such as bacteria, fungi and algae (84).

Therefore, a great effort to extend the lifespan of food and improve food quality as well as reduce packaging waste has encouraged the exploration of new packaging materials. In view of the background analyzed, it becomes evident that there is enough evidence to support the possibility of using antimicrobial agents in food packaging, especially for dried fruits, in order to improve safety and to extend the product lifespan.

Even though the use of ecologically friendly material is a worldwide demand there are no publications where the application of antimicrobial agents in industrial-level packaging is reported, especially regarding ways of attaching the agent to the packaging and, least of all, the use of packaging that is not only antimicrobial, but environmentally friendly as well. That is the challenge of the present project: to extend the lifespan of berries through the incorporation of an antimicrobial agent to an ecologically friendly material. Although there are only a few patents on antimicrobial food packaging, there are patents concerning active packaging with antioxidant capacity, not antimicrobial, however they present microencapsulation of active agents, i.e., their incorporation into the food itself, in pharmaceutical and cosmetic products.

In particular, the US patent (85) shows active packaging and/or active packaging closing in different formats, made of metal and/or synthetic and/or associated natural materials where at least one element and/or component of the packaging contains and distributes one or several food additives and supplements and/or ingredients or processing aids used to produce food and cosmetics. The additive can fulfill antimicrobial functions regarding the packaged product.

Materials can be made through cryogenic grinding, along with the substance that generates singlet oxygen, ammonium stearate or other surfactant material with properties as soup and a polymeric resin to form a consistent concentrate as a thin homogeneous powder. The concentrate is added to large batches of polymer which are then processed by conventional equipment to create plastic films, cloth, etc.

A preparation method for bioactive packaging materials made of synthetic polymers or polymeric materials from natural sources has been shown (86). It covers the packaging material with a polymeric coating which contains immobilized preservatives, where the application of a solution, or a polymer dispersion with preservatives in the packaging material is by means of spray and/or coating and/or sinking the packaging material in such solution or dispersion and/or fold or carving printing followed by evaporation of the solvent under room or increased temperature, and the preservative is a microbicide and/or bacteriostatic preservative and/or fungicides; antibiotics and/or silver compounds and/or phenol derivatives and/or bacteriocins and/or esters of *p*-hydroxybenzoic acid and/or fatty acid monoglycerides and/or benzoic acid and/or sorbic acid and/or chitosan and/or 2,4,4-trichloro-2-hydroxydiphenylether and/or lysozyme and/or nisin and/or hexamethylenetetramine. Some of these compounds are shown in Figure 1.17.

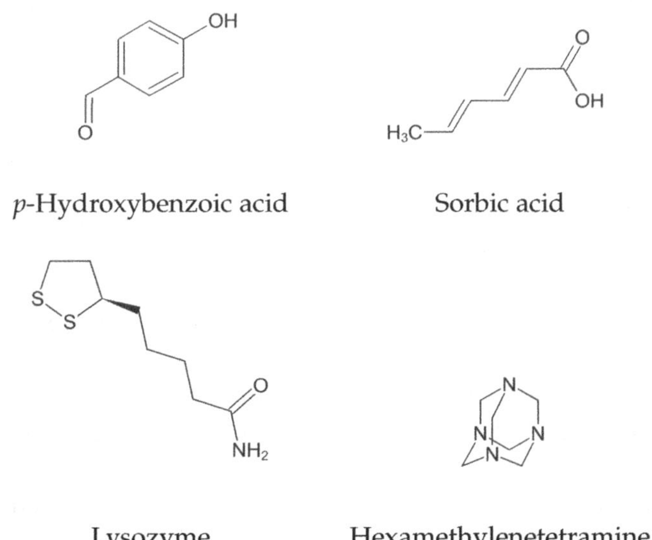

Figure 1.17 Bacteriocins.

An antimicrobial packaging material for food materials may contain between 0.05% to 1.5% weight of an essential oil. The oil can be selected from linalool and/or methyl chavicol, but also one or more from citral, generaniol, methyl cinnamate, methyl eugenol, 1,8-cineol, *trans-α*-bergamot, carvacrol and thymol, c.f. Figure 1.18. These compounds are mixed with one or more selected polymers of ethylene vinyl alcohol copolymer, polyacrylates, which include copolymers of ethyl acrylate, methyl methacrylate, ionomers, nylons and other hydrophilic polymers, or polymers with functional groups capable of partly anchoring the additives, and the mixture covers the food contact layer of a film of food grade packaging or is incorporated into a film of food grade packaging. A binding agent, such as poly(ethylene glycol), is added to the mixture to improve volatile oil retention in the polymer during the process. This material has no regulatory limitations and in the mentioned concentrations, it does not create detectable unwanted flavors.

In recent years, there has been great interest in the development of active packaging with antimicrobial capacity, to exercise this function on the food surface for as long as the food is in direct contact with the packaging. The packaging materials with antimicrobial characteristics can be divided into two major groups: those which possess antimicrobial agents that must migrate to the surface of the packaging to exercise their function and that come into direct contact with the food, and those which do not need to migrate to exercise their function, Various films with antimicrobial characteristics have been developed, mainly in Japan. Thus, Dupont has introduced MicroFree™, based on the incorporation of three kinds of silver or copper salts on a support of inorganic powder thermosetting agents and which can therefore be introduced into the polymeric resin itself during the extrusion process. At present, three kinds of MicroFree™ are being offered: Z-200 (silver on a zinc support), T-558 (silver, copper oxide and zinc silicate on a titanium dioxide support), and B-558 (silver, copper oxide and zinc silicate on a barium sulfate support), which have proved to have a beneficial effect on *Escherichia coli, Pseudomonas aeruginosa, Klebsiella pneumoniae*, and *Staphylococcus aureus*. These agents, in spite of being approved by the Environmental Protection Agency, have not been approved by the FDA for direct contact with food.

Metallic ions, copper and silver, as well as quaternary ammonium

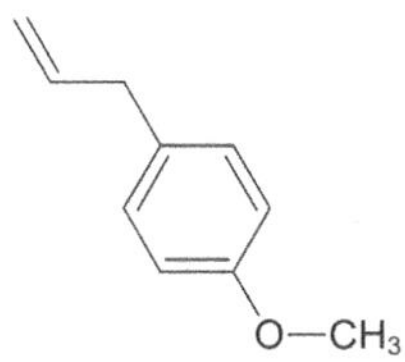

Methyl chavicol

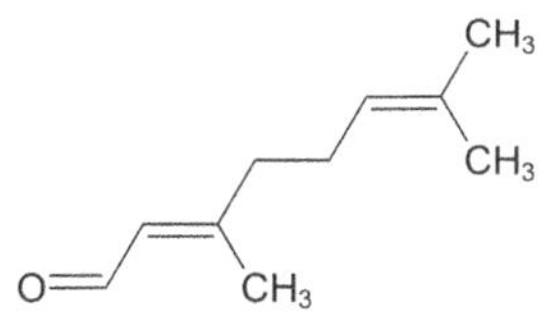

Citral

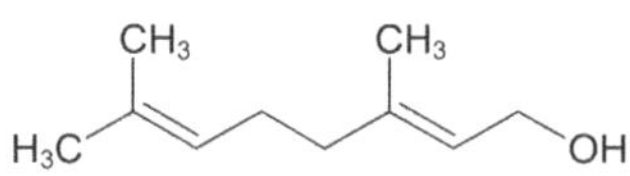

Generaniol

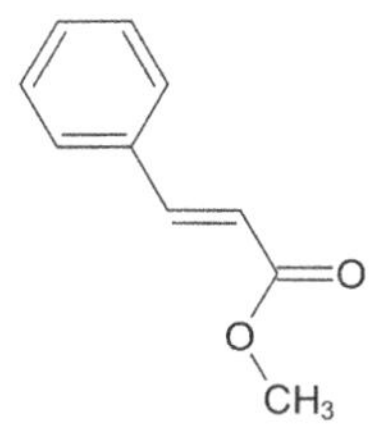

Methyl cinnamate

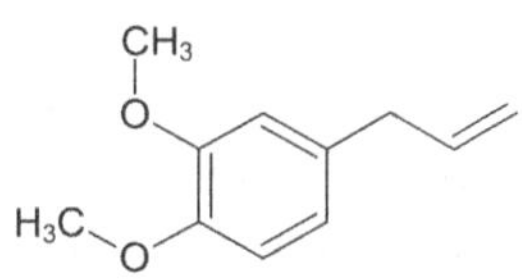

Methyl eugenol

Figure 1.18 Oil compounds.

salts and other compounds, are allowed in Japan, but not in the USA as safe antimicrobial agents. For their application in food packaging, these agents not only must be safe, but they must also migrate with difficulty towards the food. Despite the fact that copper ions can destroy microorganisms and viruses and copper is essential for living things as an enzyme constituent, it is also considered toxic and, therefore, according to US regulations is not allowed for direct contact with food.

In Japan, the regulations for the use of metallic ions are not as strict as they are in the US, and for that reason there has been plenty of work on the incorporation of silver ions, inert supports such as synthetic zeolites in many cases, such as Apacider A™, in which the silver ions chemically and physically bind with calcium phosphate on a zeolite support. Nevertheless, this agent presents problems in its incorporation as extrusion additive, requiring a large number of surfactants that must be added both during the extrusion process, to thereby enhance its dispersion in the plastic matrix and its migration towards the food service. The Japanese companies Shinanen New Ceramics along with Mitsubishi, developed Zeomic™, in which silver ions were incorporated on a zeolite support. Although they have a broad range of action that shows good performance in the extrusion process, the use of ions are still restricted due to their incompatibility with direct food contact. Other compounds have been proposed and their antimicrobial capacity has been evaluated.

For example, antimicrobial Microban™, i.e., 5-chloro-2-(2,4 dichlorophenoxy)phenol or triclosan, can be found in the market, and it is widely used in polyolefins with applications in toys, garbage bags, kitchen cutting boards, among others. However, it has to be mentioned that Microban is not accepted by US regulations as a material for direct contact with food, and it is even being questioned by the US Environmental Protection Agency (EPA). The Japanese company Sumitomo developed a thermoplastic resin for food packaging films, incorporating benomyl, one of the systemic fungicides with the broadest range of action; however, it is highly improbable that US regulating authorities will accept benomyl for direct contact with food. Some other structures with antimicrobial characteristics refer to the formation of chlorine dioxide inside the packaging. Benomyl is shown in Figure 1.19.

This invention by the Southwest Research Institute and Bertrand

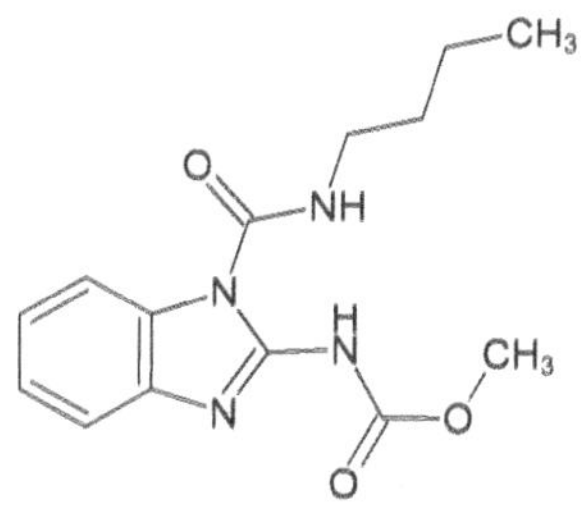

Figure 1.19 Benomyl.

Technologies, Inc., has led to the packaging system called Microatmosphere™, where the use of chlorine dioxide for food packaging has been accepted in the US. Many efforts are being made in the search for active agents from natural sources with antimicrobial capacity, many of which are being tested for their effectiveness *in-vitro* in laboratory experiments; however, there is a lack of technological development towards the ability of applying these agents in food packaging processes, and that is what will be addressed by this patent by means of the extrusion process or coating. The incorporation of antimicrobial agents both inside the polymer matrix and in the form of coating must meet the international legislation concerning packaging materials to be in direct contact with food. For instance, the US legislation defines a food additive as any substance which under use conditions may become part of the food, so all those components which intentionally or non-intentionally, may pass to the food from the packaging material must be previously admitted as food additive, except for substances recognized as GRAS (title 21 of the Federal Regulations). Therefore, in this case the antimicrobial materials to be incorporated into the structure must be approved for their use in direct contact with food by the FDA, or they must be GRAS substances, as is the case of natural origin substances (plant extracts such as thymol, carvacrol, linalool, among others).

In the case of using antimicrobial materials approved by the FDA, the legislation requires levels of maximum global migration of the extractable substances from the packaging material. These regulations and requirements concerning food packaging material are included in the form of positive lists in the Code of Federal Regulations (C.F.R.), in parts 174-179. The regulations are subdivided into sections for polymers, adhesives, coatings, adjuvants, among others, thus establishing the additive identity, safety conditions for its use and, in the case of polymer resins, they specify limits of all extractable substances of the finished article.

The use of plastic packaging with a high permeability to oxygen is being considered, as required by American legislation, in order to avoid anaerobic bacterial growth. The European Union has a very strict regulation on the necessary characteristics of packaging that is in direct contact with food, so that packaging must be ruled by Directive 2005/79/EC of 18 Nov. 2005, which rules all aspects concerning articles and materials which are in direct contact with food, and where the limits of maximum global migration (i.e., total amount of components that can pass from the packaging into the food, regardless of their nature) and specific migration (maximum amount of a specific component that can pass from the packaging into the food, subject which has a particular interest from the toxicological or sensorial point of view) for the starting substances and additives used when manufacturing plastic materials for food. Nonetheless, European legislation in Regulation 1935/2004 (87) authorizes two kinds of packaging: one that reacts *intelligently* on the product by indicating if the product has expired or is degraded, and another one that reacts *actively* by introducing beneficial chemical modifications in food, which increases its storage time.

This regulation is consistent with the latest developments in the food-packaging innovation area. In any case, the incorporation of antimicrobial agents will be performed according to and following international legislation. Food packaging can become the most evident waste source produced by the public. This has raised environmental concerns, leading to a strengthening of European legislation and boxing regulation, in order to reduce the amount of packaging waste, since solid waste recycling is a common practice in countries committed to sustainable development. The European Parliament and Council Directive 94/62/EC of 20 Dec. 1994, required that a minimum of 15% plastic from packaging and packaging waste be recycled by the end of 2005, whereas Directive 2004/12/EC of the European Parliament and Council of 11 Feb. 2004, stated that a minimum of 22.5% must be recycled by the end of 2011.

The reduction of the impact on the environment caused by traditional packaging waste is addressed. Whereas berries are currently commercialized in primary packaging made of non-biodegradable polymeric materials which produce environmental contamination after its use, changing that kind of packaging to recyclable and/or biodegradable packaging, which is in better accordance with international regulation requirements 2005/20/EC and consumer requirements internationally, will lead to the fulfillment of such requirements.

There is still a need to consolidate or improve the presence of food stuff in highly demanding markets, a goal that can be achieved as long as the food, besides being competitive and attractive, is safe, good quality and with a lifespan appropriate to the distance it is sought to bridge. Thus, the development of food preserving technologies is required in which the packaging systems play an essential role in extending the lifespan and increasing the quality and safety of the food, thereby incorporating added value to the product. To be a leader in the food industry, it is not only necessary to be a leader in the areas of production and exporting, but also in the innovation and development of packaging systems capable of safeguardong the quality of food, with special attention given to consumer requirements worldwide.

Thus, one of the requirements in developed countries, where consumers are highly aware of the environmental damage that packaging waste produces, is linked to environmentally friendly packaging, an area in which biodegradable and/or recyclable packaging plays a very important part. It is possible to develop totally recyclable or biodegradable packaging with antimicrobial capacity to inhibit microbiological growth on the surface of the packaged product, in order to be able to offer safer food, with better quality and a longer lifespan, thus reducing the losses accrued due to the rejection of fresh refrigerated fruits.

This innovation will also allow reaching new markets which increasingly require environmentally friendly packaging. The packaging incorporates a proactive and positive aptitude on the packaged food, combining the antimicrobial activity with the reduction of the environmental impact. Thus, a solution that can be easily adopted by the industry can be obtained (70, 71).

1.6 Recycling Safety

The European Food Safety Authority (EFSA) panel on Food Contact Materials, Enzymes and Processing Aids (CEP) assessed the safety of the recycling process for circular plastics (EU register number RECYC228) (88), which uses the Starlinger iV+ technology (89).

The input is hot caustic washed and dried PET flakes mainly originating from collected post-consumer PET containers, with no more than 5% PET from non-food consumer applications. The flakes are dried and crystallized in a first reactor, then extruded into pellets. These pellets are crystallized, preheated and treated in a solid-state polycondensation (SSP) reactor. Having examined the challenge test provided, the panel concluded that the drying and crystallization, extrusion and crystallization and SSP are critical in determining the decontamination efficiency of the process.

The operating parameters to control the performance of these critical steps are temperature, air flow and residence time for the drying and crystallization step, and temperature, pressure and residence time for the extrusion and crystallization step as well as the SSP step. It was demonstrated that this recycling process is able to ensure that the level of migration of potential unknown contaminants into food is below the conservatively modeled migration of 0.1 $\mu g\, kg^{-1}$ food. Therefore, the panel concluded that the recycled PET obtained from this process is not a safety concern when used at up to 100% for the manufacture of materials and articles for contact with all types of foodstuffs for long-term storage at room temperature, with or without hotfill. The final articles made of this recycled PET are not intended to be used in microwave and conventional ovens and such uses are not covered by this evaluation (88).

References

1. Y. Motarjemi, G. Moy, and E. Todd, (Eds.), *Encyclopedia of Food Safety*, Academic Press, 2013.
2. C. Matthews, F. Moran, and A.K. Jaiswal, A review on European Union's strategy for plastics in a circular economy and its impact on food safety, *Journal of Cleaner Production*, Vol. 283, p. 125263, 2021.

3. S. Jiang, F. Wang, Q. Li, H. Sun, H. Wang, and Z. Yao, Environment and food safety: A novel integrative review, *Environmental Science and Pollution Research*, Vol. 28, p. 54511, 2021.
4. J. Yates, M. Deeney, H.B. Rolker, H. White, S. Kalamatianou, and S. Kadiyala, A systematic scoping review of environmental, food security and health impacts of food system plastics, *Nature Food*, Vol. 2, p. 80, 2021.
5. I.S. Arvanitoyannis and L. Bosnea, Migration of substances from food packaging materials to foods, *Critical Reviews in Food Science and Nutrition*, Vol. 44, p. 63, 2004.
6. K. Barnes, R. Sinclair, and D. Watson, (Eds.), *Chemical Migration and Food Contact Materials*, Woodhead publishing, 2006.
7. I.S. Arvanitoyannis and K.V. Kotsanopoulos, Migration phenomenon in food packaging. Food–package interactions, mechanisms, types of migrants, testing and relative legislation. A review, *Food and Bioprocess Technology*, Vol. 7, p. 21, 2014.
8. J. Brandsch, P. Mercea, M. Rüter, V. Tosa, and O. Piringer, Migration modelling as a tool for quality assurance of food packaging, *Food Additives & Contaminants*, Vol. 19, p. 29, 2002.
9. P.V. Mercea, A. Kalisch, M. Ulrich, H. Benz, O.G. Piringer, V. Toşa, R. Schuster, S. Aranyi, and P. Sejersen, Modelling migration of substances from polymers into drinking water. Part 1 - Diffusion coefficient estimations, *Polymer Testing*, Vol. 65, p. 176, 2018.
10. R. Rafiei Nazari, S. Noorian, and M. Arabameri, Migration modelling of phthalate from non-alcoholic beer bottles by adaptive neuro-fuzzy inference system, *Journal of the Science of Food and Agriculture*, Vol. 98, p. 2113, 2018.
11. X. Ouyang, Z. Lu, Y. Hu, Z. Xie, and G. Li, Research progress on sample pretreatment methods for migrating substances from food contact materials, *Journal of Separation Science*, Vol. 44, p. 879, 2021.
12. Z. Pilevar, A. Bahrami, S. Beikzadeh, H. Hosseini, and S.M. Jafari, Migration of styrene monomer from polystyrene packaging materials into foods: Characterization and safety evaluation, *Trends in Food Science & Technology*, Vol. 91, p. 248, 2019.
13. S. Kontou, E. Dessipri, and E. Lampi, Determination of styrene monomer migrating in foodstuffs from polystyrene food contact articles using HS-SPME-GC-MS/MS: Results from the Greek market, *Food Additives & Contaminants: Part A*, Vol. 39, p. 415, 2022.
14. F. Vilarinho, R. Sendón, A. van der Kellen, M.F. Vaz, and A.S. Silva, Bisphenol A in food as a result of its migration from food packaging, *Trends in Food Science & Technology*, Vol. 91, p. 33, 2019.
15. W.A.C. Anderson and L. Castle, Benzophenone in cartonboard packaging materials and the factors that influence its migration into food, *Food Additives & Contaminants*, Vol. 20, p. 607, 2003.

16. T.H. Begley, K. White, P. Honigfort, M.L. Twaroski, R. Neches, and R.A. Walker, Perfluorochemicals: Potential sources of and migration from food packaging, *Food Additives & Contaminants*, Vol. 22, p. 1023, 2005.
17. L. Coulier, H.G. Orbons, and R. Rijk, Analytical protocol to study the food safety of (multiple-)recycled high-density polyethylene (HDPE) and polypropylene (PP) crates: Influence of recycling on the migration and formation of degradation products, *Polymer Degradation and Stability*, Vol. 92, p. 2016, 2007.
18. P.S. Mead, L. Slutsker, V. Dietz, L.F. McCaig, J.S. Bresee, C. Shapiro, P.M. Griffin, and R.V. Tauxe, Food-related illness and death in the United States, *Journal of Environmental Health*, Vol. 62, p. 9, 2000.
19. P.S. Mead, L. Slutsker, V. Dietz, L.F. McCaig, J.S. Bresee, C. Shapiro, P.M. Griffin, and R.V. Tauxe, Food-related illness and death in the United States, *Emerging Infectious Diseases*, Vol. 5, p. 607, 1999.
20. B. Tang, P. Alam, and N.L.C. Leung, AIE-active chemo sensors for amine detection and related food-safety monitoring, US Patent 11 332 449, assigned to Hong Kong University of Science and Technology, May 17, 2022.
21. P. Alam, N.L.C. Leung, H. Su, Z. Qiu, R.T.K. Kwok, J.W.Y. Lam, and B.Z. Tang, A highly sensitive bimodal detection of amine vapours based on aggregation induced emission of 1,2-dihydroquinoxaline derivatives, *Chemistry–A European Journal*, Vol. 23, p. 14911, 2017.
22. C.R. Armendáriz, Á. Daschner, E.G. Fandos, M. José, G. Muñoz, M.V. Moreno-Arribas, P.T. Oliag, and J.B.G. de Castro, Report of the Scientific Committee of the Spanish Agency for Food Safety and Nutrition (AESAN) on the presence and safety of plastics as contaminants in food, Report 007, Scientific Committee of the Spanish Agency for Food Safety and Nutrition, Madrid, 2019.
23. S. Rainieri and A. Barranco, Microplastics, a food safety issue?, *Trends in Food Science & Technology*, Vol. 84, p. 55, 2019.
24. M. Camacho, A. Herrera, M. Gómez, A. Acosta-Dacal, I. Martínez, L.A. Henríquez-Hernández, and O.P. Luzardo, Organic pollutants in marine plastic debris from Canary Islands beaches, *Science of The Total Environment*, Vol. 662, p. 22, 2019.
25. A. Llamas, A.-M. Al-Lal, M.-J. García-Martínez, M.F. Ortega, J.F. Llamas, M. Lapuerta, and L. Canoira, Polycyclic aromatic hydrocarbons (PAHs) produced in the combustion of fatty acid alkyl esters from different feedstocks: Quantification, statistical analysis and mechanisms of formation, *Science of The Total Environment*, Vol. 586, p. 446, 2017.
26. K. Bhuyan and A. Giri, Polycyclic aromatic hydrocarbon compounds as emerging water pollutants: Toxicological aspects of phenanthrene on aquatic animals in V. Shikuku, (Ed.), *Effects of Emerging Chemical*

Contaminants on Water Resources and Environmental Health, Advances in Environmental Engineering and Green Technologies, chapter 4, pp. 45–67. IGI Global, 2019.

27. Wikipedia contributors, Benz(a)anthracene — Wikipedia, the free encyclopedia, 2021. [Online; accessed 21-June-2022].
28. Wikipedia contributors, Benzo(j)fluoranthene — Wikipedia, the free encyclopedia, https://en.wikipedia.org/w/index.php?title=Benzo(j)fluoranthene&oldid=1065626450, 2022. [Online; accessed 21-June-2022].
29. S. Masala, C. Bergvall, and R. Westerholm, Determination of benzo[a]pyrene and dibenzopyrenes in a Chinese coal fly ash certified reference material, *Science of The Total Environment*, Vol. 432, p. 97, 2012.
30. C. Bergvall and R. Westerholm, Determination of highly carcinogenic dibenzopyrene isomers in particulate emissions from two diesel- and two gasoline-fuelled light-duty vehicles, *Atmospheric Environment*, Vol. 43, p. 3883, 2009.
31. I. Sadiktsis, C. Bergvall, C. Johansson, and R. Westerholm, Automobile tires? A potential source of highly carcinogenic dibenzopyrenes to the environment, *Environmental Science & Technology*, Vol. 46, p. 3326, 2012.
32. I. Fernando Tamames, Toxin binding system, US Patent 10 058 568, assigned to Special Nutrients Inc., August 28, 2018.
33. Wikipedia contributors, Aflatoxin — Wikipedia, the free encyclopedia, 2022. [Online; accessed 19-June-2022].
34. P.M. Fratamico, A.K. Bhunia, J.L. Smith, et al., (Eds.), *Foodborne Pathogens: Microbiology and Molecular Biology*, CRC Press LLC, 2005.
35. S.Z. Iqbal, S. Nisar, M.R. Asi, and S. Jinap, Natural incidence of aflatoxins, ochratoxin A and zearalenone in chicken meat and eggs, *Food Control*, Vol. 43, p. 98, 2014.
36. P. Khlangwiset, G.S. Shephard, and F. Wu, Aflatoxins and growth impairment: A review, *Critical Reviews in Toxicology*, Vol. 41, p. 740, 2011.
37. H.K. Abbas, (Ed.), *Aflatoxin and Food Safety*, CRC Press, Boca Raton, 2005.
38. L.E. Smith, A.J. Prendergast, P.C. Turner, M.N.N. Mbuya, K. Mutasa, G. Kembo, and R.J. Stoltzfus, The potential role of mycotoxins as a contributor to stunting in the SHINE trial, *Clinical Infectious Diseases*, Vol. 61, p. S733, 2015.
39. Y.M. Abdulrazzaq, N. Osman, and A. Ibrahim, Fetal exposure to aflatoxins in the United Arab Emirates, *Annals of Tropical Paediatrics*, Vol. 22, p. 3, 2002.
40. Y.M. Abdulrazzaq, N. Osman, Z.M. Yousif, and O. Trad, Morbidity in neonates of mothers who have ingested aflatoxins, *Annals of Tropical Paediatrics*, Vol. 24, p. 145, 2004.

41. Y.Y. Gong, K. Cardwell, A. Hounsa, S. Egal, P.C. Turner, A.J. Hall, and C.P. Wild, Dietary aflatoxin exposure and impaired growth in young children from Benin and Togo: Cross sectional study, *British Medical Journal*, Vol. 325, p. 20, 2002.
42. J.J. Ellis, Section liseola of fusarium, *Mycologia*, Vol. 80, p. 255, 1988.
43. Wikipedia contributors, Ochratoxin — Wikipedia, the free encyclopedia, 2022. [Online; accessed 19-June-2022].
44. A. Pfohl-Leszkowicz and R.A. Manderville, Ochratoxin A: An overview on toxicity and carcinogenicity in animals and humans, *Molecular Nutrition & Food Research*, Vol. 51, p. 61, 2007.
45. Wikipedia contributors, Zearalenone — Wikipedia, the free encyclopedia, 2021. [Online; accessed 19-June-2022].
46. J. Liu and T. Applegate, Zearalenone (ZEN) in livestock and poultry: Dose, toxicokinetics, toxicity and estrogenicity, *Toxins*, Vol. 12, p. 377, 2020.
47. C.V. Bulgaru, D.E. Marin, G.C. Pistol, and I. Taranu, Zearalenone and the immune response, *Toxins*, Vol. 13, p. 248, 2021.
48. K. Ropejko and M. Twarużek, Zearalenone and its metabolites–General overview, occurrence, and toxicity, *Toxins*, Vol. 13, p. 35, 2021.
49. C. Peillod, M. Laborde, A. Travel, A. Mika, J.D. Bailly, D. Cleva, C. Boissieu, J. Le Guennec, O. Albaric, S. Labrut, et al., Toxic effects of fumonisins, deoxynivalenol and zearalenone alone and in combination in ducks fed the maximum EU tolerated level, *Toxins*, Vol. 13, p. 152, 2021.
50. T. Kuiper-Goodman, P.M. Scott, and H. Watanabe, Risk assessment of the mycotoxin zearalenone, *Regulatory Toxicology and Pharmacology*, Vol. 7, p. 253, 1987.
51. T. Tanaka, A. Hasegawa, S. Yamamoto, U.S. Lee, Y. Sugiura, and Y. Ueno, Worldwide contamination of cereals by the Fusarium mycotoxins nivalenol, deoxynivalenol, and zearalenone. 1. Survey of 19 countries, *Journal of Agricultural and Food Chemistry*, Vol. 36, p. 979, 1988.
52. E.R. Prossnitz and M. Barton, Estrogen biology: New insights into GPER function and clinical opportunities, *Molecular and Cellular Endocrinology*, Vol. 389, p. 71, 2014.
53. D.K. Mahato, S. Devi, S. Pandhi, B. Sharma, K.K. Maurya, S. Mishra, K. Dhawan, R. Selvakumar, M. Kamle, A.K. Mishra, and P. Kumar, Occurrence, impact on agriculture, human health, and management strategies of zearalenone in food and feed: A review, *Toxins*, Vol. 13, p. 92, 2021.
54. M.D. Marazuela, M. Klaiber, E. Moreno-Gordaliza, A. Barata, and M.M. Gómez-Gómez, Safety assessment of commercial antimicrobial food packaging: Triclosan and microplastics, a closer look, *Food Packaging and Shelf Life*, Vol. 31, p. 100780, 2022.

55. Z. Li, Y. Yu, Z. Li, and T. Wu, A review of biosensing techniques for detection of trace carcinogen contamination in food products, *Analytical and Bioanalytical Chemistry*, Vol. 407, p. 2711, 2015.

56. D.L. Barry, C.J. Burnham, P.M. Desai, P.K. Joseph, H.K.-H. Leung, J.R. Masson, V.N.M. Rao, R.W. Saunders, J.W. Stalder, and M.G. Topor, Method for reducing acrylamide formation in thermally processed foods, US Patent 7 393 550, assigned to Frito-Lay North America Inc., July 01, 2008.

57. Wikipedia contributors, Maillard reaction — Wikipedia, the free encyclopedia, https://en.wikipedia.org/w/index.php?title=Maillard_reaction&oldid=1101978831, 2022. [Online; accessed 11-August-2022].

58. J. Weisburger and R.C. Jones, Method of treating a foodstuff to inhibit the development of mutagens and related product, US Patent 4 777 052, assigned to American Health Foundation, October 11, 1988.

59. Z. Hua, T. Yu, D. Liu, and Y. Xianyu, Recent advances in gold nanoparticles-based biosensors for food safety detection, *Biosensors and Bioelectronics*, Vol. 179, p. 113076, 2021.

60. H.J. Lee, K.W. Lee, K.S. Kang, D. Kim, H. Park, M. Lee, H. Kim, and I. Kwon, Theobromine with an anti-carcinogenic activity, US Patent 6 693 104, assigned to Lotte Confectionery Co. Ltd., May 29, 2003.

61. M.B. Sporn and N. Suh, Chemoprevention of cancer, *Carcinogenesis*, Vol. 21, p. 525, 2000.

62. A.S. Tsao, E.S. Kim, and W.K. Hong, Chemoprevention of cancer, *CA: A Cancer Journal for Clinicians*, Vol. 54, p. 150, 2004.

63. G.J. Kelloff, E.T. Hawk, and C.C. Sigman, (Eds.), *Cancer Chemoprevention: Volume 2: Strategies for Cancer Chemoprevention*, Vol. 2 of *Cancer Drug Discovery and Development*, Humana Press, 2008.

64. B.L. Zoumas, W.R. Kreiser, and R. Martin, Theobromine and caffeine content of chocolate products, *Journal of Food Science*, Vol. 45, p. 314, 1980.

65. M. Takahashi, S. Yanoma, Y. Yamamoto, Y. Rino, T. Amano, and T. Imada, Combined effect of CDDP and caffeine against human gastric cell line in vivo., *Anticancer Research*, Vol. 18, p. 4399, 1998.

66. F.-L. Chung, M. Wang, A. Rivenson, M.J. Iatropoulos, J.C. Reinhardt, B. Pittman, C.-T. Ho, and S.G. Amin, Inhibition of lung carcinogenesis by black tea in Fischer rats treated with a tobacco-specific carcinogen: Caffeine as an important constituent, *Cancer Research*, Vol. 58, p. 4096, 1998.

67. J.-P. Perchellet and R.K. Boutwell, Effects of 3-isobutyl-1-methylxanthine and cyclic nucleotides on the biochemical processes linked to skin tumor promotion by 12-o-tetradecanoylphorbol-13-acetate, *Cancer Research*, Vol. 41, p. 3927, 1981.

68. H.U. Gali, E.M. Perchellet, D.S. Klish, J.M. Johnson, and J.-P. Perchellet, Hydrolyzable tannins: Potent inhibitors of hydroperoxide production and tumor promotion in mouse skin treated with 12-o-tetradecanoylphorbol-13-acetate in vivo, *International Journal of Cancer*, Vol. 51, p. 425, 1992.
69. L.J. Mauer and B.F. Ozen, Food packaging in J.S. Smith and Y.H. Hui, (Eds.), *Food Processing: Principles and Applications*, chapter 5, pp. 101–132. Blackwell Publishing Hoboken, Ames, Iowa (USA), 2004.
70. M.P.J.C. Gonsalves and A.M.T. Ulloa, Method of packaging food, US Patent Application 20 150 025 144, assigned to Universidad de Santiago de Chile, January 22, 2015.
71. M.P.J.C. Gonsalves and A.M.T. Ulloa, Method of packaging food, US Patent 9 763 439, assigned to Universidad de Santiago de Chile, September 19, 2017.
72. A.D. King, Jr. and H.R. Bolin, Physiological and microbiological storage stability of minimally processed fruits and vegetables, *Food Technology (USA)*, Vol. 43, p. 132, 1989.
73. M. Arranz, A.P. Eslava, and E.P. Benito, Pathogenicity factors in Botrytis cinerea, *Revista Iberoamericana de Micologia*, Vol. 17, p. S43, 2000.
74. J. Miltz, I. Raz, and N. Passy, Shelf-life of grapefruit segments in multilayer plastic trays, *LWT - Food Science and Technology*, Vol. 28, p. 442, 1995.
75. P. Appendini and J.H. Hotchkiss, Review of antimicrobial food packaging, *Innovative Food Science & Emerging Technologies*, Vol. 3, p. 113, 2002.
76. S. Quintavalla and L. Vicini, Antimicrobial food packaging in meat industry, *Meat Science*, Vol. 62, p. 373, 2002.
77. I. Sebti and V. Coma, Active edible polysaccharide coating and interactions between solution coating compounds, *Carbohydrate Polymers*, Vol. 49, p. 139, 2002.
78. E. Willmore, A.J. Frank, K. Padget, M.J. Tilby, and C. Austin, Etoposide targets topoisomerase iiα and iiβ in leukemic cells: Isoform-specific cleavable complexes visualized and quantified in situ by a novel immunofluorescence technique, *Molecular Pharmacology*, Vol. 54, p. 78, 1998.
79. F. Debeaufort, J.-A. Quezada-Gallo, and A. Voilley, Edible films and coatings: Tomorrow's packagings: A review, *Critical Reviews in Food Science*, Vol. 38, p. 299, 1998.
80. R. Chan and V. Chen, Characterization of protein fouling on membranes: Opportunities and challenges, *Journal of Membrane Science*, Vol. 242, p. 169, 2004.

81. M.J. Kirwan and J.W. Strawbridge, Plastics in food packaging in R. Coles, D. McDowell, and M.J. Kirwan, (Eds.), *Food Packaging Technology*, Vol. 7 of *Sheffield Packaging Technology*, pp. 174–240. CRC Press LLC Boca Raton, Florida, 2003.
82. V. Siracusa, P. Rocculi, S. Romani, and M.D. Rosa, Biodegradable polymers for food packaging: A review, *Trends in Food Science & Technology*, Vol. 19, p. 634, 2008.
83. M.T. Carvalho, E. Agante, and F. Durão, Recovery of PET from packaging plastics mixtures by wet shaking table, *Waste Management*, Vol. 27, p. 1747, 2007.
84. Standard test method for deflection temperature of plastics under flexural load in the edgewise position, ASTM Standard D648-18, ASTM International, West Conshohocken, PA, 2018. https://www.astm.org/d0648-18.html.
85. M. Bonneau and M. Bonneau-Calone, Bioactive packages and package closures, US Patent 20 080 169 217, assigned to Marc Bonneau, Margaret Bonneau-Calone, July 17, 2008.
86. M. Marek, E. Kralova, V. Stastnikova, J. Dobian, and M. Voldoeich, Method of preparation of bioactive packaging materials, WO Patent 2 004 056 214, assigned to Institute of Chemical Technology Prague, Miroslav Marek, Eva Kralova, Vira Stastnikova, Jaroslav Dobian, Michal Voldoeich, October 07, 2004.
87. European Union, Materials and articles intended to come into contact with food and repealing directives 80/590/EEC and 89/109/EEC, *Official Journal of the European Union*, pp. 338/4–338/17, 2004.
88. EFSA Panel on Food Contact Materials, Enzymes and Processing Aids (CEP), C. Lambré, J.M. Barat Baviera, C. Bolognesi, A. Chesson, P.S. Cocconcelli, R. Crebelli, D.M. Gott, K. Grob, M. Mengelers, A. Mortensen, G. Rivière, I.-L. Steffensen, C. Tlustos, H.V. Loveren, L. Vernis, H. Zorn, V. Dudler, M.R. Milana, C. Papaspyrides, M. de Fátima Tavares Poças, A. Lioupis, and E. Lampi, Safety assessment of the process circular plastics, based on the Starlinger iV+ technology, used to recycle post-consumer PET into food contact materials, *EFSA Journal*, Vol. 20, p. e07019, 2022.
89. C. Lovranich, A. Pechhacker, M. Wallner, C. Weberhofer, and S. Obermüller, Automated method for packaging food, A Patent EP2 995 436, assigned to Starlinger and Co. GmbH, March 16, 2016.

2

Regulations

The issues related to global legislation for food contact materials have been presented in a monograph (1).

Food contact materials, such as packaging, storage containers and processing surfaces, can pose a substantial hazard to both food manufacturer and consumer due to the migration of chemicals or other substances from the material to the food, which can cause tainting of flavors and other sensory characteristics, or even illness. Here, the main materials used for food contact are reviewed in terms of the global legislation in place to ensure their safe and effective use. Part One provides an overview of food contact legislation issues such as chemical migration and compliance testing. Part Two looks in detail at the legislation for specific food contact materials and their advantages, hazards and use in industry. In detail, this monograph deals with (1):

* A global coverage of food contact legislation,
* Expert analysis of future trends in global food packaging regulation, and
* It focuses on specific materials such as plastic, paper and rubber materials in contact with food.

2.1 European Strategy for Plastics

In 2018, the European Union adopted a European strategy for plastics in a circular economy as a part of their action plan. Sustainability is the underlying motivation behind the plastics' strategy with a goal of addressing how plastics are designed, used and recycled in the European Union.

One of these strategies is that by 2030, all plastic packaging placed on the European Union market is either reusable or should be able to become recycled in a cost-effective manner. A large portion of food packaging is multilayer plastic that is not recyclable in a cost-effective manner.

2.2 Basic Principles of Packaging

The basic principles of packaging to preserve, merchandize, protect, market, and distribute are applied very effectively in the food industry (2). More recently, food packages have been developed which offer tamper-evident or tamper-resistant features, allow for product preparation, and provide dispensing features and many conveniences such as single serving portions. Often the food product and its package are developed to be an integrated unit such as an aerosol product.

Technological breakthroughs in metals, glass, paperboard, composites, and most especially plastics have provided a multitude of opportunities for improved food packaging. The plastics segment of the packaging industry has shown the most rapid growth for many reasons. Some of the most important are: Some plastics can be used in microwave ovens; plastics have a wide range of physical and barrier properties; and plastics offer design capabilities and features not available with other packaging materials.

However, ecological and environmental concerns are growing rapidly, along with federal, state, and local regulations and laws which will have an effect on the whole packaging industry. Furthermore, the demographic changes related to the increased percentage of older consumers will also have significant effects on food packaging in the future (2).

2.3 Colorants and Optical Brighteners

Colorants and optical brighteners explicitly cleared by the Food and Drug Administration (FDA) for use in polymers (*food contact polymers*) are listed in 21 C.F.R. §178.3297 (*Colorants for polymers*) (3). The

FDA also continues to allow those skilled in the art to determine the food additive status of colorants in food contact polymers by, for example, testing that determines that a particular color is not expected to migrate to food and is hence not a *food additive* subject to premarket regulatory clearance. Various colorants are also permitted for use in polymers by virtue of Threshold of Regulation listings or applicable food contact Notifications available from the FDA, such as those listed at Food Ingredients & Packaging (4), Packaging & Food Contact Substances (5).

2.4 Non-intentionally Added Substances in Food Contact Material

Food contact materials (FCMs) contain substances that have the potential to migrate into foodstuffs. There are methods used to assess the identity and quantities of non-intentionally added substances (NIAS) in FCMs.

Examples of NIAS are impurities present in the intentionally added substances (IAS) or degradation and reaction products created during the manufacturing of the material. The evaluation of NIAS has become of interest since they were specifically mentioned in the European legislation on plastic materials and articles intended to come into contact with food (EU Regulation 10/2011) with the following definition (6):

A non-intentionally added substance means an impurity in the substances used or a reaction intermediate formed during the production process or a decomposition or reaction product.

Different approaches to analyze NIAS in FCMs were presented and compared (7).

The approaches can give different results and it can be difficult to make meaningful comparisons (7). The approaches for preparing the sample for analysis, recommended procedures for screening, identification, and quantification of NIAS as well as the reporting requirements were outlined. Also, different analytical equipment and procedures are compared.

The identification of NIAS in FCMs and their migration is very challenging since not all NIAS are predictable. Migrating substances from FCMs can globally be subdivided into (7):

1. Intentionally added substances,
2. Known or expected/predictable non-intentionally added substances, and,
3. Unknown NIAS.

Although dedicated analytical techniques are already partially available and are in some cases standardized for IAS and known/predictable NIAS, no standardized methodology is available for non-target screenings and quantification of unknown/unpredictable NIAS in FCMs. FCMs can have very complex structures with many different layers from different materials, either coextruded or laminated, manufactured using different adhesives and different polymers as substrates, even paper or aluminum, which can also be coated by adhesives, varnishes, and printing inks. Each component of FCMs can be a source of NIAS (7).

The sources of NIAS can be classified as:

1. Degradation products,
2. Reaction products,
3. Impurities, and
4. Contaminants.

Many laboratories from industry, contract research, and academia are performing non-targeted screening tests to identify and quantify potential migrating substances in FCMs, where results are often semi-quantitative in nature. A 10 $\mu g\,kg^{-1}$ limit was set as the limit for migration from behind a functional barrier in Commission Regulation (EU) 10/2011, with exclusions of substances that are classified as carcinogenic, mutagenic, toxic to reproduction, or nanomaterials.

The 10 $\mu g\,kg^{-1}$ limit was pragmatically based on what could be reasonably achieved by available methods and what would be a meaningful limit from a risk management perspective. For *unknowns*, quantification is difficult or impossible with this $\mu g\,kg^{-1}$ limit, but it is typically the level currently used for screening NIAS.

This NIAS screening approach with the 10 $\mu g\,kg^{-1}$ migration limit is acknowledged in Recital (28) of the 15th amendment of the Plastics Regulation, but substances for which genotoxicity cannot be ruled out are excluded (8, 9).

According to Commission Regulation (EU) 2020/1245 of 2 September 2020 amending and correcting Regulation (EU) No 10/2011 on plastic materials and articles intended to come into contact with food, there is a limit of 0.15 $\mu g\, kg^{-1}$ for genotoxic substances based on the application of the threshold of toxicological concern concept (10).

The *Official Journal of the European Union* is the official gazette of record for the European Union (EU) (11). It is published daily (from Monday to Friday regularly, on Saturdays, Sundays and public holidays only in urgent cases) in the official EU languages, and in Irish when required. Only legal acts published in the Official Journal are binding.

2.4.1 *Analysis Methods*

Volatile, semi-volatile and non-volatile substances can be analyzed according to the methods listed below (7). The application ranges of these methods overlap but may differ in sensitivity.

Some substance groups, e.g., primary aromatic amines from poly(urethane) adhesives or pigments, polyaromatic hydrocarbons, require dedicated methods as the concentration level of interest is so low that general non-target screening methods cannot detect them.

The analysis of mineral oil, a common NIAS in many samples of paper, board, and polymers, both virgin and recycled materials, requires specific procedures and still represents a considerable analytical challenge (12).

Solubility is another important parameter. The expected NIAS needs to be soluble enough in the chosen extraction solvent or food simulant.

Some components react with ethanol and form artefacts by transesterification. Isocyanates react with ethanol in ethanol-water mixtures to generate ethyl urethanes. Therefore, the migration of primary aromatic amines from poly(urethane)s (PUs) is investigated in 3% acetic acid or water instead of aqueous ethanolic simulants. Artefacts can also be formed in hot injectors, e.g., a split-splitless injector of a gas chromatograph. An example is the formation of carcinogenic dichlorobenzidine, c.f. Figure 2.1, from traces of diaryl pigments in extracts at temperatures above 200°C. Another example is the re-formation of isocyanates from PU oligomers in the injector (13).

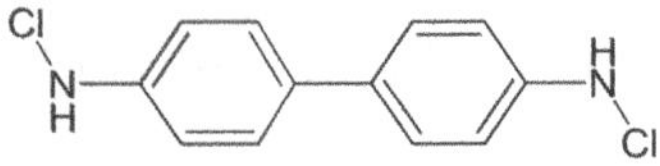

Figure 2.1 Dichlorobenzidine.

Here, a method is shown for the determination of thermal decomposition products of PU. The method is based on the collection of air samples in midget impinger flasks containing 0.01 $mol\,l^{-1}$ dibutylamine dissolved in toluene. The compounds are shown in Figure 2.2. Dibutylamine reacts with isocyanates to give urea derivatives. The amines simultaneously collected were derivatized with ethyl chloroformate to give urethane derivatives. Two isomers of toluene diisocyanate (TDI) (2,4-toluene diisocyanate and 2,6-toluene diisocyanate), toluene-2,4-diamine and toluene-2,6-diamine, three isomers of toluene aminoisocyanates and oligomers with two and three aromatic rings were determined in air containing thermal decomposition products of TDI-poly(urethane). Liquid chromatography with gradient elution at 0.5 $ml\,min^{-1}$, starting with acetonitrile-water (46 + 54) and ending after 14 *min* with acetonitrile-water (77.5 + 22.5), and atmospheric pressure ionization monitoring of positive ions demonstrated highly sensitive and selective determinations. Quantification was performed by monitoring the molecular ion [M + 1]+ of TDI-dibutylamine and toluene-diamine — ethyl chloroformate derivatives. The instrumental detection limit was 1 $nmol\,l^{-1}$ of TDI-dibutylamine and 2 $nmol\,l^{-1}$ of TDI-ethyl chloroformate. Linear calibration curves were obtained and the correlation coefficient was 0.9981-0.9996 for 50–500 $nmol\,l^{-1}$(n= 14). The overall precision for samples spiked at a concentration of 500 $nmol\,l^{-1}$ of toluene-diamine-ethyl chloroformate and 2,4-TDI and 2,6-TDI-dibutylamine was 1.7%, 1.6%, 1.8% and 1.9% (n= 11), respectively.

These effects need to be considered for the evaluation or the presence verified by another technique, such as gas chromatography with on-column injection or high-performance liquid chromatography.

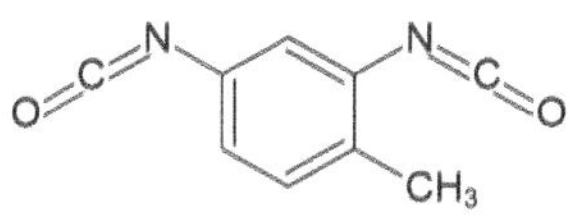

2,4-Toluene diisocyanate

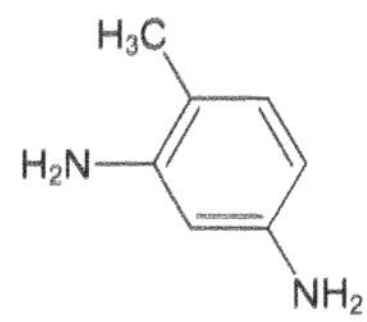

Toluene-2,4-diamine

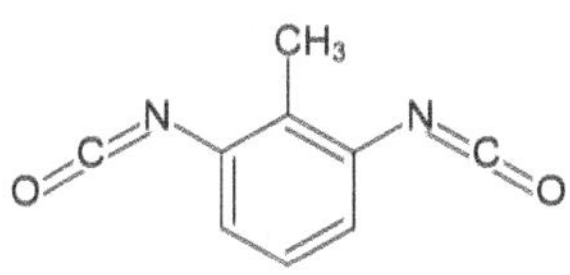

2,6-Toluene diisocyanate

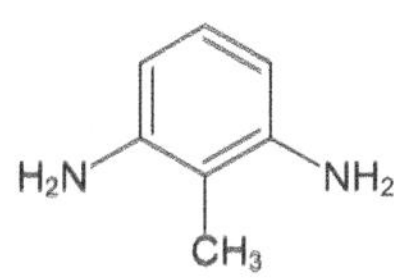

Toluene-2,6-diamine

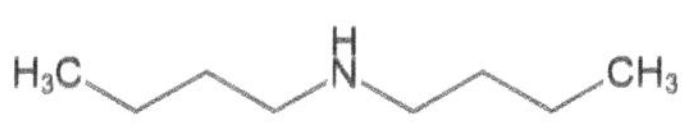

Dibutylamine

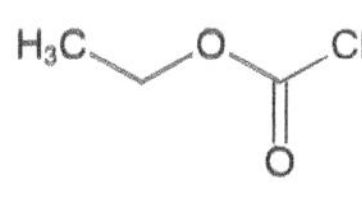

Ethyl chloroformate

Figure 2.2 Amines and Isocyanates.

To assess isocyanate exposures of people in a variety of processes involving the manufacture and use of surface coatings, PU foams, adhesives, resins, elastomers, binders, and sealants, it is important to be able to measure airborne reactive isocyanate-containing compounds (14).

Choosing the correct methodology can be difficult. Isocyanate species, including monomers, prepolymers, oligomers, and poly(isocyanate)s, are capable of producing irritation of the skin, eyes, mucous membranes, and the respiratory tract. The most common adverse health effect is respiratory sensitization, and to a lesser extent dermal sensitization and hypersensitivity pneumonitis.

Isocyanate species that are formed during the production of PU or thermal degradation may also produce adverse health effects. Isocyanate measurement is complicated by the fact that (14):

1. Isocyanates may be in the form of vapors or aerosols of various particle size,
2. The species of interest are reactive and therefore unstable,
3. Only a few pure analytical standards exist, and
4. A high analytical sensitivity is needed.

There are numerous points in the sampling and analytical procedures at which errors can be introduced. The factors to be considered for selecting the most appropriate methodology for a given workplace include collection, derivatization, sample preparation, separation, identification, and quantification (14).

2.5 Recycling

2.5.1 Recycled Plastics in Food Packaging

There is an emphasis throughout the United States on increasing the uses of post-consumer recycled materials, including plastic (15). The FDA is involved when industry collects used polymeric materials (usually food containers) and proposes to recycle these materials to make new food containers. The main safety concerns with the use of post-consumer recycled plastic materials in food contact articles are that:

1. Contaminants from the post-consumer recycled material may appear in the final food contact product made from the recycled material,
2. Post-consumer recycled material that may not be regulated for food contact use may be incorporated into the food contact article, and
3. Adjuvants in the post-consumer recycled plastic may not comply with the regulations for food contact use.

To address these concerns, the FDA considers each proposed use of recycled plastic on a case-by-case basis and issues informal advice as to whether the recycling process is expected to produce post-consumer recycled plastic of suitable purity for food contact applications. The FDA has prepared a document entitled Guidance for Industry: Use of Recycled Plastics in Food Packaging (Chemistry Considerations) that will assist manufacturers of food packaging in evaluating processes for post-consumer recycled plastic into food packaging.

If a manufacturer would like the FDA to consider the use of recycled plastic for a food contact application, the following information should be submitted (15):

1. A complete description of the recycling process, including a description of the source of the post-consumer recycled plastic and a description of any source controls in place intended to ensure that only plastic that initially complied with the applicable regulations is recycled. Also, a description of any steps that are taken to ensure that the recyclable plastic is not contaminated at some point, either before collection for recycling, or during the recycling process.
2. The results of any tests performed to show that the recycling process removes possible incidental contaminants. For use of the recycled material as a substitute for plastic made from virgin materials, it would be necessary to either show that there has been no possibility of contamination with substances other than food, such as strict source control on the input material, or to demonstrate the cleaning efficiency of the recycling process through surrogate contaminant testing. If appropriate, additional migration testing or migration modeling may be conducted to show that the recy-

cling process successfully removes possible incidental contaminants to a level that does not allow each contaminant to migrate from recycled plastic to food that would result in a dietary concentration (DC) exceeding 0.5 *ppb*, which the FDA equates to a negligible exposure level for the use of recycled plastics for food packaging. However, surrogate contaminant testing is no longer considered necessary to demonstrate that post-consumer recycled poly(ethylene terephthalate) (PET) or poly(ethylene naphthalate) (PEN) produced by a tertiary recycling process is suitable for food contact use. Because the FDA has determined that tertiary recycling processes produce post-consumer recycled-PET or post-consumer recycled-PEN of suitable purity for food contact use, the Agency no longer evaluates tertiary recycling processes for post-consumer recycled-PET or post-consumer recycled-PEN or issues individual opinion letters for those processes.

3. A description of the proposed conditions of use of the plastic, e.g., information on intended temperature of use, type of food with which the plastic will come into contact, the duration of the contact, and whether the food contact plastic will be for repeated or single-use applications PET.

2.5.2 *Planning and Decision-Making Process*

To underpin the planning and decision-making process involved in determining future European community legislation on materials and articles in contact with food, a strong science base is required (16).

Significant progress has been made in European funded work in this area, with many developments contributing to a much better understanding of the migration process, and better and simpler approaches to food control (16).

2.5.3 *Mechanical Recycled Plastics*

An overview of the legislative requirements for the use of mechanical recycled plastics in articles placed on the EU market has been presented (17). These issues are seen from the perspective of a plastics recycler.

First, the main principles included in the overarching legislation on Registration, Evaluation, Authorization and Restrictions of Chemicals (REACH) are given, and also to what extent these are applicable for mechanical recyclers of plastics. The interactions between REACH and the Waste Framework Directive (WFD) are discussed, as well as the difficulties for recyclers to comply with certain REACH requirements.

In a second part, the focus moves to the use of recycled plastics as FCM. The scope of the different applicable EU FCM regulations is inventorized as well as the key legislative principles involved. A final section is dedicated to a discussion on the authorization of recycling processes under FCM regulation and the practical challenges involved for the effective introduction of FCMs containing recycled plastics.

All in all, it could be concluded that the complexity of the different legal perspectives, a lack of communication and transparency within the plastic value chain together with technical challenges related to recycling processes have been hindering the effective uptake of recycled plastic FCM (with the exception of bottle PET). The development of targeted solutions across the entire value chain, taking into account different perspectives in terms of legislation and health protection, economic growth and technical innovations, will be crucial in achieving a circular economy for plastics, including recycled plastics for FCM (17).

2.6 Standards

Some standards are collected in Table 2.1 and others are also shown in Table 3.3 in Chapter 3, Section 3.4.

Table 2.1 Standards.

Standard	Ref.	Name
ASTM D7329-07	(18)	Food Service Gloves
ISO 17088:2021	(19)	Specifications for Compostable Plastics
ISO 13106:2014	(20)	Containers for Packaging of Liquid Foodstuffs

2.6.1 *Gloves*

The specification ASTM D7329-07 covers the properties necessary for thin film, unlined polymer gloves to be used in food preparation and food handling. It is also intended to serve as a referee and a guide to permit obtaining gloves of a consistent performance. The materials shall include the following:

1. Natural (latex), and
2. Synthetic poly(chloroprene) and poly(nitrile) rubbers,
3. Vinyl, and
4. Low-density, high-density, and cast poly(ethylene).

The gloves shall be sampled using a sampling plan derived from ISO 2859-1 (21) or its equivalent, or other suitable statistical rationale and shall be tested for the following performance requirements:

1. Freedom from holes,
2. Dimensions and tolerances,
3. Physical properties such as tensile strength, yield strength, and ultimate elongation before and after accelerated aging,
4. Residual powders for powder-free gloves,
5. Powder amount,
6. Aqueous extractable protein content, and
7. Antigenic protein content.

2.6.2 *Specifications for Compostable Plastics*

The standard ISO 17088:2021 specifies procedures and requirements for plastics, and products made from plastics, that are suitable for recovery through organic recycling (19). The four following aspects are addressed:

1. Disintegration during composting,
2. Ultimate aerobic biodegradation,
3. No adverse effects of compost on terrestrial organisms, and
4. Control of constituents.

These four aspects are suitable to assess the effects on the industrial composting process.

This standard is intended to be used as the basis for systems of labeling and claims for compostable plastic materials and products.

The recovery of compostable plastics through composting can be carried out under the conditions found in well-managed industrial composting processes, where the temperature, water content, aerobic conditions, carbon/nitrogen ratio and processing conditions are optimized. Such conditions are generally obtained in industrial and municipal composting plants. Under these conditions, compostable plastics disintegrate and biodegrade at rates comparable to yard trimmings, kraft paper bags and food scraps.

2.6.3 *Containers for Packaging of Liquid Foodstuffs*

The standard ISO 13106:2014 provides the requirements for poly(propylene) resins intended for use in blow-molded, round containers with capacities up to, and including, two liters intended for the packaging of liquids for human consumption. It also provides tolerances on mass, dimensions, methods of sampling, testing, and performance requirements (20).

2.6.4 *Williams-Landel-Ferry Equation*

The Williams-Landel-Ferry (WLF) equation is a useful tool in the prediction of temperature-induced physical changes in foods that are exposed to specific food processing and storage conditions, because so many deteriorative processes are diffusion controlled (22).

The general form of the WLF equation is (23):

$$\log \frac{\varepsilon(T)}{\varepsilon(T_g)} = \log a_T = \frac{-C_1(T - T_g)}{C_2 + T - T_g}. \tag{2.1}$$

In Eq. 2.1, $\varepsilon(T)$ is the viscosity at temperature T and $\varepsilon(T_g)$ is the viscosity at the glass transition temperature T_g. C_1 and C_2 are WLF constants for a given relaxation process.

In food polymers, viscoelastic properties are a strong function of the key plasticizers present in foods, such as water (22). However, WLF constants as a function of water content and water activity are not available in the food literature.

The WLF properties of soy flour were investigated as a function of water activity and elucidated the effect of water on the magnitude of WLF constants. Based on a time-temperature superposition principle, shift factors (*aT*) were obtained, which were used to estimate WLF constants. WLF constants for soy flour were shown to be material specific.

References

1. J.S. Baughan, (Ed.), *Global Legislation for Food Contact Materials*, Woodhead Publishing Series in Food Science, Technology and Nutrition, Elsevier, 2nd edition, 2022.
2. J.K. Cage, Introduction to food packaging, in *Introduction to Food Packaging*, pp. 3–12, Fullerton, CA, 1991. ASTM, Beatrice/Hunt-Wesson, Inc. https://www.astm.org/stp14837s.html.
3. Colorants for Polymers CFR 178.3297, Regulation, U.S. Food & Drug Administration, Silver Spring, Maryland, US, 2021. https://www.govinfo.gov/app/details/CFR-2000-title21-vol3/CFR-2000-title21-vol3-sec178-3297/context.
4. Food Ingredients & Packaging, Regulation, U.S. Food & Drug Administration, Silver Spring, Maryland, US, 2021. https://www.fda.gov/food/food-ingredients-packaging.
5. Packaging & Food Contact Substances, Regulation, U.S. Food & Drug Administration, Silver Spring, Maryland, US, 2021. https://www.fda.gov/food/food-ingredients-packaging/packaging-food-contact-substances-fcs.
6. E. Union, Commission Regulation (EU) No 1282/2011 of 28 November 2011 amending and correcting Commission Regulation (EU) No 10/2011 on plastic materials and articles intended to come into contact with food, *Official Journal of the European Union*, Vol. 50, p. 328, 2011.
7. C. Nerín, S. Bourdoux, B. Faust, T. Gude, C. Lesueur, T. Simat, A. Stoermer, E.V. Hoek, and P. Oldring, Guidance in selecting analytical techniques for identification and quantification of non-intentionally added substances (NIAS) in food contact materials (FCMS), *Food Additives & Contaminants: Part A*, Vol. 0, p. 1, 2022.
8. E. Commission, Commission Regulation (EU) 2016/1416 of 24 August 2016 amending and correcting Regulation (EU) No 10/2011 on plastic materials and articles intended to come into contact with food (Text with EEA relevance), *Official Journal of the European Union*, Vol. 230, p. 22, 2016.

9. E. Commission, Commission Regulation (EU) No 10/2011 of 14 January 2011 on plastic materials and articles intended to come into contact with food, *Official Journal of the European Union*, Vol. 12, p. 1, 2011.
10. E.S. Committee, Guidance on the use of the threshold of toxicological concern approach in food safety assessment, *EFSA Journal*, Vol. 17, p. e05708, 2019.
11. Wikipedia contributors, Official journal of the european union — Wikipedia, the free encyclopedia, https://en.wikipedia.org/w/index.php?title=Official_Journal_of_the_European_Union&oldid=1069718981, 2022. [Online; accessed 25-March-2022].
12. A. Hochegger, S. Moret, L. Geurts, T. Gude, E. Leitner, B. Mertens, S. O'Hagan, F. Poças, T.J. Simat, and G. Purcaro, Mineral oil risk assessment: Knowledge gaps and roadmap. Outcome of a multi-stakeholders workshop, *Trends in Food Science & Technology*, Vol. 113, p. 151, 2021.
13. H. Tinnerberg, M. Spanne, M. Dalene, and G. Skarping, Determination of complex mixtures of airborne isocyanates and amines. Part 2. Toluene diisocyanate and aminoisocyanate and toluenediamine after thermal degradation of a toluene diisocyanate–polyurethane, *Analyst*, Vol. 121, p. 1101, 1996.
14. R.P. Streicher, C.M. Reh, R.J. Key-Schwartz, P.C. Schlecht, M.E. Cassinelli, and P.F. O'Connor, Determination of airborne isocyanate exposure: Considerations in method selection, *AIHAJ - American Industrial Hygiene Association*, Vol. 61, p. 544, 2000.
15. Recycled Plastics in Food Packaging, Regulation, U.S. Food & Drug Administration, Silver Spring, Maryland, US, 2022. https://www.fda.gov/food/packaging-food-contact-substances-fcs/recycled-plastics-food-packaging.
16. J. Gilbert and L. Rossi, European priorities for research to support legislation in the area of food contact materials and articles, *Food Additives & Contaminants*, Vol. 17, p. 83, 2000.
17. E. De Tandt, C. Demuytere, E. Van Asbroeck, H. Moerman, N. Mys, G. Vyncke, L. Delva, A. Vermeulen, P. Ragaert, S. De Meester, and K. Ragaert, A recycler's perspective on the implications of REACH and food contact material (FCM) regulations for the mechanical recycling of FCM plastics, *Waste Management*, Vol. 119, p. 315, 2021.
18. Standard specification for food preparation and food handling (food service) gloves, ASTM Standard D7329-07, ASTM International, West Conshohocken, PA, 2018. https://www.astm.org/d7329-07r18.html.
19. Plastics – Organic recycling – Specifications for compostable plastics, ISO Standard 17088, International Organization for Standardization, Geneva, Switzerland, 2021. https://www.iso.org/standard/74994.html.

20. Plastics – Blow-moulded polypropylene containers for packaging of liquid foodstuffs, ISO Standard 13106, International Organization for Standardization, Geneva, Switzerland, 2014. https://www.iso.org/en/contents/data/standard/05/28/52836.html.
21. Sampling procedures for inspection by attributes – Part 1: Sampling schemes indexed by acceptance quality limit (AQL) for lot-by-lot inspection, ISO Standard 2859, International Organization for Standardization, Geneva, Switzerland, 1999. https://www.iso.org/standard/35236.html.
22. M.E. Yildiz and J.L. Kokini, Determination of Williams-Landel-Ferry constants for a food polymer system: Effect of water activity and moisture content, *Journal of Rheology*, Vol. 45, p. 903, 2001.
23. M.L. Williams, R.F. Landel, and J.D. Ferry, The temperature dependence of relaxation mechanisms in amorphous polymers and other glass-forming liquids, *Journal of the American Chemical Society*, Vol. 77, p. 3701, 1955.

3

Testing Methods

A monograph that focuses on packaging material testing and quality assurance has been published (1). It provides information to find practical solutions to packaging defects and to develop innovative packaging materials for food products.

Material properties, such as mechanical and other physical properties, permeability, sealing, and migration of substances upon food contact, are determining factors for food quality, shelf life, and food safety. Therefore, food packaging materials have to be tested to ensure that they have correct properties in terms of permeability for gases, water vapor, and contaminants; of mechanical and other physical properties (1).

An extensive knowledge of the packaging material is crucial for the design of a packaging material for particular products (2). The properties of the packaging materials, such as mechanical strength, gas permeability, and sealing ability. determine the suitability of packaging material for the intended purpose.

Thus, it is necessary to determine the properties of a packaging material to select the suitable material for any product. Testing of packaging materials generally involves the determination of physical properties. But it is also important to study the chemical stability of a packaging material, particularly when it will be used for food-based applications, so as to ensure the safety of the product. Several testing methods for packaging materials have been described (2).

3.1 Risk Assessment

A fast and automated approach has been developed for the tentative identification and risk assessment of unknown substances in plastic food contact materials (FCMs) by quadrupole-Orbitrap gas chromatography/high-resolution mass spectrometry (GC-HRMS) (3). This approach combines GC-HRMS full scan data acquisition coupled with the Compound Discoverer™ 3.2 software for automated data processing and compound identification.

To perform the tentative identification of the detected features, a restrictive set of identification criteria was used, including matching with the NIST Mass Spectral Library, exact mass of annotated fragments, and retention index calculation. After the tentative identification, a risk assessment of the identified substances was performed by using the threshold of toxicological concern (TTC) approach.

This strategy has been applied to a recycled low-density poly(ethylene) (LDPE), which could be used as FCMs, as a proof-of-concept demonstration. In the analyzed sample, 374 features were detected, of which 83 were tentatively identified after examination of the identification criteria. Most of these were additives, such as plasticizers, used in a wide variety of plastic applications, oligomers of LDPE, and substances with chemical, industrial, or cosmetic applications. Tentatively identified substances in the analyzed recycled LDPE sample are shown in Table 3.1. Some of these compounds are shown in Figure 3.1.

The risk assessment was performed and, according to the TTC approach, the obtained results showed that there was no risk associated with the release of the identified substances (3).

3.2 Antimicrobial Packaging

There are many *in-vitro* antimicrobial testing methods for antimicrobial food packaging. A practical guide to the different aspects of antimicrobial testing has been presented (4).

Antimicrobial packaging is a type of active packaging that can be developed by incorporating different kinds and concentrations of antimicrobial substances into packaging polymers in order to increase the shelf life of the food product.

Table 3.1 Substances in the analyzed recycled LDPE sample.

Aldehydes and Ketones	
Benzophenone	
2,6-Di-*tert*-butyl-1,4-benzoquinone	
3,5-Di-*tert*-butyl-4-hydroxybenzaldehyde	
7,9-Di-*tert*-butyl-1-oxaspiro(4,5)deca-6,9-diene-2,8-dione	
Dehydroabietic aldehyde	
Cyclic and aromatic hydrocarbons	
2,6-Diisopropylnaphthalene	
2,2′,5,5′-Tetramethyl-1,1′-biphenyl	
Undecylcyclohexane	
1-Ethyldecylbenzene	
7-Isopropyl-1-methyl-1,2,3,4-tetrahydrophenanthrene	
7-Isopropyl-1,4-dimethyl tetradecahydrophenanthrene	
m-Camphorene	
1,3,5-Triphenylcyclohexane	
Alkenes	Phenol derivatives
1-Hexadecene	1,2-Diphenoxyethane
9-Nonadecene	2,4-Di-*tert*-butylphenol
1-Docosene	2,6-Di-*tert*-butylphenol
9-Tricosene	Butylated hydroxytoluene
1-Tetracosene	2,4-Di-*tert*-pentylphenol
1-Hexacosene	Metilox
Squalene	Irganox 1076
Phthalates	Primary alcohols
Dibutyl phthalate	3-Nonenol
Diisobutyl phthalate	Octadecanol
Bis(2-ethylhexyl) phthalate	Nonadecanol
Bis(2-ethylhexyl) terephthalate	Eicosanol
	Docosanol
	Tetracosanol

Table 3.1 (cont) Substances in the analyzed recycled LDPE sample.

Fatty acid methyl esters	Other fatty acid esters
Methyl laurate	Isopropyl myristate
Methyl palmitate	Ethyl palmitate
Methyl heptadecanoate	Butyl palmitate
Methyl linolelaidate	Butyl stearate
Methyl elaidate	Hexadecyl palmitate
Methyl stearate	
Methyl erucate	
Methyl isopimarate	
Methyl icosanoate	
Linear and branched PE oligomers	
Tetradecane	Tetraiacontane
Hexadecane	3-Methylpentadecane
Heptadecane	3-Methyl heptadecane
Octadecane	3-Methylnonadecane
Eicosane	2,6,10,15-Tetramethylheptadecane
Tetracosane	3-Methylheneicosane
Pentacosane	5-Methylheneicosane
Heptacosane	11-Methyltricosane
Octacosane	2-Methyloctacosane
	5-Methylnonacosane
Acetate esters	**Other compounds**
7-Tetradecen-1-yl acetate	1-Chlorohexadecane
Hexadecyl acetate	*N,N*-Dimethylpalmitylamine
Hexadecyl trifluoro acetate	Galaxolide
Octadecyl trifluoro acetate	Methyl dehydroabietate
Tri-*n*-butyl acetyl citrate	Methyl abietate
	Diphenyl sulfone
	Irgafos 1076

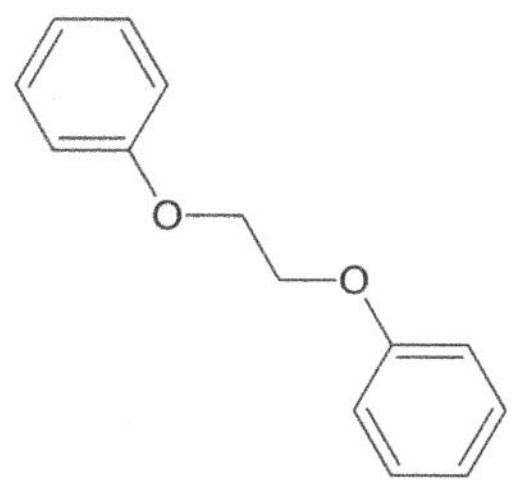

1,2-Diphenoxyethane

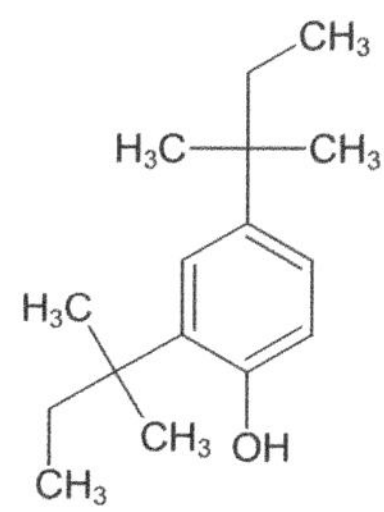

2,4-Di-*tert*-pentylphenol

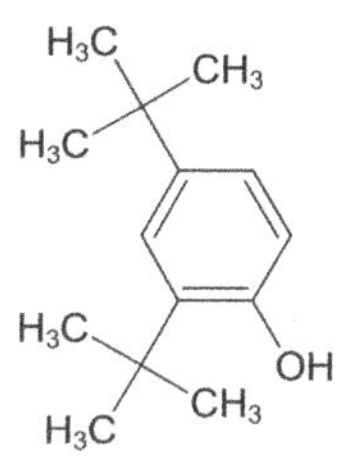

2,4-Di-*tert*-butylphenol

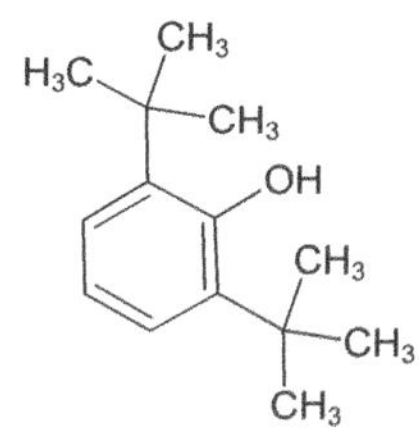

2,6-Di-*tert*-butylphenol

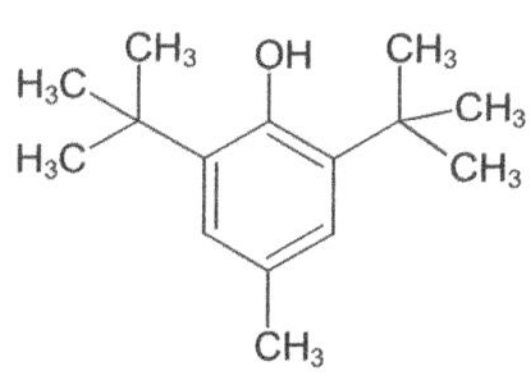

Butylated hydroxytoluene

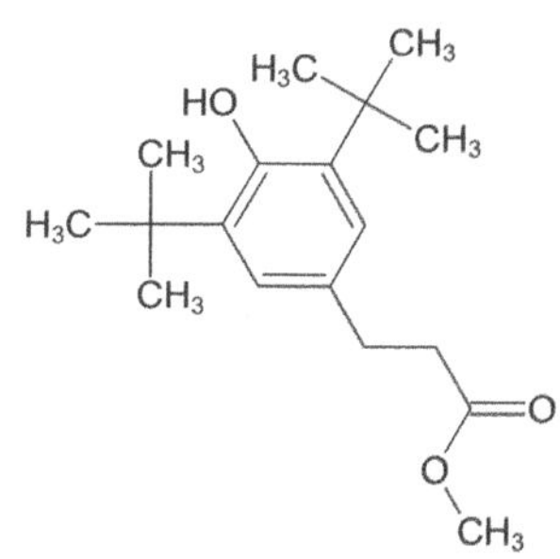

Metilox

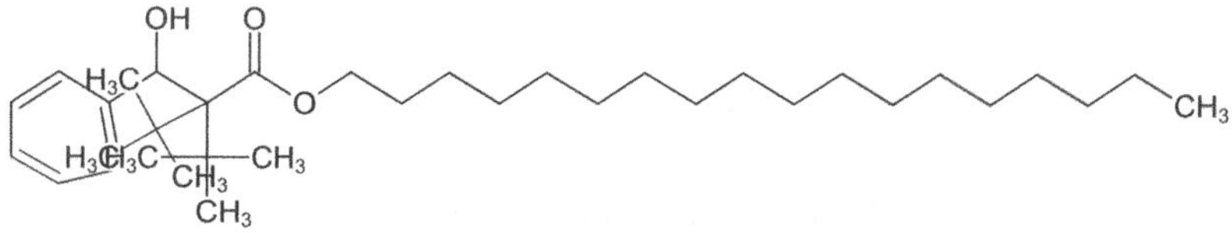

Irganox 1076

Figure 3.1 Phenol derivatives.

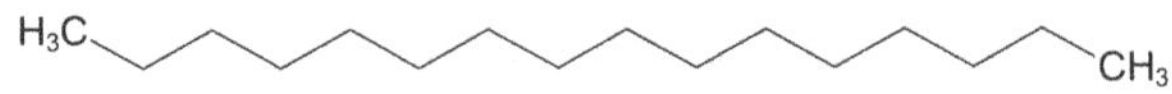

Hexadecane

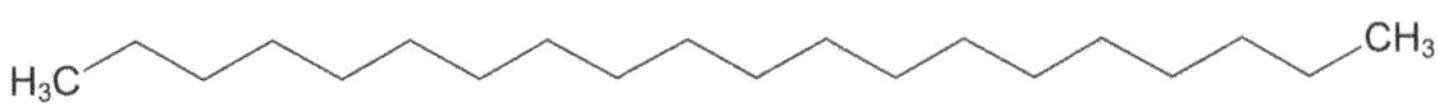

Eicosane

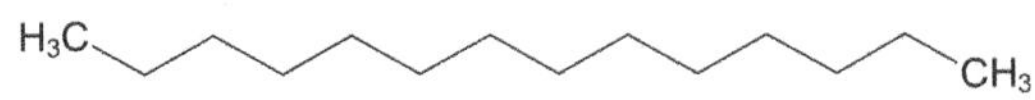

Tetradecane

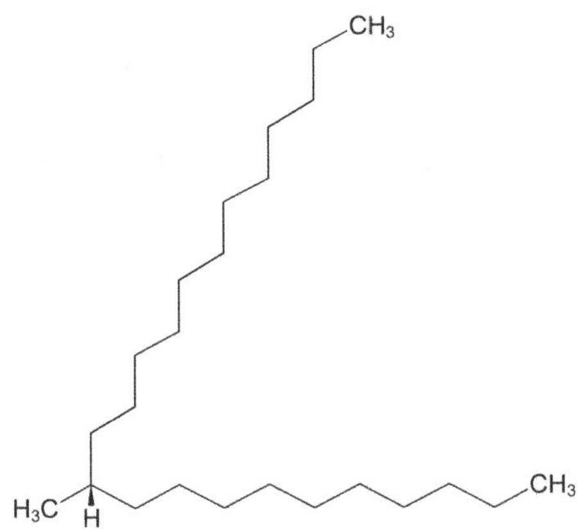

11-Methyltricosane

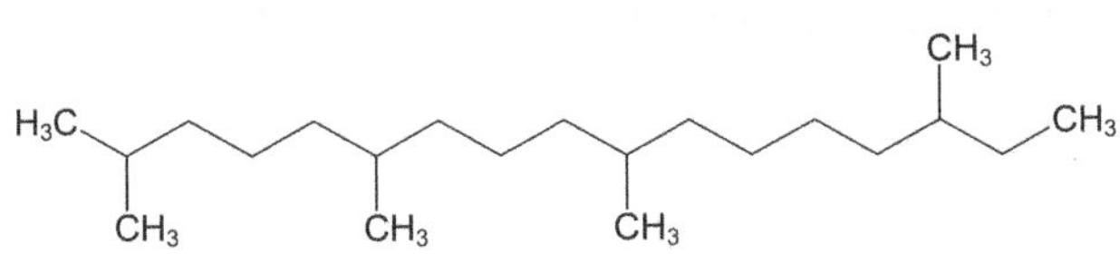

2,6,10,15-Tetramethylheptadecane

Figure 3.1 (cont) Linear and branched PE oligomers.

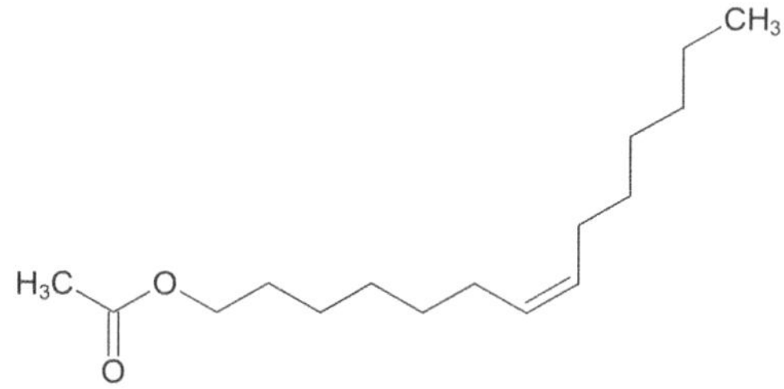

7-Tetradecen-1-yl acetate

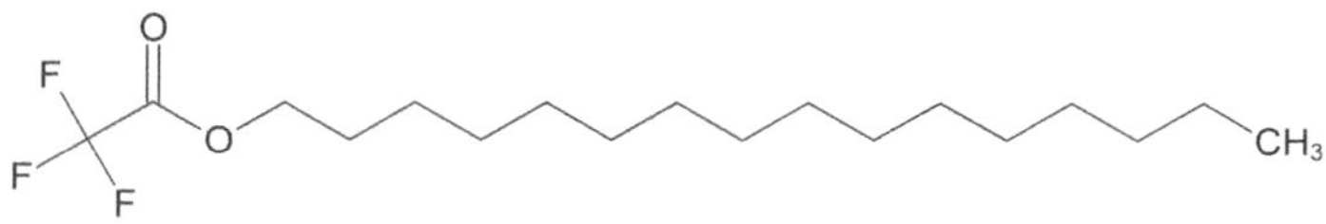

Hexadecyl trifluoro acetate

Figure 3.1 (cont) Acetate esters.

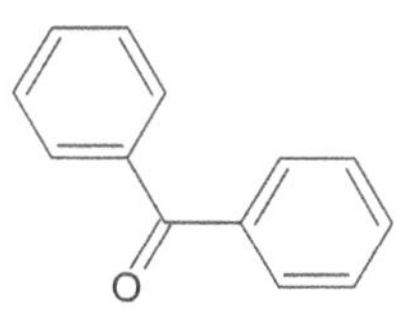

Benzophenone

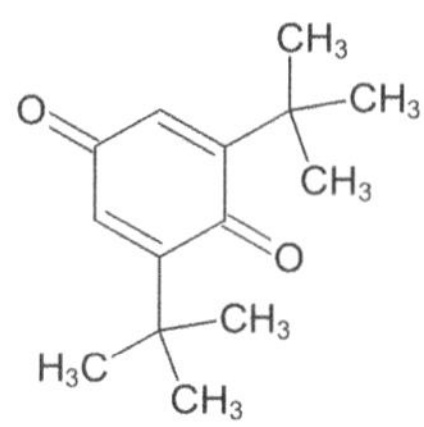

2,6-Di-*tert*-butyl-1,4-benzoquinone

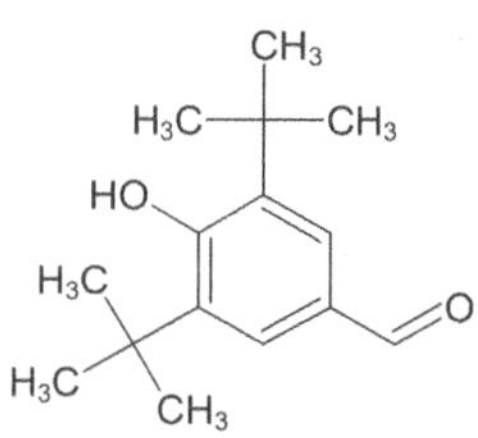

3,5-Di-*tert*-butyl-
4-hydroxybenzaldehyde

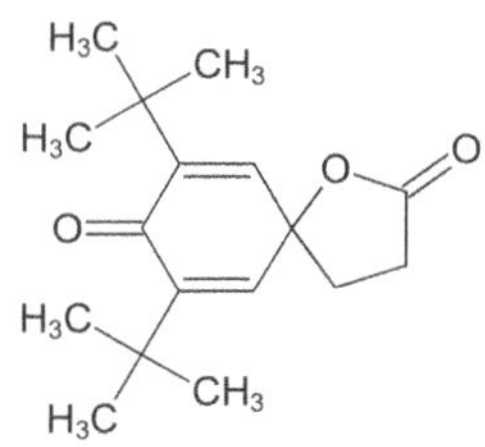

7,9-Di-*tert*-butyl-1-oxaspiro(4,5)deca-
6,9-diene-2,8-dione

Figure 3.1 (cont) Aldehydes and Ketones.

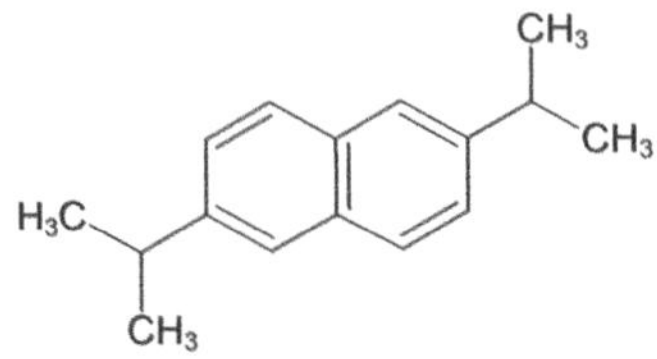

2,6-Diisopropylnaphthalene

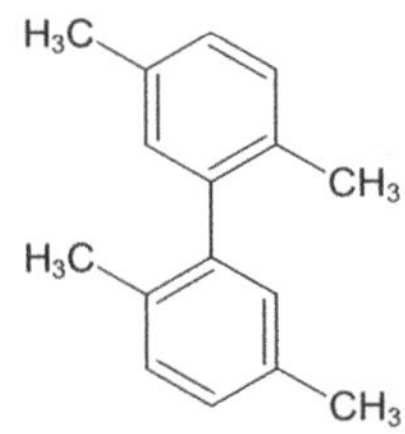

2,2′,5,5′-Tetramethyl-1,1′-biphenyl

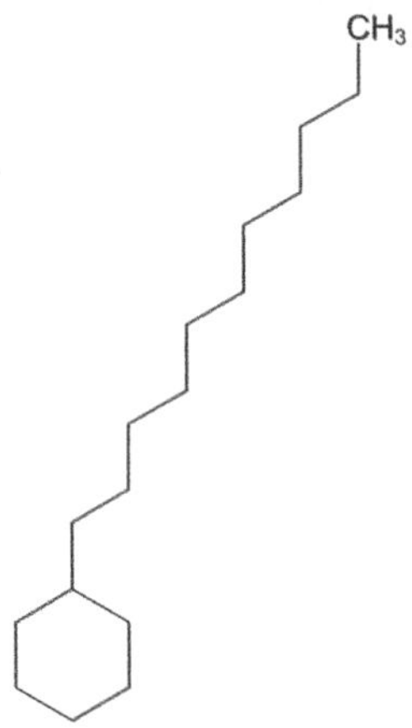

Undecylcyclohexane

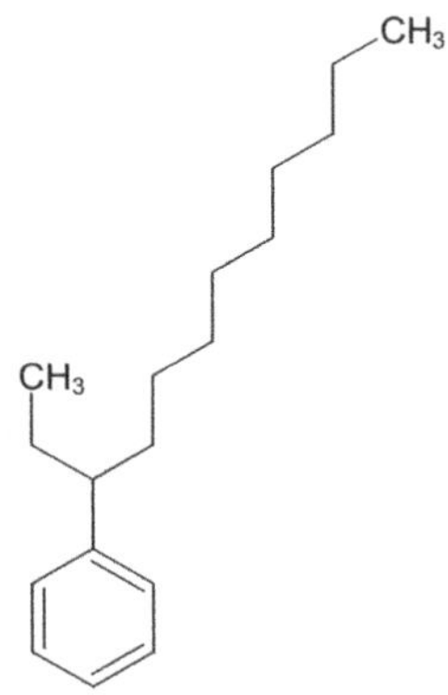

1-Ethyldecylbenzene

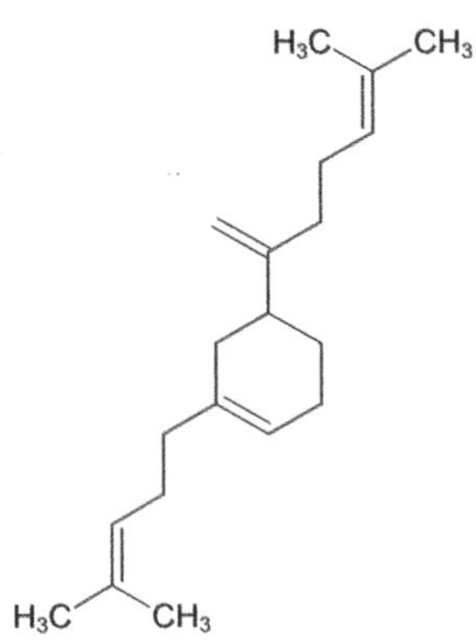

m-Camphorene

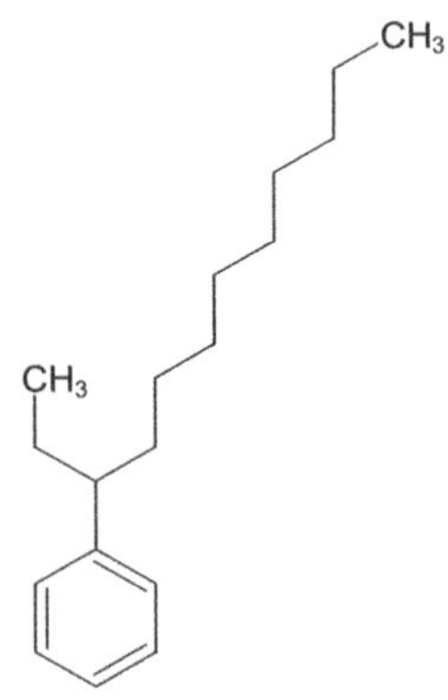

1-Ethyldecylbenzene

Figure 3.1 (cont) Cyclic and aromatic hydrocarbons.

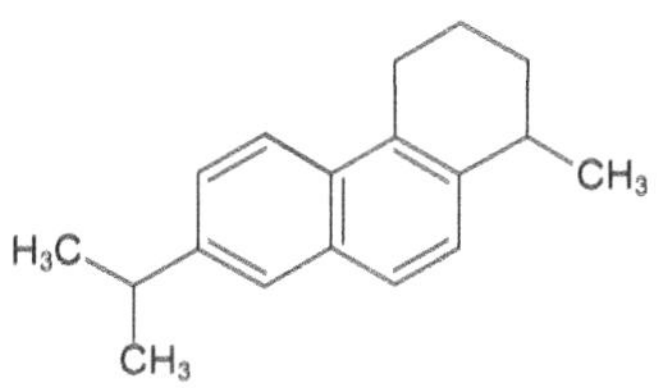

Undecylcyclohexane

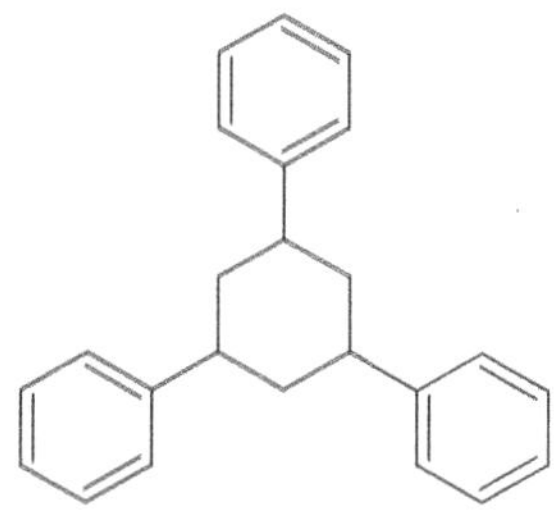

1,3,5-Triphenylcyclohexane

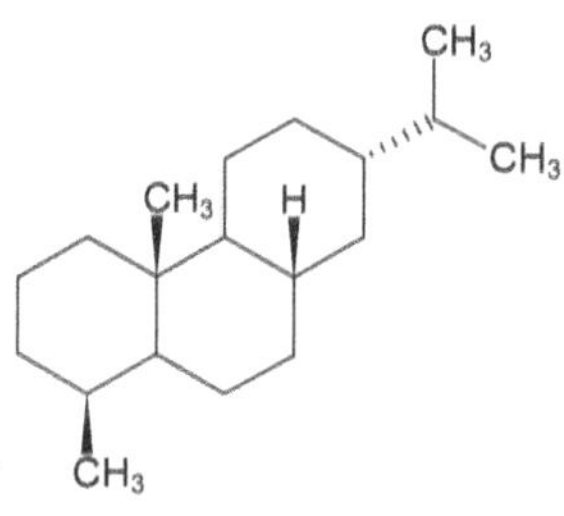

7-Isopropyl-1,4-dimethyl tetradecahydrophenanthrene

Figure 3.1 (cont) Cyclic and aromatic hydrocarbons.

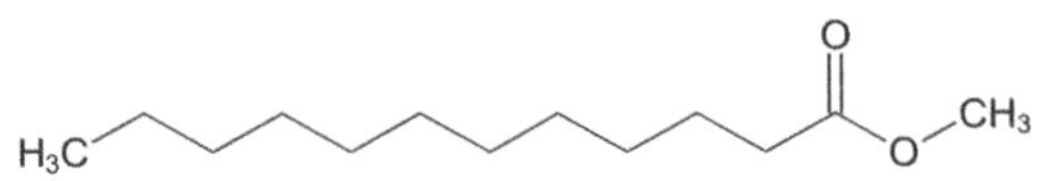

Methyl laurate

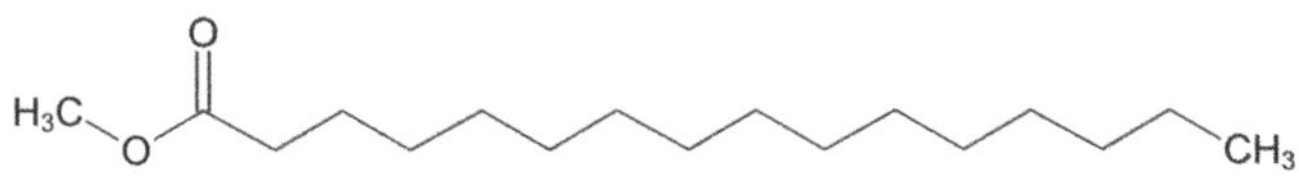

Methyl palmitate

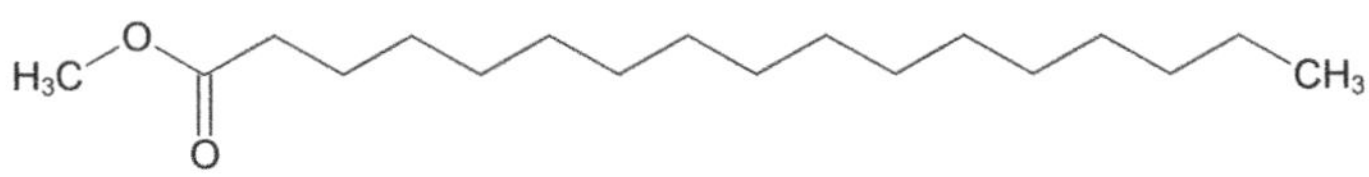

Methyl heptadecanoate

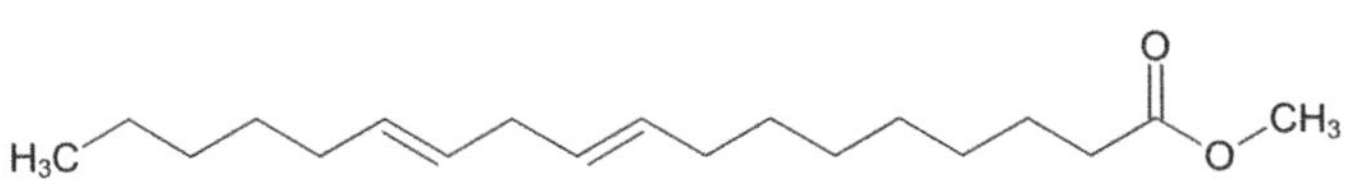

Methyl linolelaidate

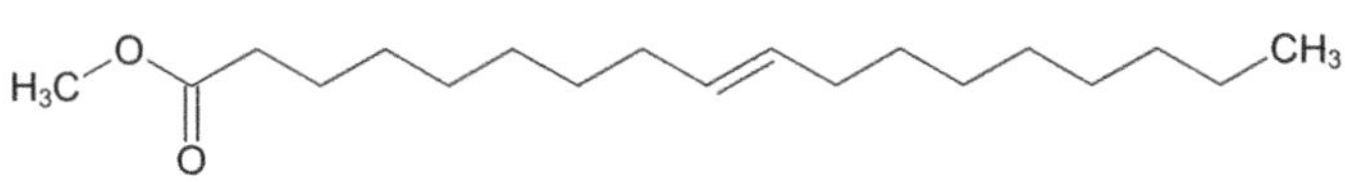

Methyl elaidate

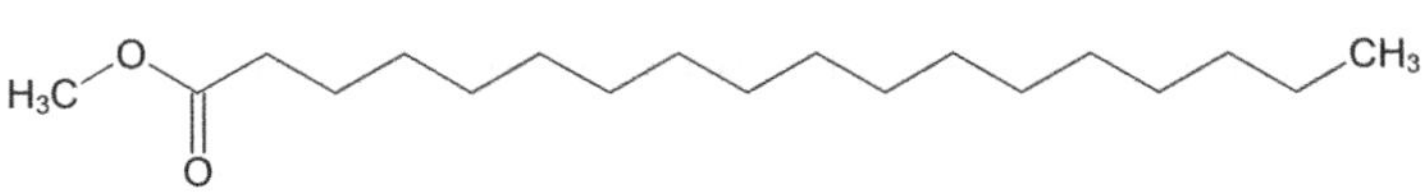

Methyl stearate

Figure 3.1 (cont) Fatty acid methyl esters.

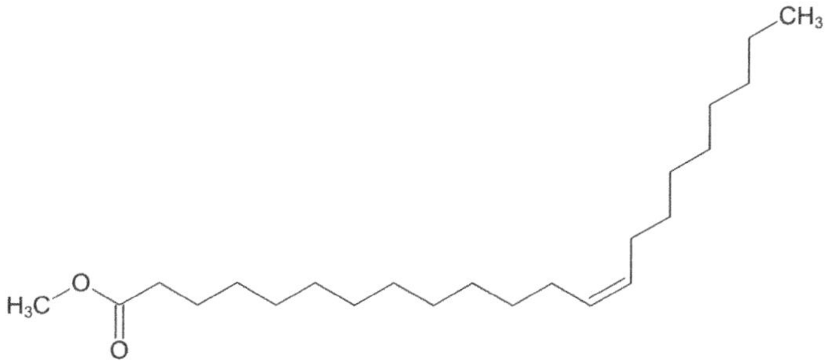

Methyl erucate

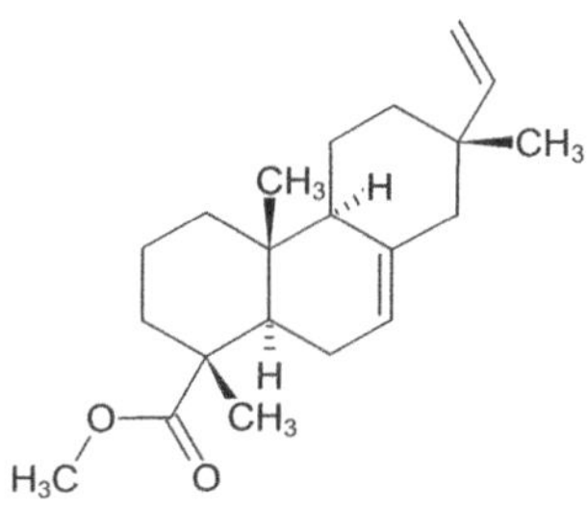

Methyl isopimarate

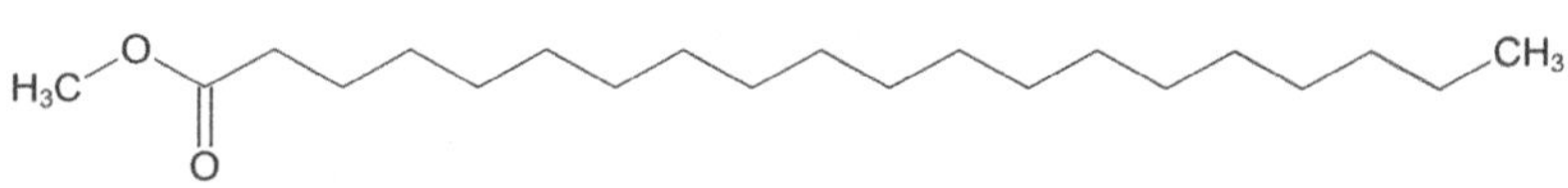

Methyl icosanoate

Figure 3.1 (cont) Fatty acid methyl esters.

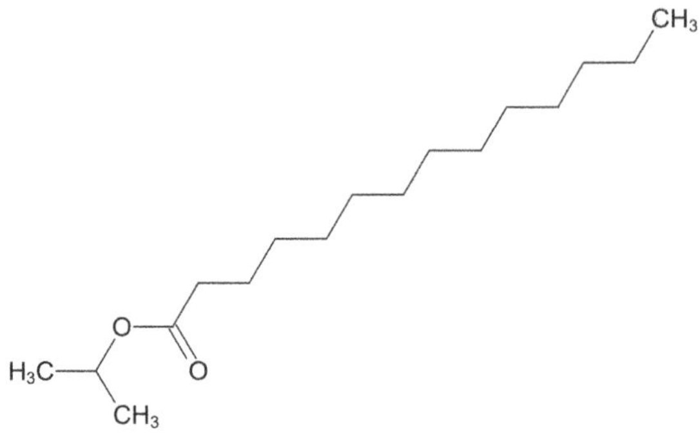

Isopropyl myristate

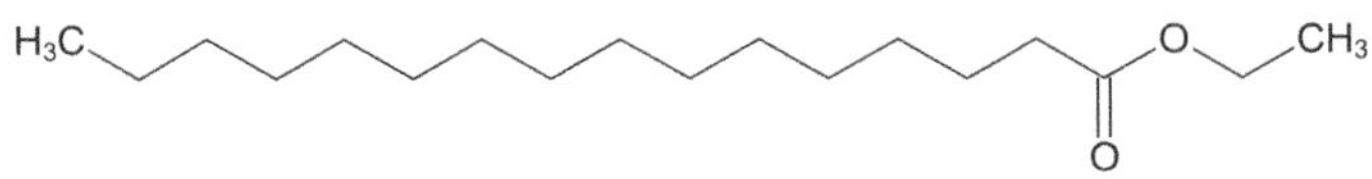

Ethyl palmitate

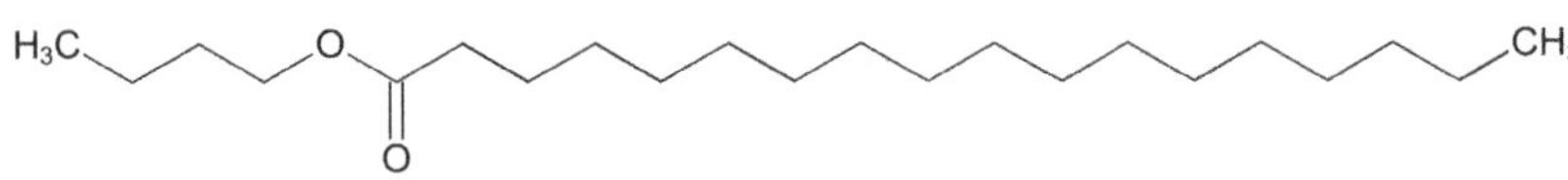

Butyl stearate

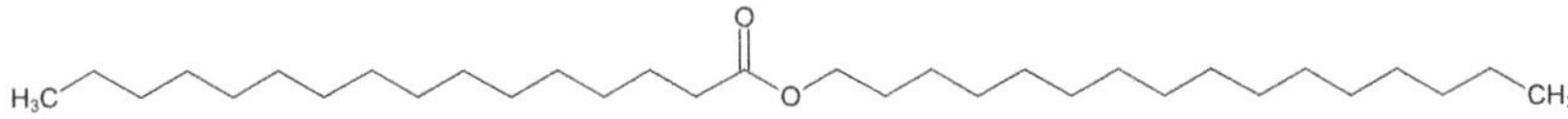

Hexadecyl palmitate

Figure 3.1 (cont) Other fatty acid esters.

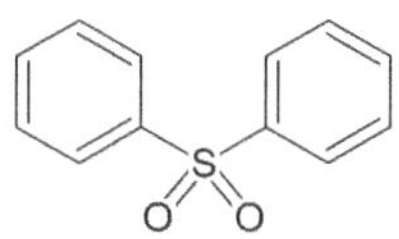

Diphenyl sulfone

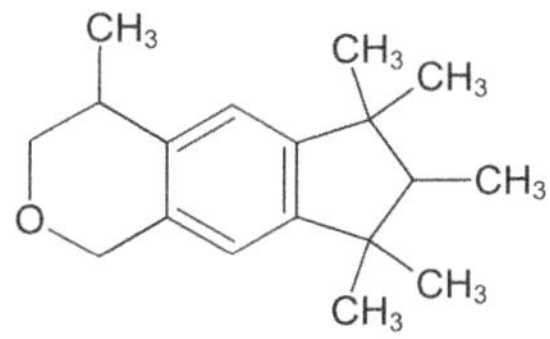

Galaxolide

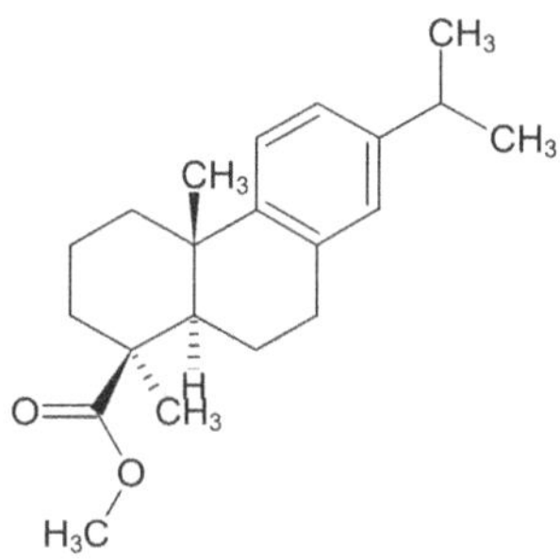

Methyl dehydroabietate

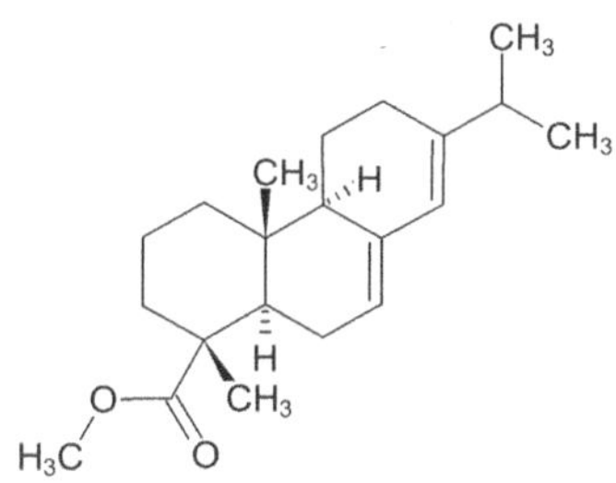

Methyl abietate

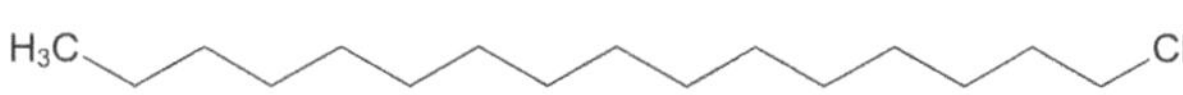

1-Chlorohexadecane

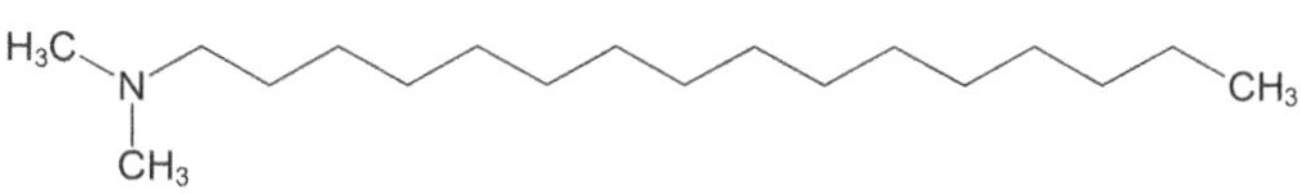

N,N-Dimethylpalmitylamine

Figure 3.1 (cont) Other compounds.

The determination of the hygiene status of the packaging material is of great importance. With the aim of achieving a maximum recovery rate at low contamination levels, an improved experimental design was developed for the enumeration of the total colony count, yeasts and molds and Enterobacteriaceae on the surface of roll stock packaging materials (5).

Two different types of paper laminates were selected and exemplarily used as objects of testing. The performance of different growth media was compared for each microbiological parameter. This approach was followed by a method of validation using a selection of quantitative reference materials of representative microorganisms, including resistant forms of microbes such as bacterial endospores and fungal spores (5).

3.3 Food Colorants

The state of a food can be shown in smart packaging materials (6). Here, the quality or freshness of the packaged food can be evaluated in real time. As a color indicator, a natural colorant, anthocyanin, drew a lot of attention due to its various colors as well as useful functional properties such as antioxidant activity and anticarcinogenic and anti-inflammatory effects, prevention of cardiovascular disease, obesity, and diabetes.

In particular, the pH-responsive color-changing function of anthocyanins is useful for making smart color indicator packaging films. The recent information on the use of natural pigment anthocyanins for intelligent and active food packaging applications has been reviewed. So, recent studies on eco-friendly biodegradable polymer-based color indicator films incorporated with anthocyanins have been addressed. Also, studies on the use of smart packaging films to monitor the freshness of foods such as milk, meat, and fish were reviewed. This review highlights the potential and challenges for the use of anthocyanins as pH-responsive color-changing films for intelligent food packaging applications, which may be beneficial for further development of smart color indicator films for practical use (6).

The chemical form of anthocyanins is based on the glycosides of flvylium salts, i.e., 2-phenyl benzo pyrylium salts. Cyanidin is a purple-reddish colored pigment mainly found in berries and vegetables having a red color like purple corn, sweet potato, and others (7, 8). Delphinidin is a red-blue pigment found in plants, and usually, the blue color of flowers is attributed to this pigment (9). Pelargonidin is a red-orange color pigment, mainly found in red fruits and orange flowers (10). Peonidin is a methylated anthocyanidin that exhibits magenta color in herbs such as berries and grapes (10). Malvidin is a purple-colored methylated anthocyanidin, mainly found in blue colored flowers such as Summer Wave blue and red wine (11). Petunidin is a water-soluble dark red colored methylated anthocyanidin (12).

Also, some other types of anthocyanidins are known, such as aurantindin, europinidin, luteolinidin, and rosinidin.

The color stability of the anthocyanin compounds varies depending on the pigment type, pH, light, temperature, co-pigment, metal ions, enzymes, oxygen and the structure of the anthocyanidin. The pH-dependent color-changing property of anthocyanins arises from the ionic nature of anthocyanin. In general, anthocyanins are reddish in acidic conditions, pink in neutral conditions, and blue in basic conditions.

The pH-dependent color-changing properties of anthocyanins were used to develop indicator films for monitoring the pH change of packaged foods in real time.

Polymer-based color indicator films were developed by adding anthocyanins. Biopolymers are mainly used rather than synthetic plastics as solid supports of anthocyanins because biopolymers are biodegradable, biocompatible, nontoxic and environmentally friendly (13, 14).

Biopolymers have many advantages over synthetic-petroleum based non-biodegradable polymers. Biopolymers are annually renewable and abundantly available, emit fewer greenhouse gases, thereby reducing carbon dioxide emissions, and are readily biodegradable in the environment.

Among the poly(saccharide)-based biopolymers, chitosan has received the most attention due to its abundance, biocompatibility, biodegradability, and functional properties such as antimicrobial and antioxidant activities. Compositions of chitosan-based intelligent packaging films are listed in Table 3.2.

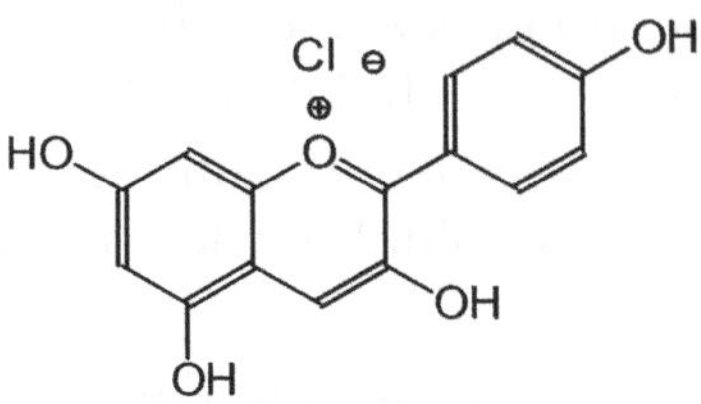

Pelargonidin

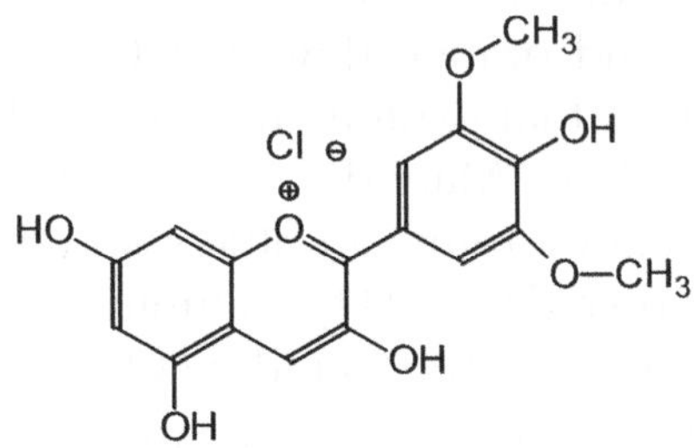

Malvidin

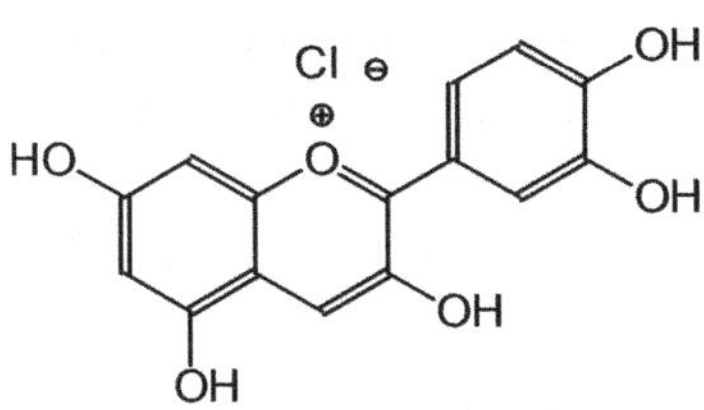

Cyanidin

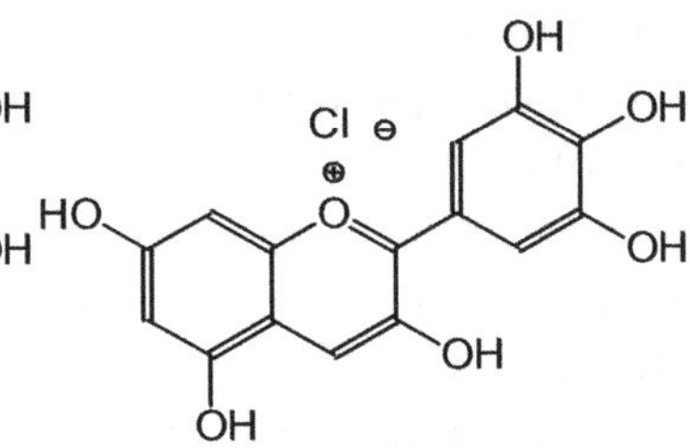

Delphinidin

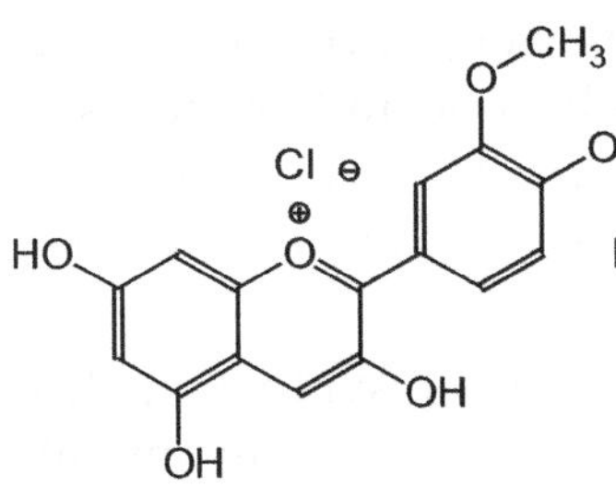

Peonidin

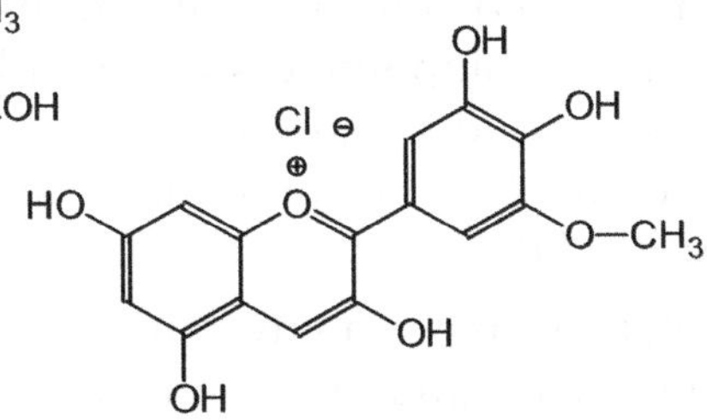

Petunidin

Figure 3.2 Color-changing compounds.

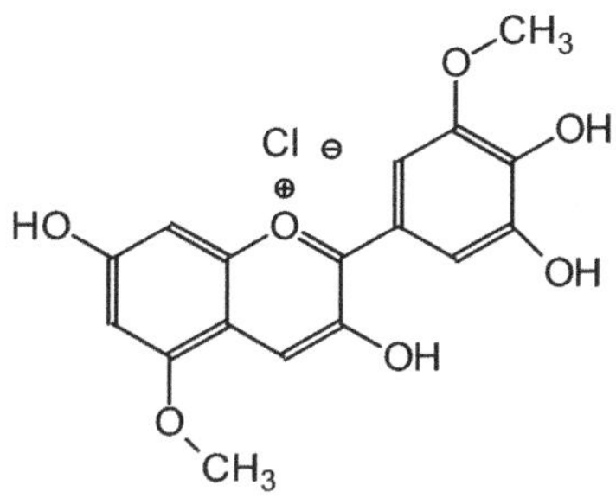

Europinidin

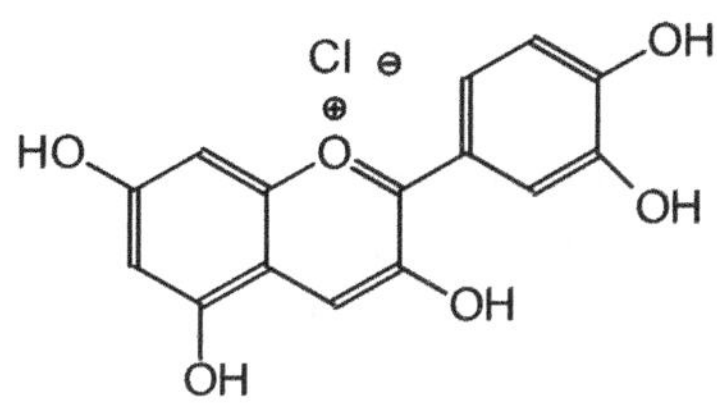

Luteolinidin

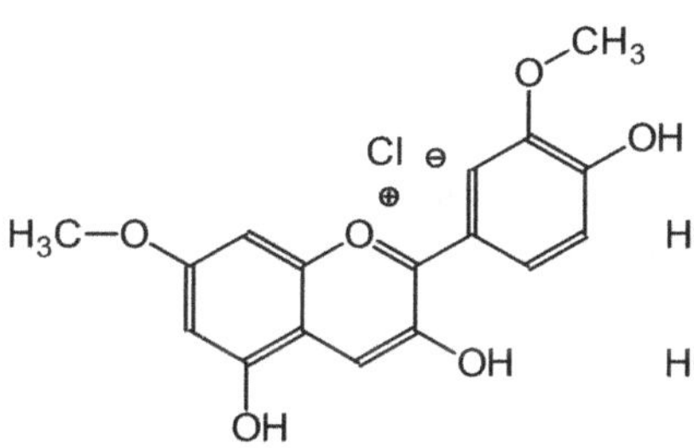

Rosinidin

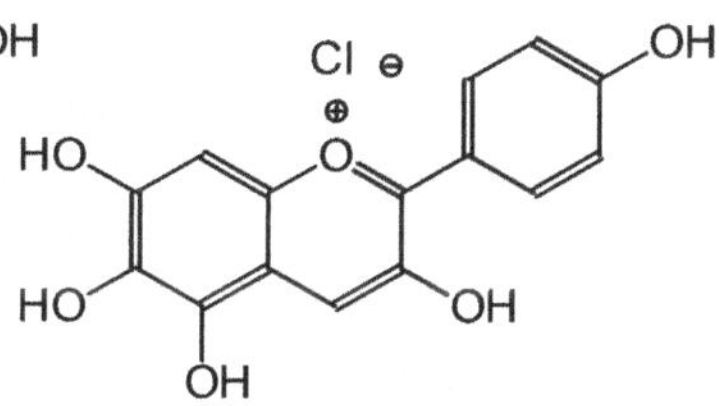

Aurantindin

Figure 3.2 (cont) Color-changing compounds.

Table 3.2 Chitosan-based intelligent packaging films (6).

Anthocyanins	Applications	Ref.
Red cabbage	Smart packaging for monitoring pork belly slices spoilage	(15)
Black soybean seed coat extract	Antioxidant and pH-sensing intelligent food packing	(16)
Purple corn extract	Intelligent food packaging films	(17)
Roselle anthocyanins	Checking pork freshness	(18)
Black chokeberry pomace extract	Colorimetric pH indicator film application	(19)
Grapes anthocyanin	Smart food packaging films	(20)
Aqueous hibiscus extract	Intelligent food packaging films	(21)
Blueberry and blackberry pomace extracts	Smart food packaging films	(22)
Red cabbage	Intelligent food packaging films	(23)
Brassica oleraceae (red cabbage)	Indicator of fish fillet deterioration	(23)
Purple-fleshed sweet potato extract	Intelligent pH-sensing films	(24)
Black carrot anthocyanin	Smart food packaging films	(25)
Purple potato extractions	Smart food packaging films	(24)
Black rice bran anthocyanins	Seafood (fish and shrimp) spoilage monitoring	(26)
Purple rice extract and black rice extract	Pork spoilage monitoring	(27)
Purple and black eggplant extracts	Milk spoilage monitoring	(28)
Blueberry and red grape skin pomace extracts	Monitoring freshness of chicken meat	(29)
Black plum peel extract	pH-sensitive food packing	(18)
Eggplant	Intelligent food packaging films	(30)
Carrot anthocyanin	Monitoring milk spoilage	(31)
Jambolan (*Syzygium cumini*)	Monitoring freshness of shrimp	(32)
Pomegranate peel extract	Monitoring cream cheese spoilage	(33)

Anthocyanin-based pH-responsive smart packaging films for packaging applications were tested for various food products. Mostly, the indicator films were tested for sensing the deterioration of milk, meat, and fish products.

3.3.1 *Milk Freshness Testing*

A smart packaging was tested with a pH indicator in pasteurized milk (34). The evaluation of the pH indicator function was performed using an intelligent composite film immersed in milk at room temperature for 48 *h*. After 48 *h*, the color of the indicator film changed from purple to red to indicate that the deterioration of milk occurred.

Grape skins extract incorporated tara gum (*Caesalpinia spinosa*) and cellulose-based packaging film for the detection of milk spoilage were tested (35). It was found that the pH of milk was initially 6.5 at room temperature, but after 48 *h* of storage, it changed to 2.94, so the change in pH became apparent from a color change in the film. As the pH of milk changes due to deterioration of milk and growth of microorganisms, the degradation of milk quality can be evaluated using a pH-sensitive smart packaging film.

Rosemary anthocyanin incorporated PVA/starch/propolis-based composite film was prepared and tested for monitoring the quality of pasteurized milk (36).

Here it was observed that the color of the film changed with the content of the propolis. When the packaging film contained lower content of propolis (10%), it changed from brown to reddish pink after 48 *h* of storage, while when the packaging film contained 20% of propolis, the color did not change after the storage. The color change of the former film was due to the decreased pH of milk, but milk packed with the latter film did not change the pH of milk due to the antibacterial activity of the film preventing the growth of lactic acid bacteria. The color change of the smart color indicator film was due to the production of lactic acid by the growth of lactic acid bacteria.

3.3.2 *Meat Freshness Testing*

A double-layer indicator film was prepared. The indicator layer was made of carrageenan, curcumin and/or anthocyanin. The emulsified layer was made of konjac glucomannan and camellia oil. These compounds are shown in Figure 3.3.

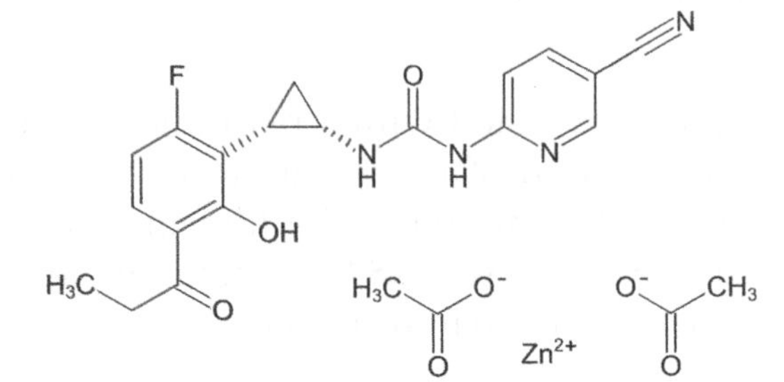

Carrageenan

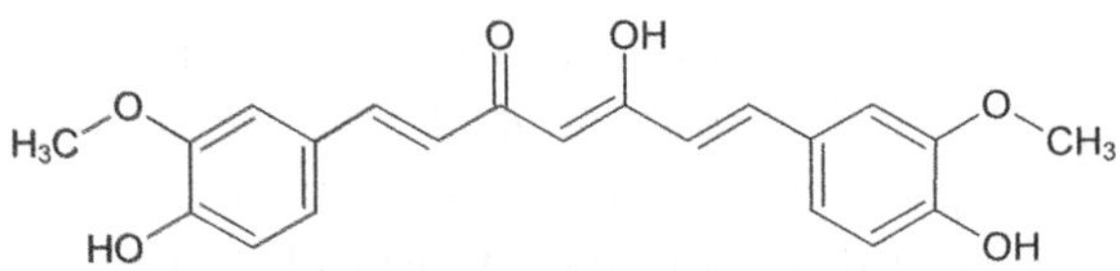

Curcumin

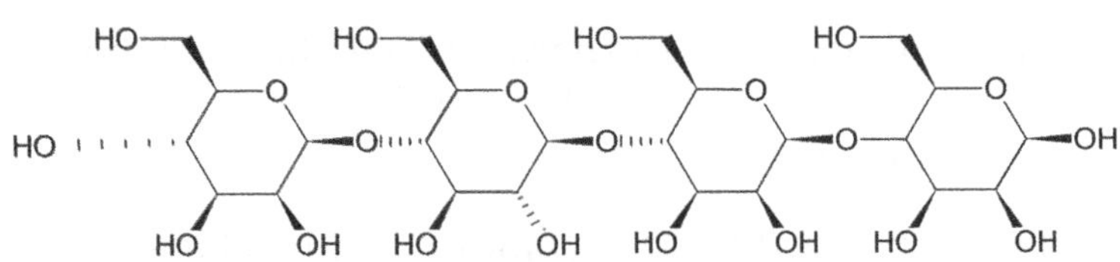

Glucomannan

Figure 3.3 Compounds for a double-layer indicator film.

The film was fabricated to monitor the freshness of chicken meat. All three kinds of the double-layer indicators were tested. The results of a Fourier transform infrared study demonstrated that different components were bound together by hydrogen bond.

The double-layer structure films were clearly observed by scanning electron microscopy images. The presence of the layer made from konjac glucomannan and camellia oil significantly improved the thermal stability, the water resistance and the UV resistance, and also the mechanical properties of the indicator layer.

It was found that the emulsified layer with carrageenan and anthocyanin and the emulsified layer with carrageenan, curcumin and anthocyanin could be used to monitor the chicken meat freshness at 25°C with a more obvious color change than the emulsified layer with carrageenan and curcumin (37).

3.3.3 Fish Freshness Testing

The pH indicator-based smart packaging films have also been used for the packaging of shrimp and fish fillets. The freshness of fish (grass carp) was monitored using a smart packaging film made of carboxymethyl cellulose/starch and purple-fleshed sweet potato (38). It was found that this intelligent film was able to monitor the freshness of grass carp at 20°C as it changed its color from red to blue.

The freshness of shrimp was tested using a smart film made of PVA/okra mucilage poly(saccharide) film blended with rose anthocyanins (39).

It was found that the color of the shrimp-packed smart film first changed from purple to blue at 18 *h*, then to dark green at 24 *h*, and finally to yellow at 32 *h* of the test, which is very effective in indicating the quality of the packaged shrimp. The authors also showed that the increased tendency of the total volatile basic nitrogen in packaged shrimp depends on the color change of the film (39).

In another study, mulberry anthocyanin incorporated ethylene-vinyl alcohol/montmorillonite-based multilayer film was tested for monitoring the quality of packaged shrimp (40). In this study, the authors showed that the color of the packaging film changed from rose red to light blue after 24 *h*.

3.4 Standards

Several testing methods for packaging materials are summarized in a chapter of a monograph (2). Some standards are collected in Table 3.3.

Table 3.3 ASTM Standards (2).

Number	Year	Name
B479-19	2019	Standard specification for annealed aluminum and aluminum-alloy foil for flexible barrier, food contact, and other applications.
D1653-13	2021	Standard test methods for water vapor transmission of organic coating films.
D2457-21	2021	Standard test method for specular gloss of plastic films and solid plastics.
D3078-02	2021	Standard test method for determination of leaks in flexible packaging by bubble emission.
D3418-21	2021	Standard test method for transition temperatures and enthalpies of fusion and crystallization of polymers by differential scanning calorimetry.
D3985-17	2017	Standard test method for oxygen gas transmission rate through plastic film and sheeting using a coulometric sensor.
D4003-98	2019	Standard test methods for programmable horizontal impact test for shipping containers and systems.
D523-14	2018	Standard test method for specular gloss.
D5277-92	2015	Standard test method for performing programmed horizontal impacts using an inclined impact tester.
D6604-00	2017	Standard practice for glass transition temperatures of hydrocarbon resins by differential scanning calorimetry.
D6868-21	2021	Standard specification for labeling of end items that incorporate plastics and polymers as coatings or additives with paper and other substrates designed to be aerobically composted in municipal or industrial facilities.
D880-92	2021	Standard test method for impact testing for shipping containers and systems.
E1356-08	2014	Standard test method for assignment of the glass transition temperatures by differential scanning calorimetry.
E1870-11	2019	Standard test method for odor and taste transfer from polymeric packaging film.

Table 3.3 (cont) ASTM Standards (2).

Number	Year	Name
E2602-22	2022	Standard test methods for the assignment of the glass transition temperature by modulated temperature differential scanning calorimetry.
E398-20	2020	Standard test method for water vapor transmission rate of sheet materials using dynamic relative humidity measurement.
E460-21	2021	Standard practice for determining effect of packaging on food and beverage products during storage.
E96M-22	2022	Standard test methods for gravimetric determination of water vapor transmission rate of materials.
F1249-20	2020	Standard test method for water vapor transmission rate through plastic film and sheeting using a modulated infrared sensor.
F1307-20	2020	Standard test method for oxygen transmission rate through dry packages using a coulometric sensor.
F1356-22	2022	Standard guide for irradiation of fresh, frozen or processed meat and poultry to control pathogens and other microorganisms.
F1640-21	2021	Standard guide for selection and use of contact materials for foods to be irradiated.
F1886-16	2016	Standard test method for determining integrity of seals for flexible packaging by visual inspection-
F1921	2018	Standard test methods for hot seal strength (hot tack) of thermoplastic polymers and blends comprising the sealing surfaces of flexible webs.
F1927-20	2020	Standard test method for determination of oxygen gas transmission rate, permeability and permeance at controlled relative humidity through barrier materials using a coulometric detector.
F2096-11	2010	Standard test method for detecting gross leaks in packaging by internal pressurization (bubble test).
F2338-09	2020	Standard test method for nondestructive detection of leaks in packages by vacuum decay method.
F2391-05	2016	Standard test method for measuring package and seal integrity using helium as the tracer gas.
F2714-08	2021	Standard test method for oxygen headspace analysis of packages using fluorescent decay.

Table 3.3 (cont) ASTM Standards (2).

Number	Year	Name
F3039-16	2015	Standard test method for detecting leaks in non-porous packaging or flexible barrier materials by dye penetration.
F3169-16	2017	Standard test method for leak detection in blister packaging by vacuum deflection method by laser measurement.
F372-99	2010	Standard test method for water vapor transmission rate of flexible barrier materials using an infrared detection technique (withdrawn 2009).

3.4.1 *Odor and Taste Transfer from Packaging Film*

The ASTM standard E1870-11 (41) deals with a method for examining the odor and taste properties of a polymeric film that is intended for use as flexible packaging materials. This test method can be used for single layers, coextruded, and laminate materials. The focus of this test method is the evaluation of the film in terms of its perceived inherent odor and the transfer of package-related odors, or flavors, to water and other model systems, e.g., bland food simulants.

This test method provides two methods of evaluating the film performance: the Score Method, which allows the comparison of any film sample to another, and the Ranking Method, which allows the comparison of samples within a set.

3.4.2 *Annealed Aluminum and Aluminum-Alloy Foil for Food Contact*

The ASTM standard B479-19 (42) deals with annealed aluminum and aluminum-alloy foil used in packaging, which includes food handling and processing applications, requiring flexible barrier materials, and other general applications. The chemical composition shall be in conformity with provisions in the standard, and shall be determined through suitable chemical and spectrochemical methods. Tensile breaking load limits shall be as agreed, and shall be tested twice with test methods for tension testing of metallic foil using two samples from different rolls of foil.

3.4.3 Detection of Leaks in Packages

3.4.3.1 Vacuum Loss Measurement

The ASTM standard F2338-09 (43) can detect package leaks by measuring the rise in pressure (vacuum loss) in an enclosed evacuated test chamber containing the test package. Vacuum loss results from leakage of test package headspace gases and/or volatilization of test package liquid contents located in or near the leak. When testing for leaks that may be partially or completely plugged with the package's liquid contents, the test chamber is evacuated to a pressure below the vaporization pressure of the liquid. All methods require a test chamber to contain the test package and a leak detection system designed with one or more pressure transducers. The sensitivities can be determined using specific product-package systems selected for the precision and bias studies.

Trays sealed with porous barrier lidding material (headspace gas leakage)-hole or crack defects in the wall of the tray/cup of at least 100 μm in diameter can be detected. Also, channel defects in the seal area (made using wires of 125 μm in diameter) can be detected.

Severe seal bonding defects in both continuous adhesive and dot matrix adhesive package systems can be detected. Slightly incomplete dot matrix adhesive bonding defects can also be detected. All porous barrier lidding material packages were tested at -400 *mbar*. The sensitivity of the test for porous lidded packages is approximately 10^{-2} $Pa\,m^3 s^{-1}$ using a calibrated volumetric airflow meter.

3.4.3.2 Bubble Emission

The ASTM standards F2096-11 (44) and ASTM D3078-02 (45) can detect leaks in flexible packaging by bubble emission.

These test methods cover the determination of gross leaks in flexible packaging containing a headspace gas. The test sensitivity is limited to $1{\times}10^{-6}$ $Pa^{m}3s^{-1}$.

Small leaks may not be detected by this procedure. Viscoelastic effects on the products, or entrapped air, become significant and prevent passage through small openings. A positive pressure inside the pouch after the vacuum is drawn may force the product to plug small leaks. The size of the leak that can be detected is dependent upon the products contained, the nature of the packaging material, and the test parameters selected (45).

Traditional and recently proposed test methods for the measurement of hermeticity in packages with very small cavity volumes have been detailed (46).

Closed-form expressions of the minimum and maximum true leak rates achievable are provided for the helium fine leak test method. These expressions are shown to provide practical guidelines for the accurate testing of hermeticity for ultra-small packages. A portfolio of hermeticity test methods was also presented outlining the limitations and advantages of each method (46).

3.4.3.3 Gas Transmission Rates

The assessment of the transport properties of gases in polymeric materials is of importance for the evaluation of the performance of packaging materials in order to protect packed food products and ensure a prolonged shelf life (47). The most relevant substances for such application can be identified in O_2, CO_2 and C_2H_4, together with N_2 as control, as permanent gases, and water vapor as condensable species.

All these penetrants are able to significantly alter the properties of food in terms of sensory characteristics, textures, and quality.

Relevant techniques for the experimental characterization of such mass transport properties in polymeric films or packages were critically reviewed (47). Standard methods were described In detail. The following methods are available for the characterization of gas transport (47):

- Static permeation with pure penetrants,
- Static permeation with concentration analysis,
- Dynamic permeation with concentration analysis,
- Dynamic permeation with specific sensor,
- Gas permeation in packages,
- Effect of temperature or moisture on gas permeability,
- Water permeation by a gravimetric method, and
- Water permeation in dynamic systems.

Oxygen Gas Transmission Rate. The ASTM standard D3985-17 (48) is a test method for oxygen gas transmission rate through plastic film and sheeting using a coulometric sensor.

The effects of methyl cellulose coating containing carvacrol or menthol, c.f. Figure 3.4, on the physicochemical, mechanical, and antimicrobial activity of poly(ethylene) (PE) films were investigated (49).

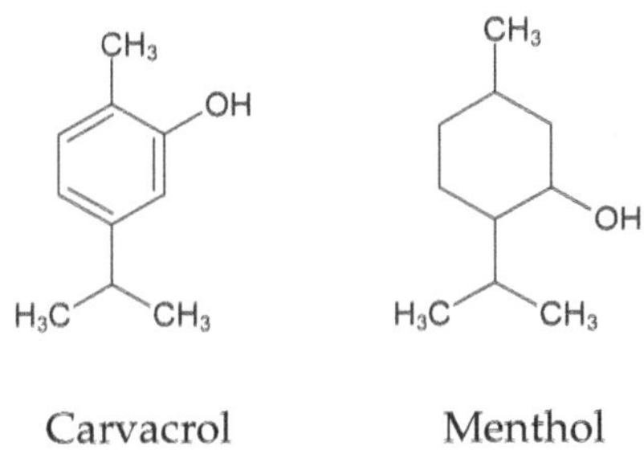

Figure 3.4 Coating materials.

The oxygen permeability of PE films, the ASTM Standard Method D3985-17 and a Mocon OXTRAN 2/21 system (Minneapolis, USA) were used. Film samples were conditioned at a temperature of 25°C, atmospheric pressure, and a relative humidity of 55% for 48 h. Then, their thickness was measured and mounted into the diffusion cell of the equipment. By using the convergent method, the oxygen permeability of the films was estimated by WinPerm® software (49).

The ASTM standard F1307-14 (50) is a test method for the gas transmission rate of oxygen through dry packages using a coulometric sensor.

3.4.3.4 *Integrity of Seals for Flexible Packaging*

The ASTM standard F1886-16 (51) is a test method for the integrity of seals for flexible packaging by visual inspection.

The issues of sealing are most problematic in food packaging for controlling the moisture and gas ingress and preserving the product quality (52).

Heat sealing is a widely used method in flexible packaging. It is critical for maintaining the product quality throughout the storage and preventing food waste. Understanding the factors causing the leak formation in the seal interface helps to avoid failures and to increase the integrity for all seal types produced by heat sealing methods.

During a heat sealing process, multiple factors affect seal strength and integrity development. These factors can be grouped as (52):

1. Process parameters,
2. Material properties,
3. Contaminants, and
4. Further processing.

The sealing quality is strongly dependent on the main process parameters, i.e., temperature, time and pressure (53, 54).

By using the correlations between these parameters and precise controlling, different types of seals, such as lock seal or peel seal, can be obtained. The energy consumption can be decreased or high-speed processing can be achieved without even changing the material type (55).

However, a poor optimization of these parameters may cause a lot of problems, which eventually results in poor sealing and leakages. Even though the effects of some process parameters, such as time, temperature and pressure on the seal quality, are well studied, the roles of other parameters, such as contaminants, seal bar design and cooling rate on the formation of gaps and channels at the sealing area, are still unknown. As a result, understanding the factors/mechanisms behind the formation of gaps and channels that result in leakages is critical to prevent food waste (52).

The seal strength is an important quality parameter for packaged products such as biaxially oriented poly(propylene) (PP) films (54). The seal quality can be insured and maintained under optimum heat sealing process parameters. This article (54) shows that heat seal strength can be optimized through simultaneous study of dwell time, pressure, and temperature.

A heat seal strength optimization trial for 18 μm thick untreated, both sides sealable biaxially oriented PP film, utilized response surface methods with a replicated central composite statistical experimental design. This gives a second-order model for the heat seal strength. This experiment showed that dwell time, temperature, and their interaction have significant effects and that high seal strength is attained at a dwell time of 0.9 *s* and a temperature of 120°C (54).

3.4.3.5 *Dye Penetration*

The ASTM standard F3039-15 (56) is a test method for detecting leaks in nonporous packaging or flexible barrier materials by dye penetration.

3.4.3.6 *Package Integrity*

The ASTM standard F2391-05 (57) is a test method for measuring the integrity of a package and the seal using helium as tracer gas.

3.4.3.7 *Leak Detection*

The ASTM standard F3169-16 (58) is a test method for leak detection in blister packaging by a vacuum deflection method by laser measurement.

3.4.4 *Changes in Sensory Attributes*

The ASTM standard E460-21 has been designed to detect the changes in sensory attributes of foods and beverages stored in various packaging materials or systems (59).

This practice may be used for testing a wide variety of materials associated with many kinds of products. There are many ways in which a packaging material may influence a product during storage. First, the packaging material may contaminate the product with off-flavors by direct transfer of packaging component compounds to the product. Second, the packaging material may adsorb components from the product which may then be further transferred to the atmosphere, thus reducing aroma intensity in the product. Third, external contaminants may permeate the package and possibly be transferred to the product. In addition to flavor influences, packaging materials may allow color or textural changes, or both, and many other measurable sensory effects.

3.4.5 Irradiated Food Contact Materials

Food irradiation is a promising technology for food processing and is considered as a safe, clean and eco-friendly process (60). The effects of ionizing radiation (γ-electron-beam and X-rays) on natural matrices depend on the food product type, size, state and temperature, and on the irradiation conditions.

The effects of this technology on food bioactives have been detailed in a chapter of a monograph (60). The bioactive compounds extracted from the irradiated food can be suitable alternatives to use in food, cosmetic and pharmaceutical industries, promoting the circular economy.

The ASTM standard F1640-21 (61) provides a format to assist producers in selecting food contact materials that have the desirable characteristics for their intended use and that comply with applicable standards or government authorizations. It outlines some parameters that should be considered when selecting food contact materials intended for use during irradiation of prepackaged foods and examines the criteria for fitness for their use.

This standard identifies known regulations and regulatory frameworks worldwide pertaining to food contact materials for holding foods during irradiation. However, it does not address all regulatory issues associated with the selection and the use of packaging materials for foods to be irradiated.

Meats are an ideal medium for the proliferation of microorganisms. So the extended shelf life is important for safety (62). This can be done by using food irradiation. Several issues, like radiation source, dose ranges and dose rate are discussed, together with the logistics and operational considerations (63).

In the ASTM standard F1356-22 (64), the issues of irradiation of fresh, frozen or processed meat and poultry to control pathogens and other microorganisms are presented.

The effect of the sources of irradiation, i.e., γ-ray, electron-beam, and X-ray, and the dose levels on the physicochemical, organoleptic and microbial properties of cooked beef patties and pork sausages was studied (65). The experiments were performed during 10 d of storage at 30±1°C. The processed meat products were irradiated at 0, 2.5, 5, 7.5, and 10 kGy by the above-mentioned three different irradiation sources.

The pH of cooked beef patties and pork sausages was unaffected by irradiation sources or their doses. The redness of beef patties linearly decreased with increasing dose level ($P<0.05$), obviously by electron-beam irradiation compared to γ-ray and X-ray ($P<0.05$). The redness of pork sausages was increased by γ-ray irradiation, whereas it decreased by electron-beam irradiation depending on absorbed dose level. No significant changes in overall acceptability were observed for pork sausages regardless of irradiation source ($P>0.05$), while γ-ray irradiated beef patties showed significantly decreased overall acceptability in a dose-dependent manner ($P<0.05$) (65).

The lipid oxidation of samples was accelerated by the irradiation depending on irradiation sources and dose levels during storage at 30°C. The electron-beam reduced the total aerobic bacteria of beef patties more effectively, while γ-ray considerably decreased microbes in pork sausages as the irradiation dose increased.

So, the quality attributes of meat products, in particular color, lipid oxidation, and microbial properties, are significantly influenced by the irradiation sources (65).

The combined effect of low-dose γ-irradiation (1.0 kGy), antioxidants (rosemary extract, c.f. Figure 3.5, and butylated hydroxytoluene, c.f. Figure 6.3) and antimicrobial materials (nisin and sodium nitrate) on physicochemical, oxidative and microbial quality of mutton meat emulsions was evaluated (66).

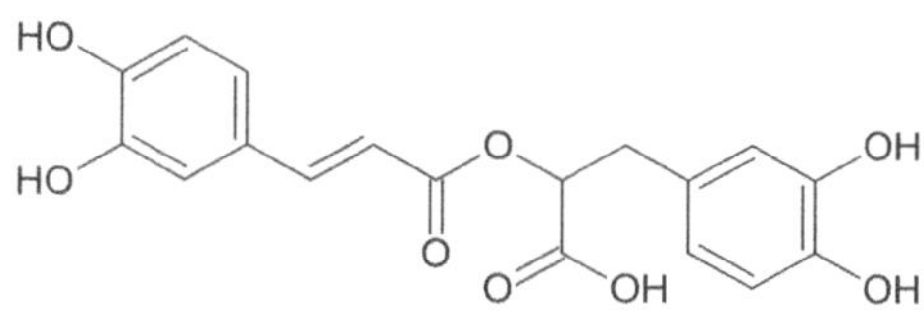

Figure 3.5 Rosemarinic acid.

Irradiated samples treated with rosemary extract and butylated hydroxytoluene exhibited a significantly ($P = 0.05$) higher emulsion stability, cooking yield and penetration force and displayed non-significant variation in protein and moisture content (66).

The results revealed that peroxide value, thiobarbituric acid reactive substances (TBARs) value (c.f. Figure 3.6), carbonyl content and loss of --S–H sulfydryl groups of irradiated samples treated with rosemary extract were comparable to that of butylated hydroxytoluene, which indicated the protective effect of natural antioxidant against irradiation-induced oxidation.

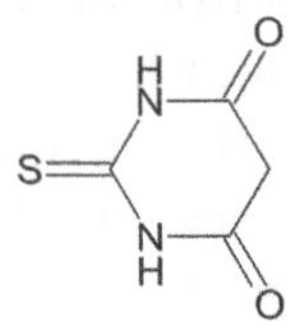

Figure 3.6 Thiobarbituric acid.

Irradiation and a nisin treatment significantly reduced bacterial growth as compared to irradiated sodium nitrate treatments. Therefore, the application of low-dose γ-irradiation in combination with rosemary extract and nisin could be employed for a better quality retention of meat emulsions (66).

3.4.6 Oxygen Headspace Analysis

The ASTM standard F2714-08 (67) is a procedure for the determination of the oxygen concentration in the headspace within a sealed package without opening or compromising the integrity of the package.

This test method requires that chemically coated components are placed on the inside surface of the package before closing. The package must be either transparent, translucent, or a transparent window must be affixed to the package surface without affecting the package's integrity.

Since this test method determines the oxygen headspace over time, the oxygen permeability can easily be calculated as ingress per unit time as long as the volume of the container is known.

References

1. P. Singh, A.A. Wani, and H.-C. Langowski, (Eds.), *Food Packaging Materials: Testing & Quality Assurance*, CRC Press, 2017.

2. H. Jayan, J.A. Moses, and C. Anandharamakrishnan, Testing methods for packaging materials in *Bio-Based Materials for Food Packaging*, pp. 57–79. Springer, 2018.
3. P. Miralles, V. Yusà, A. Pineda, and C. Coscollà, A fast and automated strategy for the identification and risk assessment of unknown substances (IAS/NIAS) in plastic food contact materials by GC-Q-Orbitrap HRMS: Recycled LDPE as a proof-of-concept, *Toxics*, Vol. 9, p. 283, 2021.
4. M. Moradi, S.A. Kousheh, R. Razavi, Y. Rasouli, M. Ghorbani, E. Divsalar, H. Tajik, J.T. Guimarães, and S.A. Ibrahim, Review of microbiological methods for testing protein and carbohydrate-based antimicrobial food packaging, *Trends in Food Science & Technology*, Vol. 111, p. 595, 2021.
5. M. Feichtinger, U. Zitz, H. Fric, W. Kneifel, and K.J. Domig, An improved method for microbiological testing of paper-based laminates used in food packaging, *Food Control*, Vol. 50, p. 548, 2015.
6. S. Roy and J.-W. Rhim, Anthocyanin food colorant and its application in pH-responsive color change indicator films, *Critical Reviews in Food Science and Nutrition*, Vol. 61, p. 2297, 2021.
7. N.P. Seeram, R.A. Momin, M.G. Nair, and L.D. Bourquin, Cyclooxygenase inhibitory and antioxidant cyanidin glycosides in cherries and berries, *Phytomedicine*, Vol. 8, p. 362, 2001.
8. B.A. Cevallos-Casals and L. Cisneros-Zevallos, Stoichiometric and kinetic studies of phenolic antioxidants from Andean purple corn and red-fleshed sweetpotato, *Journal of Agricultural and Food Chemistry*, Vol. 51, p. 3313, 2003.
9. Y. Katsumoto, M. Fukuchi-Mizutani, Y. Fukui, F. Brugliera, T.A. Holton, M. Karan, N. Nakamura, K. Yonekura-Sakakibara, J. Togami, A. Pigeaire, G.-Q. Tao, N.S. Nehra, C.-Y. Lu, B.K. Dyson, S. Tsuda, T. Ashikari, T. Kusumi, J.G. Mason, and Y. Tanaka, Engineering of the rose flavonoid biosynthetic pathway successfully generated blue-hued flowers accumulating delphinidin, *Plant and Cell Physiology*, Vol. 48, p. 1589, 2007.
10. L. Jaakola, New insights into the regulation of anthocyanin biosynthesis in fruits, *Trends in Plant Science*, Vol. 18, p. 477, 2013.
11. G. Mazza and F.J. Francis, Anthocyanins in grapes and grape products, *Critical Reviews in Food Science & Nutrition*, Vol. 35, p. 341, 1995.
12. H.E. Khoo, A. Azlan, S.T. Tang, and S.M. Lim, Anthocyanidins and anthocyanins: Colored pigments as food, pharmaceutical ingredients, and the potential health benefits, *Food & Nutrition Research*, Vol. 61, p. 1361779, 2017.
13. T. Hoffmann, D. Peters, B. Angioletti, S. Bertoli, L. Péres, M. Reiter, and C. De Souza, Potentials nanocomposites in food packaging, *Chem. Eng. Trans.*, Vol. 75, 2019.

14. S. Kumar, A. Mukherjee, and J. Dutta, Chitosan based nanocomposite films and coatings: Emerging antimicrobial food packaging alternatives, *Trends in Food Science & Technology*, Vol. 97, p. 196, 2020.
15. T.-V. Vo, T.-H. Dang, and B.-H. Chen, Synthesis of intelligent pH indicative films from chitosan/poly (vinyl alcohol)/anthocyanin extracted from red cabbage, *Polymers*, Vol. 11, p. 1088, 2019.
16. X. Wang, H. Yong, L. Gao, L. Li, M. Jin, and J. Liu, Preparation and characterization of antioxidant and pH-sensitive films based on chitosan and black soybean seed coat extract, *Food Hydrocolloids*, Vol. 89, p. 56, 2019.
17. Y. Qin, Y. Liu, L. Yuan, H. Yong, and J. Liu, Preparation and characterization of antioxidant, antimicrobial and pH-sensitive films based on chitosan, silver nanoparticles and purple corn extract, *Food Hydrocolloids*, Vol. 96, p. 102, 2019.
18. J. Zhang, X. Zou, X. Zhai, X. Huang, C. Jiang, and M. Holmes, Preparation of an intelligent pH film based on biodegradable polymers and roselle anthocyanins for monitoring pork freshness, *Food Chemistry*, Vol. 272, p. 306, 2019.
19. K. Halász and L. Csóka, Black chokeberry (Aronia melanocarpa) pomace extract immobilized in chitosan for colorimetric pH indicator film application, *Food Packaging and Shelf Life*, Vol. 16, p. 185, 2018.
20. C.M. Yoshida, V.B.V. Maciel, M.E.D. Mendonça, and T.T. Franco, Chitosan biobased and intelligent films: Monitoring pH variations, *LWT-Food Science and Technology*, Vol. 55, p. 83, 2014.
21. J. Peralta, C.M. Bitencourt-Cervi, V.B. Maciel, C.M. Yoshida, and R.A. Carvalho, Aqueous hibiscus extract as a potential natural pH indicator incorporated in natural polymeric films, *Food Packaging and Shelf Life*, Vol. 19, p. 47, 2019.
22. M. Kurek, I.E. Garofulić, M.T. Bakić, M. Ščetar, V.D. Uzelac, and K. Galić, Development and evaluation of a novel antioxidant and pH indicator film based on chitosan and food waste sources of antioxidants, *Food Hydrocolloids*, Vol. 84, p. 238, 2018.
23. J. Pereira, Valdir Aniceto, I.N.Q. de Arruda, and R. Stefani, Active chitosan/PVA films with anthocyanins from Brassica oleraceae (red cabbage) as time–temperature indicators for application in intelligent food packaging, *Food Hydrocolloids*, Vol. 43, p. 180, 2015.
24. Y. Li, Y. Ying, Y. Zhou, Y. Ge, C. Yuan, C. Wu, and Y. Hu, A pH-indicating intelligent packaging composed of chitosan-purple potato extractions strength by surface-deacetylated chitin nanofibers, *International Journal of Biological Macromolecules*, Vol. 127, p. 376, 2019.
25. M. Koosha and S. Hamedi, Intelligent chitosan/PVA nanocomposite films containing black carrot anthocyanin and bentonite nanoclays with improved mechanical, thermal and antibacterial properties, *Progress in Organic Coatings*, Vol. 127, p. 338, 2019.

26. C. Wu, J. Sun, P. Zheng, X. Kang, M. Chen, Y. Li, Y. Ge, Y. Hu, and J. Pang, Preparation of an intelligent film based on chitosan/oxidized chitin nanocrystals incorporating black rice bran anthocyanins for seafood spoilage monitoring, *Carbohydrate Polymers*, Vol. 222, p. 115006, 2019.
27. J. Liu, H. Wang, M. Guo, L. Li, M. Chen, S. Jiang, X. Li, and S. Jiang, Extract from Lycium ruthenicum Murr. incorporating κ-carrageenan colorimetric film with a wide pH–sensing range for food freshness monitoring, *Food Hydrocolloids*, Vol. 94, p. 1, 2019.
28. H. Yong, X. Wang, X. Zhang, Y. Liu, Y. Qin, and J. Liu, Effects of anthocyanin-rich purple and black eggplant extracts on the physical, antioxidant and pH-sensitive properties of chitosan film, *Food Hydrocolloids*, Vol. 94, p. 93, 2019.
29. M. Kurek, L. Hlupić, M. Ščetar, T. Bosiljkov, and K. Galić, Comparison of two pH responsive color changing bio-based films containing wasted fruit pomace as a source of colorants, *Journal of Food Science*, Vol. 84, p. 2490, 2019.
30. S. Bilgiç, E. Söğüt, and A.C. Seydim, Chitosan and starch based intelligent films with anthocyanins from eggplant to monitor pH variations, *Turkish Journal of Agriculture - Food Science and Technology*, Vol. 7, p. 61, 2019.
31. F.E. Tirtashi, M. Moradi, H. Tajik, M. Forough, P. Ezati, and B. Kuswandi, Cellulose/chitosan pH-responsive indicator incorporated with carrot anthocyanins for intelligent food packaging, *International Journal of Biological Macromolecules*, Vol. 136, p. 920, 2019.
32. B. Merz, C. Capello, G.C. Leandro, D.E. Moritz, A.R. Monteiro, and G.A. Valencia, A novel colorimetric indicator film based on chitosan, polyvinyl alcohol and anthocyanins from jambolan (Syzygium cumini) fruit for monitoring shrimp freshness, *International Journal of Biological Macromolecules*, Vol. 153, p. 625, 2020.
33. S. Pirsa, I. Karimi Sani, M.K. Pirouzifard, and A. Erfani, Smart film based on chitosan/melissa officinalis essences/pomegranate peel extract to detect cream cheeses spoilage, *Food Additives & Contaminants: Part A*, Vol. 37, p. 634, 2020.
34. B. Liu, H. Xu, H. Zhao, W. Liu, L. Zhao, and Y. Li, Preparation and characterization of intelligent starch/PVA films for simultaneous colorimetric indication and antimicrobial activity for food packaging applications, *Carbohydrate Polymers*, Vol. 157, p. 842, 2017.
35. Q. Ma and L. Wang, Preparation of a visual pH-sensing film based on tara gum incorporating cellulose and extracts from grape skins, *Sensors and Actuators B: Chemical*, Vol. 235, p. 401, 2016.
36. P. Mustafa, M.B.K. Niazi, Z. Jahan, G. Samin, A. Hussain, T. Ahmed, and S.R. Naqvi, PVA/starch/propolis/anthocyanins rosemary extract

composite films as active and intelligent food packaging materials, *Journal of Food Safety*, Vol. 40, p. e12725, 2020.

37. X. Zhou, X. Yu, F. Xie, Y. Fan, X. Xu, J. Qi, G. Xiong, X. Gao, and F. Zhang, pH-responsive double-layer indicator films based on konjac glucomannan/camellia oil and carrageenan/anthocyanin/curcumin for monitoring meat freshness, *Food Hydrocolloids*, Vol. 118, p. 106695, 2021.
38. G. Jiang, X. Hou, X. Zeng, C. Zhang, H. Wu, G. Shen, S. Li, Q. Luo, M. Li, X. Liu, et al., Preparation and characterization of indicator films from carboxymethyl-cellulose/starch and purple sweet potato (Ipomoea batatas (L.) lam) anthocyanins for monitoring fish freshness, *International Journal of Biological Macromolecules*, Vol. 143, p. 359, 2020.
39. S. Kang, H. Wang, L. Xia, M. Chen, L. Li, J. Cheng, X. Li, and S. Jiang, Colorimetric film based on polyvinyl alcohol/okra mucilage polysaccharide incorporated with rose anthocyanins for shrimp freshness monitoring, *Carbohydrate Polymers*, Vol. 229, p. 115402, 2020.
40. S. Kang, H. Wang, M. Guo, L. Zhang, M. Chen, S. Jiang, X. Li, and S. Jiang, Ethylene-vinyl alcohol copolymer–montmorillonite multilayer barrier film coated with mulberry anthocyanin for freshness monitoring, *Journal of Agricultural and Food Chemistry*, Vol. 66, p. 13268, 2018.
41. Standard test method for odor and taste transfer from polymeric packaging film, ASTM Standard E1870-11, ASTM International, West Conshohocken, PA, 2019. https://www.astm.org/e1870-11r19.html.
42. Standard specification for annealed aluminum and aluminum-alloy foil for flexible barrier, food contact, and other applications, ASTM Standard B479-19, ASTM International, West Conshohocken, PA, 2019. https://www.astm.org/b0479-19.html.
43. Standard test method for nondestructive detection of leaks in packages by vacuum decay method, ASTM Standard F2338-09, ASTM International, West Conshohocken, PA, 2020. https://www.astm.org/f2338-09r20.html.
44. Standard test method for detecting gross leaks in packaging by internal pressurization (bubble test), ASTM Standard F2096-11, ASTM International, West Conshohocken, PA, 2011. https://www.astm.org/F2096-11.html.
45. Standard test method for determination of leaks in flexible packaging by bubble emission, ASTM Standard D3078-02, ASTM International, West Conshohocken, PA, 2013. https://www.astm.org/d3078-02.html.
46. S. Costello, M.P.Y. Desmulliez, and S. McCracken, Review of test methods used for the measurement of hermeticity in packages containing small cavities, *IEEE Transactions on Components, Packaging and Manufacturing Technology*, Vol. 2, p. 430, 2012.

47. M. Giacinti Baschetti and M. Minelli, Test methods for the characterization of gas and vapor permeability in polymers for food packaging application: A review, *Polymer Testing*, Vol. 89, p. 106606, 2020.
48. Standard test method for oxygen gas transmission rate through plastic film and sheeting using a coulometric sensor, ASTM Standard D3985-17, ASTM International, West Conshohocken, PA, 2010. https://www.astm.org/D3985-17.html.
49. E. Jahdkaran, S.E. Hosseini, A. Mohammadi Nafchi, and L. Nouri, The effects of methylcellulose coating containing carvacrol or menthol on the physicochemical, mechanical, and antimicrobial activity of polyethylene films, *Food Science & Nutrition*, Vol. 9, p. 2768, 2021.
50. Standard test method for oxygen transmission rate through dry packages using a coulometric sensor, ASTM Standard F1307-14, ASTM International, West Conshohocken, PA, 2014. https://www.astm.org/F1307-14.html.
51. Standard test method for determining integrity of seals for flexible packaging by visual inspection, ASTM Standard F1886/F1886M-16, ASTM International, West Conshohocken, PA, 2016. https://www.astm.org/f1886_f1886m-16.html.
52. I. Ilhan, D. Turan, I. Gibson, and R. ten Klooster, Understanding the factors affecting the seal integrity in heat sealed flexible food packages: A review, *Packaging Technology and Science*, Vol. 34, p. 321, 2021.
53. K. Hishinuma, *Heat Sealing Technology and Engineering for Packaging: Principles and Applications*, DEStech Publications, Inc., Lancaster, Pennsylvania, US, 2009.
54. R. Aiyengar and J. Divecha, Experimental and statistical analysis of the effects of the processing parameters on the seal strength of heat sealed, biaxially oriented polypropylene film for flexible food packaging applications, *Journal of Plastic Film & Sheeting*, Vol. 28, p. 244, 2012.
55. G. Clough, Development of integrity seal sealing technology development of integrity seal hermetic sealing technology for flexible packaging and application into the food industry, *WRAP*, pp. 1–37, 2007.
56. Standard test method for detecting leaks in nonporous packaging or flexible barrier materials by dye penetration, ASTM Standard F3039-15, ASTM International, West Conshohocken, PA, 2015. https://www.astm.org/F3039-15.html.
57. Standard test method for measuring package and seal integrity using helium as the tracer gas, ASTM Standard F2391-05, ASTM International, West Conshohocken, PA, 2016. https://www.astm.org/f2391-05r16.html.
58. Standard test method for leak detection in blister packaging by vacuum deflection method by laser measurement, ASTM Standard

F3169-16, ASTM International, West Conshohocken, PA, 2016. https://www.astm.org/F3169-16.html.
59. Standard practice for determining effect of packaging on food and beverage products during storage, ASTM Standard E460-21, ASTM International, West Conshohocken, PA, 2021. https://www.astm.org/e0460-21.html.
60. J. Madureira, L. Barros, F.M.A. Margaça, C. Santos-Buelga, I.C.F.R. Ferreira, and S.C. Verde, Effects of irradiation on food bioactives in S.M. Jafari and E. Capanoglu, (Eds.), *Retention of Bioactives in Food Processing*, pp. 429–465. Springer International Publishing, Cham, 2022.
61. Standard guide for selection and use of contact materials for foods to be irradiated, ASTM Standard F1640-21, ASTM International, West Conshohocken, PA, 2021. https://www.astm.org/f1640-21.html.
62. C.H. Sommers, Irradiation of minimally processed meats in V.K. Juneja, J.S. Novak, and G.M. Sapers, (Eds.), *Microbial Safety of Minimally Processed Foods*, chapter 14, pp. 301–318. CRC Press, Boca Raton, 2003.
63. J. Mittendorfer, Food irradiation facilities: Requirements and technical aspects, *Radiation Physics and Chemistry*, Vol. 129, p. 61, 2016. Special issue on Food Irradiation commemorating József Farkas.
64. Standard guide for irradiation of fresh, frozen or processed meat and poultry to control pathogens and other microorganisms, ASTM Standard F1356-22, ASTM International, West Conshohocken, PA, 2022. https://www.astm.org/F1356-22.html.
65. Y.-K. Ham, H.-W. Kim, K.-E. Hwang, D.-H. Song, Y.-J. Kim, Y.-S. Choi, B.-S. Song, J.-H. Park, and C.-J. Kim, Effects of irradiation source and dose level on quality characteristics of processed meat products, *Radiation Physics and Chemistry*, Vol. 130, p. 259, 2017.
66. R. Akhter, F.A. Masoodi, T.A. Wani, S.A. Rather, and P.R. Hussain, Synergistic effect of low dose γ-irradiation, natural antimicrobial and antioxidant agents on quality of meat emulsions, *Radiation Physics and Chemistry*, Vol. 189, p. 109724, 2021.
67. Standard test method for oxygen headspace analysis of packages using fluorescent decay, ASTM Standard F2714-08, ASTM International, West Conshohocken, PA, 2021. https://www.astm.org/f2714-08r21.html.

4

Food Packaging

4.1 Automated Method

In restaurants, especially fast-food restaurants, fast, consistent, efficient and safe food preparation is essential for a successful operation (1).

The quality of the prepared food largely depends on the consistency of food preparation. The food must be cooked under correct conditions for a proper amount of time.

The consistency in food preparation can vary as a result of many factors. For example, people engaged in food preparation often must perform multiple tasks at frequencies that vary with time because of constantly varying customer demands throughout the day. For example, lunchtime and dinnertime may be extremely busy while other periods may be relatively slow. The product mix can vary from hour to hour and day to day. As a result, the consistency and quality of food may vary. Difficulties in proper scheduling of food production during peak and non-peak periods can cause customer delays and/or stale, wasted or unusable food.

Also, food preparation can be labor intensive, and thus, the labor cost can be a large portion of the total cost of the prepared food. An additional problem is that in sparsely populated and other areas where quick service restaurants are located, such as along interstate highways, for example, recruiting sufficient numbers of suitable employees is difficult.

A need exists for an automated, commercially suitable food dispensing, cooking and packaging device. So, an automated food processing system and method has been developed (1).

This method allows food to be dispensed, fried and packaged in a suitable container or alternatively dispensed to a food holding area for subsequent processing by a human operator.

The system includes an automated dispensing module capable of dispensing a desired quantity of food to be fried, an automated fry module adjacent to the dispensing module to receive and fry the quantity of food dispensed from the dispensing module and to produce and dispense a quantity of fried food, and an automated packaging module adjacent to the fry module to receive and package the fried food from the fry module into an individual portion-sized container.

Any one of the modules can be deactivated and a human operator can manually perform the function of the deactivated module with manually operated equipment. Optionally, an automated seasoning device is present to apply seasoning to the food. Typically, the automated dispensing module is capable of dispensing one or more of uncooked or unheated french fries, chicken nuggets, hash browns, chicken patties and fish filets or similar types of food items to be cooked and/or heated. Furthermore, the automated dispensing module includes a freezer, a storage container located in the freezer for containing food to be dispensed, a structure for dispensing a predetermined quantity of food from the storage container into a secondary or dump container, with the structure for dispensing and the secondary or dump container being located in the freezer, and with the structure for dispensing the quantity of food from the secondary or dump container to a location outside of the freezer.

The fry module of the automated modular system includes a fry vat for containing and heating cooking oil, at least one circular fry wheel having at least a generally circular perimeter in a plurality of compartments, each compartment having an opening towards the perimeter, the fry wheel mounted for rotational movement relative to the radial axis of the fry wheel, with the radial axis disposed above the normal operating level of the frying oil or the cooking oil in the fry vat. A drive mechanism is provided for rotating the fry wheel.

Also, a control system is provided for causing the drive mechanism to periodically rotate the fry wheel back and forth through a relatively small amount of angular rotation to simulate shaking of a fry basket. Such a control can be accomplished electronically.

Food is delivered from the fry module to a cooked food holding device, which can consist of a heated holding bin or bins. The automated packaging module includes a rotatable food dispensing member having an inlet location to receive a quantity of cooked food at a discharge location to discharge cooked food, the packaging module also including a food dispensing chute position to receive cooked food from the discharge location of the rotatable food dispenser, the food dispensing chute having a discharge location.

The automated modular system further includes a carton holding device for holding the individual portion-sized carton or container in position to receive food from the discharge location of the dispensing chute. The packaging module may further include a rotatable food collecting member disposed to collect food from the discharge location of a dispensing chute that is not deposited into the individual portion-sized food container. The so collected food may be subsequently deposited into the food dispensing chute for delivery to a container or alternatively to the rotatable food dispensing member or to a waste receptacle or chute.

The automated packaging device includes a conveyor system for transporting filled individual portion-sized food containers from an adjacent filling location to a filled food container holding area, for subsequent pick-up by a human operator, for example. In one embodiment, any suitable automated packaging device can be utilized. An automated food carton-retrieving device is provided for retrieving and grasping individual portion-sized food containers. The automated retrieving device consists of a moveable member for selectively grasping and releasing the food container.

The retrieving device is capable of grasping and releasing an unerected food container on one side and the device further includes a second device for selectively grasping the unerected food container on the other side with a structure for moving the retrieving device and the second device relatively apart when grasping the sides of the container to erect or partially erect the container.

An automated urging structure is provided for urging the container bottom upwardly relative to the sides of the container when the sides of the container are moved relatively apart. The automated modular system includes an electronic control system that receives current customer order information and the electronic control system causes the selection of a container from a plurality of different container sizes and further causes filling of food with the size of food container in response to a customer order.

The electronic control system can receive customer order information and controls the dispensing rate of food dispensed from the food dispensing module to the fry module which dispensing automatically determines the amount of food being fried without further intervention by the electronic control. In one aspect, the control system can include a separate control system for each of the dispensing, fry and packaging systems or modules, each of which interface with a central control system, which in turn optionally interfaces with a POS (point-of-sale) system.

The automated modular system is suitable for dispensing, frying and packaging french fries into individual portion-sized containers. An automated method of dispensing, frying and packaging food into individual portion-sized containers is provided that includes dispensing a desired quantity/portion of food to be fried from an automated dispensing module to an automated fry module and thereafter frying the portion of food dispensed from the dispensing module in the automated fry module adjacent to the dispensing module to produce a quantity of fried food. Thereafter, the quantity of fried food is dispensed from the fry module to a packaging module where the fried food dispensed from the fry module is packaged into individual portion-sized containers with an automated packaging module.

The automated method further consists of seasoning the quantity of fried food with a seasoning device. The dispensing includes dispensing a predetermined quantity of food from the storage container into a secondary container located in a freezer and dispensing the quantity of food from the secondary container to a location outside of the freezer. The frying consists of a rotating fry wheel having at least a generally circular perimeter and a plurality of compartments, each compartment having an opening towards the perimeter, the food being contained in at least one of the compartments during the frying, the fry wheel being mounted for rotational movement relative to the radial axis of the fry wheel in a fry vat with the radial axis being disposed above a normal operating level of the cooking oil in the fry vat.

The automated method further includes containing a drive mechanism to periodically rotate the fry wheel back and forth through a relatively small amount of angular rotation to simulate shaking of a fry basket during frying. The packaging includes rotating a rotatable food dispensing member having an inlet location to receive a quantity of cooked food in a discharge location to discharge cooked food, the food dispensing member being rotated to dispense food into a food dispensing chute positioned to receive cooked food from the discharge location of the rotatable food dispenser and thereafter dispensing said food from the dispensing chute to a container to be filled.

The method further includes holding an individual portion-sized carton or container positioned to receive food from the dispensing chute with an automated carton holding device. The method further includes collecting food dispensed from the discharge location of the dispensing chute that is not deposited into the individual portion-sized food container with a rotatable food collecting member disposed to collect such not deposited food. In accordance with another aspect of the method, the method includes electronically coordinating the operation of the three modules or devices within an electronic control system.

In one embodiment, the method further includes electronically received current customer order information by the electronic control system which causes selection of a container from a plurality of different sized containers and filling the container with food of the ordered size of food container in response to a customer order by the packaging module. In accordance with another aspect of the method, customer order information is electronically received and the dispensing rate of food dispensed from the food dispensing module to the fry module is controlled, with the dispensing module automatically determining the amount of food being fried without further intervention by the electronic control system.

An automated dispensing device for dispensing a quantity of food to be subsequently cooked is provided. In one embodiment, the automated dispensing device includes a freezer or refrigerated compartment, a storage container located in the freezer for containing food to be dispensed, a structure for dispensing a predetermined quantity of food from the storage container into a secondary or dump container, a structure for dispensing the predetermined quantity of food being located in the freezer, and a structure provided for dispensing the quantity of food from the secondary or dump container in the freezer to a location outside of the freezer (1).

4.2 Multilayer Packaging

Food packaging often consists of multilayers that are made from several different plastic polymer types. It was reported that multilayer packaging can consist of up to 11 ultra-thin layers offering different options such as oxygen barriers (2).

Multilayer packaging is sustainable: it does not require much material and can now also be produced from chemically recycled material.

Multilayer packaging consists of up to 11 individual, ultra-thin layers, making it considerably lighter and thinner than comparable packaging. In addition to reducing the amount of raw materials used, it also contributes to considerably lowering CO_2 emissions during transport. All in all, multilayer packaging is clearly more resource efficient than its alternatives.

However, multilayer packaging can only be recycled to a limited extent, because mechanical recycling processes are not suitable for composite materials. Within the ChemCycling project, BASF is working on advancing the chemical recycling of plastic waste, because this will make it possible to process and reuse previously non-recyclable plastics such as mixed plastics.

The raw materials for poly(amide) (PA) and poly(ethylene) (PE) were produced in very small quantities as part of the ChemCycling project. The pyrolysis oil derived from plastic waste was supplied by a partner and fed into BASF's Verbund production site in Ludwigshafen as feedstock. According to the certified mass balance method, both plastics have an allocated 100% share of recycled materials.

These packaging solutions keep the packaging volume to a minimum, reducing raw material use and lowering CO_2 emissions while offering maximum protection to the food, extending shelf life and reducing food waste. However, Dilkes-Hoffman *et. al.* pointed out that these multilayer plastics pose big challenges for recycling, with either high costs, technological difficulties in separating the different plastic polymers or the inability to recycle mixed polymers (3). It was also explained that plastic is one of the most complex materials to recycle because plastics are made up of such a large group of polymers (4).

4.3 Flexible Packaging

Flexible packaging has long been used for food, non-food, and pharmaceutical applications, and may extend beyond these areas for other uses (5). Plastic films, paper and metalized films are used in various combinations to make laminates that are then used to form different types of packaging depending on the types and conditions of use required.

Flexible packaging is economical and utilizes a lower carbon footprint as compared to rigid packaging. This trend is driving consumers and companies to move more aggressively towards the use of flexible packaging that is recyclable. Rigid packaging includes packaging made from metals, glass and rigid plastics, e.g., bottles and cans, some of which are not recyclable. As a result, the flexible packaging industry is growing faster than other packaging segments.

Polymer-based multilayer flexible packaging is popular because this approach provides a combined performance of the different polymers used to manufacture these products. These products are popular because the combination of several layers of different materials improves the mechanical and physical properties of the packaging film, e.g., improving heat and moisture resistance, oxygen barrier properties, antibacterial and antiviral properties, puncture or tear resistance, etc.

These properties have improved the packaged product quality and shelf life, thereby encouraging the big players in the field to move towards flexible multilayered packaging. At the same time, there is a global urgency to create more recyclable products and materials. Unfortunately, the current technologies are frequently not recyclable, or otherwise difficult and expensive to recycle. As a consequence of this poor recyclability, these multilayered packaging materials create mostly unwanted non-biodegradable waste.

Another emerging global issue is the increasing awareness of plastics in our landfills, oceans and rivers, and the resulting demands to reduce our global carbon footprint. Companies like P&G and Unilever are feeling pressure to have sustainable packaging mandates that include reduced packaging, reduced carbon footprints, and recyclable materials. The electron beam (EB) curing of inks and lacquers for packaging has been discussed (6–8).

The materials used in most of the currently available flexible packaging structures are not considered readily recyclable because of the presence of two or more dissimilar films or layer materials used to manufacture the final package. For packaging to be considered recyclable, the top film and the subsequent layers/films, including, for example, a laminated film (e.g., a sealant film), must be of the same or similar material and be electron-beam crosslinkable. By way of example, only polyethylene packaging containing the same or similar materials can be pelletized and recycled. Provided that the material requirements noted above are followed, many other frequently used materials in flexible packaging can be made recyclable.

A cross-sectional view of a recyclable flexible package is shown in Figure 4.1.

Electron beam treated poly(ethylene) film
Reverse printed electron beam ink layer
Laminating layer
Sealant poly(ethylene) film

Figure 4.1 Recyclable flexible package (5).

The recyclable flexible package contains an electron-beam treated film, oriented crosslinked PE film, at least one reverse printed electron-beam ink layer disposed over said electron-beam treated oriented crosslinked PE film, a laminating layer disposed over said at least one reverse printed electron-beam ink layer, and a sealant polyethylene film disposed over the laminating layer.

The curing step can be done by using electron-beam radiations, wherein the voltage is controlled to crosslink the electron-beam treated oriented crosslinked PE film, cure the reverse printed electron-beam ink layer, and cure the laminating electron-beam layer.

The voltage control restricts penetration of the sealant film to prevent problems with downstream processing (5).

4.4 Multicomponent Package

A multicomponent package has been provided which enables simultaneous heat cooking of a food product which does not require a long cooking time and a food product which requires a longer cooking time (9).

This package includes a lower container including a lower bottom container main body for containing a first food product. The container further includes an upper container including an upper container main body for containing a second filled food product. A flange of the lower container main body and a flange of the upper container main body are engaged with each other so that the lower container and the upper container are combined integrally. A vapor vent is provided in a lower container lid material, whereby water vapor generated upon heating of the first food product heats the second food product in the upper container.

A perspective view showing the lower container of the food package is shown in Figure 4.2.

The lower container 1 consists of a lower body portion 10 having an opening 11, and flange 12 provided on the peripheral edge of the opening 11. The lower body portion 10 is divided by ribs 14 to provide four well-like portions 25 (also referred to as wells) for containing the filled food product. Although the lower container is illustrated with four well-like portions, the lower container can be provided in other configurations, such as with a single well or two or more well-like portions, if desired. By increasing the number of well portions, the surface area of the container in contact with the food product is also increased, which allows the food product to be heated more evenly and quickly. It is generally more difficult for microwaves emitted from the microwave oven to warm the center part of the filled food product than to warm the periphery thereof. Therefore, the filled food product can be more uniformly warmed by dividing the containing portion of the container into more than one well. A first filled food product (not shown) can be provided in the lower container main body 10, particularly in the wells 25. In one

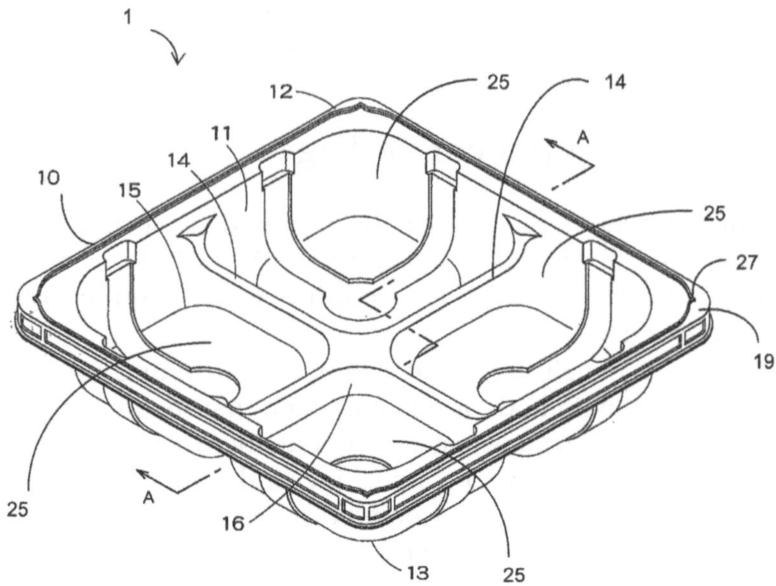

Figure 4.2 Perspective view of a lower container (9).

aspect, the first filled food product includes, but is not limited to, liquid foods, semi-liquid foods, paste foods and solids. For liquid foods, semi-liquid foods, paste foods or the like, that are generally eaten with a spoon or the like, it is preferable to provide the wells of the container main body 10 with rounded or curved corners 13. In one aspect, when the containing portion 32 of the container main body 10 is divided by the ribs 14, interior corners 15 (two in each well) where the interior of the container main body 10 and the rib 14 intersect and corners 16 (one in each well) where the ribs 14 intersect are preferably provided with a curved surface at least in the horizontal direction.

In another aspect, the interior corners 15 and 16 are provided with a curved surface in at least the vertical direction. In one aspect, interior corners 15 and 16 have the same or similar curvature as curved corners 13 and 17. The curved corners of the respective wells allow the filled food product to be warmed more uniformly and with reduced or no generation of hot spots. As a result, the time required for microwave heat cooking can be shortened. Furthermore, in one aspect and as shown in Figure 4.4, which is a cross-sectional view of Figure 4.3 taken along the line A-A, the rib 14 is provided so that a bottom surface 17 of the wells located adjacent to the center of the container has a convex shape. Since it is generally difficult for microwaves to reach the center part of the container as described above, the bottom surfaces of the wells at the center of the container have a convex shape to permit greater amounts of microwave radiation to reach those surfaces and heat the food product in the lower container. The flange 12 of the lower container main body 10 is configured to engage a flange of an upper container main body, so that the lower container and the upper container can be releasably coupled.

The side surface 18 on the peripheral edge of the flange 12 of the lower container main body 10 is configured to engage and/or couple to a complementary surface on the peripheral edge of the flange of the upper container main body, whereby the upper and lower containers can be coupled together. In one aspect, side surface 18 has a convex (or concave) configuration and the peripheral edge of the flange of the upper container main body has a complementary concave (or convex) configuration. In one aspect, the flanges of the lower container main body and upper container main body can mate, snappingly engage, releasably lock, or otherwise cooperate depending on the desired configuration and coupling of the flanges.

The lower container is made from a material with a heatproof temperature of not less than 100°C, that is permeable to microwaves emitted from a microwave oven. Such materials include poly(ethylene terephthalate) (PET), inorganic substance-containing poly(propylene) (PP), and multilayer inorganic substance-containing PP.

A perspective view of an upper container is shown in Figure 4.3.

Now turning to upper container 2 of the package provided herein, Figure 4.3 shows an upper container main body 30 consisting of an opening 31, a bottom containing portion 32, an upper surface 40, and side walls 41 which extend downward from upper surface 40 to flange 33. Flange 33 extends around the bottom periphery of the upper container main body 30. The upper container 2 does not include a bottom face other than interior facing surface 44 of containing portion 32. Instead, flange 33 extends above opening 39.

The side walls 41 extend below the bottom surface 29 of the containing portion 32. The shape of the opening 31 is not especially

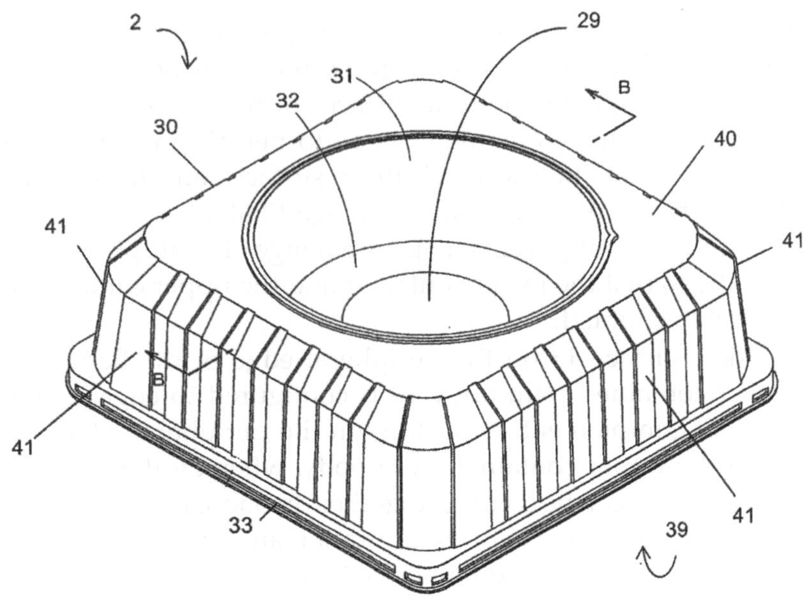

Figure 4.3 Perspective view of an upper container (9).

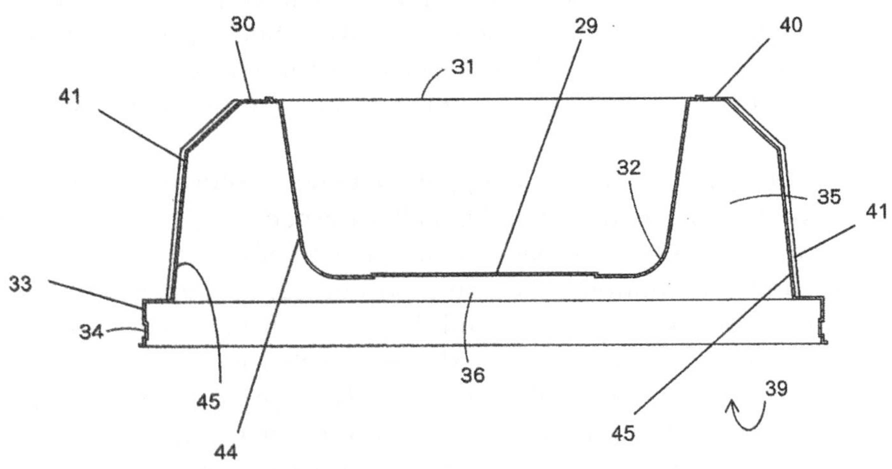

Figure 4.4 Cross-sectional view of an upper container (9).

limited and may be, for example, rectangular shape, substantially rectangular shape, circular shape, substantially circular shape, oval shape, substantially oval shape, elliptical shape and substantially elliptical shape. Other shapes may also be used, if desired. The shape of the containing portion 32 is also not especially limited so long as the containing portion 32 has a bottom surface 29 and side walls within which to contain the food product.

Figure 4.4 is a cross-sectional view of the upper container main body in Figure 4.3 taken along the line B-B, the flange 33 of the upper container main body 30 is configured to engage the flange 12 of the lower container main body 10 so that the lower and upper containers can be releasably coupled. Flange 33 is provided with a side surface 34 on the peripheral edge of the flange 33 that is configured to engage and/or couple to a complementary surface on the peripheral edge of the flange of the lower container main body, whereby the upper and lower containers can be coupled together to provide the package 100.

The upper container main body 30 is configured to provide interior space 35 between interior-facing surface 44 of containing portion 32 that faces the interior space of the container and the inner surface 45 of side walls 41. Furthermore, since the side walls 41 and flange 33 extend below the bottom surface 29 of the containing portion 32, a space 36 is formed between the lid material 20 of the lower container 1 and the bottom surface of the containing portion 32 when the lower container 1 and the upper container 2 are coupled together.

As can be seen in Figure 4.4, side walls 41 extend below the bottom surface 29 of the containing portion 32. The shape of the opening 31 is not especially limited and may be, for example, rectangular shape, substantially rectangular shape, circular shape, substantially circular shape, oval shape, substantially oval shape, elliptical shape and substantially elliptical shape. Other shapes may also be used, if desired. The shape of the containing portion 32 is also not especially limited so long as the containing portion 32 has a bottom surface 29 and side walls within which to contain the food product.

A food package can be formed as follows (9):

Preparation 4–1: A PP sheet having a thickness of 800 μm was vacuum-formed and pneumatically formed to prepare an upper container as shown in Figure 4.4.

A mixture obtained by kneading 30 *g* of a bread mix and 30 *g* of water is filled in the resulting upper container, and the container opening was sealed by a film constituted from a PET layer/an aluminum foil layer/a heat seal layer. Next, a PP sheet having a thickness of 800 *μm* was vacuum-formed and pneumatically formed to prepare a lower container as shown in Figure 4.2.

Then, 300 *cc* of soup was filled in the resulting lower container. A film constituted from a PET layer/ONY (biaxially oriented nylon)/a heat seal layer was prepared, and vapor vent holes having a diameter of 10 *mm* were provided in the film so that the holes were positioned on the four corners of the lower container. This film was sealed on a seal flange of the lower container. The upper and lower containers sealed by the film were combined integrally, and the coupled containers were stored in a freezer at -19°C for 24 *h*. Thereafter, the coupled containers were subjected to heat cooking for 8 *min* using an 1100-W turntable-type microwave oven. After completion of heating, the coupled containers were removed from the microwave oven, and then separated into the upper and lower containers. Then, the seal films of the respective containers were peeled off to check the filled products. As a result, soft bread having good texture was produced in the upper container. Furthermore, the soup contained in the lower container had a liquid temperature of 70°C and was heated uniformly.

4.4.1 *Barrier Films*

4.4.1.1 *Polyolefins*

Plastic films are widely used as packaging materials for foods. Flexible films, including multilayer films, are used to prepare bags, wrappers, pouches and other thermoformed materials. The permeability of these plastic films to gases (especially oxygen) and moisture is an important aspect for the design of a suitable food package (10).

Films prepared from thermoplastic ethylene vinyl alcohol (EVOH) copolymers are commonly employed as an oxygen barrier and also for resistance to oils. However, EVOH films are quite permeable to moisture. Conversely, polyolefins, especially high-density poly(ethylene) (HDPE), are resistant to moisture transmission but comparatively permeable to oxygen. The permeability of a linear PE film to moisture is typically described by a water vapor transmission rate (WVTR).

In certain applications some vapor transmission is desirable, for example, to allow moisture out of a package which contains produce. The use of linear low-density poly(ethylene) which may be filled with calcium carbonate (to further increase vapor transmission) is common for this purpose. Conversely, for packages which contain crispy foods such as breakfast cereals or crackers, it is desirable to limit WVTR to very low levels to prevent the food from going stale. The use of HDPE to prepare a barrier film is common for this purpose.

A review of plastic films and the WVTR behavior has been given (11). Here, barrier films were prepared from HDPE, i.e., films with low moisture vapor transmission rate (MVTR). As will be appreciated from the description of EVOH films, it is also known to prepare multilayer barrier films to produce a structure which is resistant to moisture and oxygen.

Multilayer structures may also contain additional layers to enhance packaging quality; for example, additional layers may be included to provide impact resistance or tear strength. It will also be appreciated by those skilled in the art that tie layers may be used to improve the adhesion between structural layers. In such multilayer structures, the HDPE barrier layer may either be used as an internal (core) layer or external (skin) layer.

The manufacture of barrier food packaging from plastic resins involves two basic operations (10). The first operation involves the manufacture of plastic film from the plastic resin. Most barrier films are prepared by a blown film extrusion method, in which the plastic is melted in an extruder, then forced through an annular die. The extrudate from the annular die is subjected to blown air, thus forming a plastic bubble. The use of multiple extruders and concentric dies permits multilayer structures to be co-extruded by the blown film process. The product from this operation is barrier film which is collected on rolls and shipped to the manufacturers of food packaging. The manufacturer of the food packaging generally converts the rolls of blown film into packaged foods. This typically involves three basic steps (10):

1. Forming the package,
2. Filling the package, and
3. Sealing the food in the finished package.

Although the specific details will vary from manufacturer to manufacturer, it will be readily appreciated that the film needs to have a balance of physical properties in order to be suitable for food packaging. In addition to low MVTR, it is desirable for the film to seal well and to have sufficient impact strength and stiffness (or film modulus) to allow easy handling of the package. Multilayer coextrusions are often used to achieve this balance of properties, with 3 and 5 layer coextrusions being well known. Sealant layers may be prepared with ethylene vinyl acetate ionomers, very low density PE copolymers having a density of less than 0.910 $g\,cm^{-3}$, and blends with small amounts of poly(butene). It is known to use sealant compositions in both skin layers of a coextrusion or in only one of the skin layers (10).

A first blend component of the PE composition has a comparatively high melt index. A second blend component is a HDPE, with a density from 0.955 $g\,cm^{-3}$ to 0.965 $g\,cm^{-3}$. The blends may be made by any blending process, such as (10):

1. Physical blending of particulate resin,
2. Co-feed of different HDPE resins to a common extruder,
3. Melt mixing in any conventional polymer mixing apparatus,
4. Solution blending, or
5. A polymerization process which employs 2 or more reactors.

In addition, nucleating agents are added. A nucleating agent changes the crystallization behavior of a polymer when the polymer melt is cooled.

Nucleating agents for polyolefins have been detailed (12–14). Examples are collected in Table 4.1. The organic materials without the metal ions are shown in Figure 4.5.

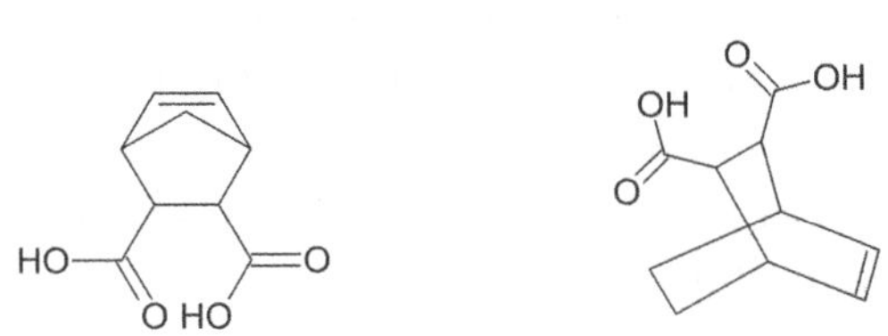

Bicyclo[2.2.1]hept-5-ene-2,3-dicarboxylic acid   Bicyclo[2.2.2]oct-5-ene-2,3-dicarboxylic acid

Figure 4.5 Organic substances for nucleating agents.

Table 4.1 Nucleating agents.

Compound	Ref.
Dilithium bicyclo[2.2.1]hept-5-en-2,3-dicarboxylate	(12)
2-Methyl-3-sodium bicyclo[2.2.1]hept-5-en-2,3-dicarboxylate	(12)
Octadecyl bicyclo[2.2.1]hept-5-en-3-carboxy-2-carboxamide	(12)
Dilithium bicyclo[2.2.2]oct-5-en-2,3-dicarboxylate	(12)
2-Methyl-3-sodium bicyclo[2.2.2]oct-5-en-2,3-dicarboxylate	(12)
trans-bicyclo [2.2.1]hept-5-en-2,3-dicarboxylic acid	(12)
Disodium bicyclo[2.2.1]hept-5-en-2,3-dicarboxylate	(12)
Disodium bicyclo[2.2.1]heptane-2,3-dicarboxylate	(13)
Calcium Bicyclo[2.2.1]heptane-2,3-dicarboxylate	(13)

4.4.1.2 Shelf Life

Packed foods that are sensitive to moisture or oxygen, water vapor transmission rate (WVTR) and oxygen transmission rate (OTR) place high demands on defining adequate desired shelf life to maintain food quality. For a maximum allowable H_2O_{max} and $O_{2\ max}$, the shelf life (t_s) can be estimated as the equations 4.1 and 4.2:

$$t_s = \frac{H_2O_{max}}{WVTR} \tag{4.1}$$

$$t_s = \frac{O_{2\ max}}{OTR} \tag{4.2}$$

The t_s is the shelf life in days, and H_2O_{max} and $O_{2\ max}$ are calculated in $g\,m^{-2}$, which can be determined by sensory evaluation (15). For calculating t_s, WVTR is generally measured at a temperature of 38°C and 90% relative humidity conditions.

There are some mathematical and computational models that can predict the required barrier properties of food packaging for the desired shelf life. A computational model to design plastic food packaging has been proposed (16). Using a heuristic optimization algorithm, the model minimizes the cost of the multilayer structure satisfying the specific product requirements.

The product requirements are defined by the expected shelf life, the storage conditions, the water sorption isotherms of foods and the maximum allowable gain or loss of gases, such as O_2, CO_2, N_2, and the moisture for the packaged food. In order to assure the food shelf life, these product requirements should be fulfilled to estimate the maximum permeance values of the plastic package.

The computational algorithm automatically generates different multilayer film structures that satisfy the product requirements (16). This algorithm combines different polymeric materials taking into account the barrier properties and cost of each layer, the compatibility between layers, the maximum number of layers and the minimum and maximum film thickness for each layer. Also, temperature and relative humidity corrections for the permeance calculations are considered.

The permeance calculations of several barrier films were compared with oxygen transmission rate and WVTR measurements. The optimization model algorithm was evaluated by means of standard numerical routines and by numerical benchmarking (16).

4.4.1.3 *Paperboard Materials*

The ongoing shift of barrier coatings from traditional synthetic polymers to sustainable breakthrough materials for paper-based packaging and films was detailed in a review (17). The advantages, challenges and adapting feasibility of these materials were described, highlighting the implications of selecting different materials and processing options.

The progress in methods of coating technologies were also briefly described. Finally, the end fate of the barrier materials was classified depending on the packaging type, coating materials used and sorting facility availability (17).

The barrier properties of petroleum and bio-based paper coatings are shown in Table 4.2.

A carboxymethyl chitosan/carboxymethyl cellulose plus poly(lactic acid) (PLA) and zinc oxide (ZnO) nanoparticles multilayer coated paperboard was prepared to explore its potential use in food package applications. The water vapor barrier properties increased by a factor of 3 even at 38°C and 90% relative humidity. The barrier performance of the ZnO nanoparticle coating was further improved. The oxygen, water vapor, and heptane vapor barrier properties all doubled, and the oil resistance time reached 235 *h* (18).

Table 4.2 Barrier properties of petroleum and bio-based paper coatings.

Coating material	Water vapor permeability/ $[g\,mm\,m^{-2}d^{-1}]$			Oxygen permeability/ $[cm^3\,mm\,m^{-2}d^{-1}atm^{-1}]$		
HDPE	0.1	–	0.24	26.3	–	453
LDPE	0.39	–	0.59	98	–	453
PET	0	–	.28	55	–	
Metalized PET	0.04	–	0.10	0.16	–	1.7
EVOH	0.8	–	2.4	0.01	–	0.15
PVC	0.94	–	0.95	3.28	–	394
PVDC	0.025	–	0.913	0.00425	–	0.57
Saran PVDC	0.009	–	0.34	0.00425	–	0.00625
PP	3.9	–	6.2	35	–	377
Metalized OPP	0.01	–	0.03	0.019	–	0.16
Nylon-6	0.24	–	125	0.394	–	2.50
Metalized Nylon-6	0.1	–	0.15	0.00078	–	
PS	09	–	155	4350	–	6200
PTFE	0.0045	–	0.30	222	–	38712
Cellulose acetate	4.5	–	77.85	0.239	–	1.301
Soy Protein	3.5	–	4.5	0.94	–	2.56
Pectin	113	–	2.4		–	
Chitosan	1.28	–	21	0.2	–	9.8
PLA	1.34	–		0.038	–	0.042
PVOH	41.904	–		0.1	–	45
PHB	0.23	–		0.42	–	
CNF	17.0	–		0.003	–	
CNC-MMT-Soy	20	–		0.28	–	

4.5 Chlorine Dioxide Gas-Releasing Package

Fresh produce consumption is associated with foodborne illnesses (19). Washing fruits and vegetables with aqueous sanitizing solutions, such as chlorine, is currently used by the industry to reduce microorganisms on the surface of these products and to improve their safety. However, these sanitizers in solution cannot penetrate inaccessible areas such as pores, channels and crevices where microorganisms are attached due to the hydrophobicity of these regions and surface tension.

Gaseous chlorine dioxide has the ability to penetrate and inactivate human pathogens attached to hard-to-reach sites. So, chlorine dioxide gas is an effective biocide over a wide range of pH levels from 3 to 8. This gas can be generated by the reaction of an acid with sodium chlorite salt in the presence of moisture.

Bio-based package labels/inserts have been developed which can release chlorine dioxide gas (19). They contain at least one layer of pectin embedded with citric acid and at least one layer of gelatin embedded with sodium chlorite, wherein each layer of pectin with citric acid is adjacent to each layer of gelatin with sodium chlorite.

Optionally, a gelatin barrier layer (without sodium chlorite) is inserted between each layer of pectin with citric acid and each layer of gelatin with sodium chlorite. Gelatin barrier layers are used to modulate the release rate and/or concentration of chlorine dioxide produced by the package label.

The chlorine dioxide gas produced by the package labels kills or inhibits the growth of microorganisms, including human pathogens, on food surfaces inside the package (19).

The design of a bio-based label/insert package is shown in Figure 4.6.

Upon activation of the label:

1. Moisture starts diffusing from the gelatin functional layer toward upper and lower pectin functional layers, causing polymer relaxation of these matrices,
2. Higher mobility and diffusion of protons and chlorite ions moves down their gradient of concentration,
3. Moisture released from respiration of fresh produce acts further on the label by causing swelling of the corresponding

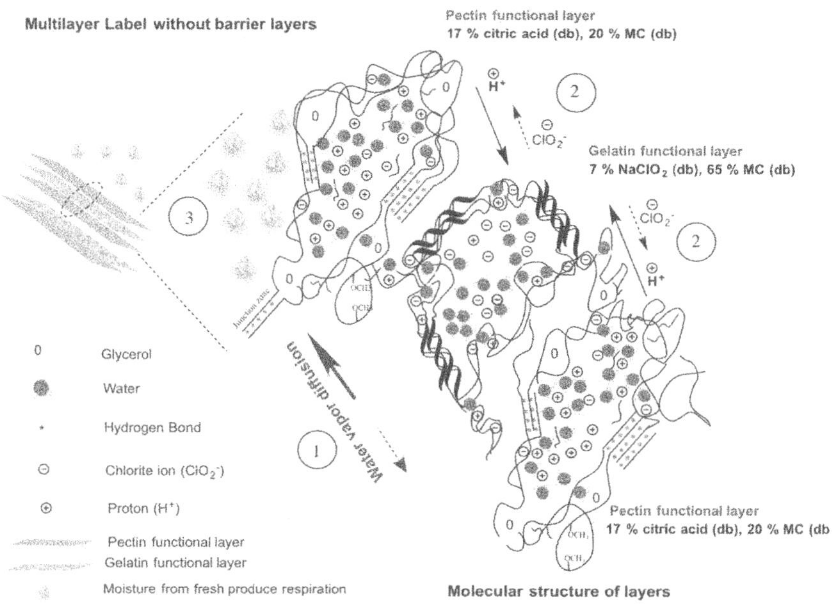

Figure 4.6 Bio-based label/insert package (19).

layers, starting with the outermost pectin functional layer and the release of the bulk of chlorine dioxide over time. Moisture content of the gelatin functional layer acts as a short- term trigger for chlorine dioxide release whereas moisture released from fresh produce respiration constitutes the long-term trigger.

The technical feasibility of the tested package label composed of alternating gelatin and pectin layers depends on the ability to design and manufacture the label, activate it, and have it generate and release chlorine dioxide in a controlled manner at levels adequate for microbial inactivation. Both release kinetics and microbial inactivation were in good agreement and support the technical feasibility of bio-based package labels in terms of label design and manufacture, chlorine dioxide generation and release after controlled activation, and antimicrobial effectiveness against *Salmonella* growth (19).

4.6 Sustainable Food Packaging

The issues of sustainable food packaging technology have been detailed in a monograph (20).

Sustainable food packaging technology deals with packaging solutions that use engineered biopolymers or biocomposites that have suitable physicochemical properties for food contact and protection and originate both from renewable or nonrenewable resources, but in both cases are compostable or edible. Modified paper and cardboard with increased protective properties towards food while keeping their compostability are presented as well. The book also covers natural components that can make the packaging functional, e.g., by providing active protection to the food by indicating food spoilage.

The major parts of this book are (20):

1. Review on biopolymers for food protection,
2. Food packaging based on individual biopolymers and their composites,
3. Biocomposites of cellulose and biopolymers in food packaging,
4. Natural principles in active and intelligent food packaging for enhanced protection and indication of food spoilange or pollutant presence, and
5. Technological developments in the engineering of biocomposite materials for food packaging applications.

4.7 Ventable Food Packages

It is a common practice in the packaging of foods for heating/cooking in a microwave oven to provide the package with a vent or some other mechanism or means to be opened to enable steam produced from the heating of the food within the package to vent (21).

4.7.1 Ventable Microwave-Safe Food Package

A package and a method for storing, venting, and heating consumable goods has been developed (22). This package includes a container and a valve for venting gas during storage and heating.

Here, a microwave-safe valve with rigid valve body is used. A membrane secures the rigid valve body for venting gas from the interior. So, the valve prevents a package rupture or damage. The valve has a well-defined flow resistance and a dimensionally stable valve gate at a vapor pressures of up to 3 *bar* and at temperatures of up to 150°C.

The container must have sufficient stability to withstand heating, including microwave energy, vapor pressure, and a temperature of 150°C.

Here, consumable goods are disposed in the interior of the container and consist of at least two different fresh, raw foods that require different cooking times in water, with said consumable goods having an amount of water content of at least 30% (22).

4.7.2 *Self-Venting Microwave-Safe Food Package*

A flexible package for food products, to enable any steam resulting from heating the product in the package to automatically vent, has been developed (21).

The packages are particularly suitable for holding a food product hermetically sealed within the hollow interior of the package. A gas can be included in the package to enhance the shelf life or sterility of the food product.

This package has a hollow interior space, in which the product is located and is made up of a flexible sheet material. Portions of the flexible sheet material are sealed to each other along an elongated seal line, which forms a portion of the periphery of the hollow interior space.

Furthermore, a portion of the seal line is in the form of a peelable interface. At least one vent is coupled to the peelable interface. The peelable interface automatically peels open upon the creation of steam within the package resulting from the heating of the product to allow the steam or other vapor to automatically exit through at least one vent (21).

4.7.3 *Smart and Active Food Packaging*

Smart and active food packaging is an innovative technology to prevent food contamination and ensure food quality and safety (23).

Active packaging protects the food from microbial contamination, while smart or intelligent packaging enables monitoring the freshness of the food or quality change in real time.

4.7.3.1 *Nanotechnology*

Integration of nanotechnology with food packaging systems is a promising advancement in the shelf life of foods which introduces novel food packaging systems. In a study, common classification, functionalities, employed nanotechnologies, and the biomaterials used are discussed (24).

According to this survey, food packaging systems are classified as active food packaging and smart food packaging systems. The functionality of both systems can be manipulated by employing nanotechnologies, such as metal nanoparticles and nanoemulsions, and appropriate biomaterials like synthetic polymers and biomass-derived biomaterials. "Degradability and antibacterial" and "indicating and scavenging" are the well-known functions for active food packaging and smart food packaging, respectively (24).

The presence of polymers, such as poly(styrene) (PS), poly(vinyl chloride) (PVC), PE, PET, PP, and PA, in food packaging systems provides more satisfaction from the customers' perspective due to their ease of use, appearance, and transparency (25).

Furthermore, these polymers showed a good availability, lower price, good water vapor permeability, good mechanical properties (tensile strength and shear strength), and good gas barrier properties (oxygen and carbon dioxide) (25, 26).

They can also be natural (e.g., gelatin-chitosan) or synthetic (e.g., polycaprolactone). In spite of this, contamination and environmental pollution issues still remain unsolved (27, 28).

Biodegradable packaging helps the environment remove waste materials left from food packaging. Active food packaging has a high potential of decomposition in comparison with common packaging due to the employed biodegradable materials and embedded specific ingredients (24).

The incorporation of natural antimicrobial-loaded nanocarriers (metal oxide, essential oil, herbal extracts, etc.) is the most effective way to create active food packaging in food packaging systems, which has become feasible by employing nanoencapsulated antimicrobial materials in coating or film structures, instead of interpolation of antibacterial agents (29–33).

This approach has further been able to create many benefits, including release control, protection against environmental stresses, and solubility improvement and natural antimicrobial absorption in active food packaging, which are the main achievements in overcoming the obstacles of using natural antimicrobial materials in food packaging (34).

4.7.3.2 Curcumin

Curcumin is one of the most well-known natural colorants (23). It has received a lot of attention due to its excellent functional properties and ability to change color with changes in pH.

Curcumin, the golden component of turmeric, a spice widely used in food since ancient times, is a cost-effective and abundant biomaterial with various biological properties such as antioxidant, antibacterial, antiviral, antitumor, and anti-inflammatory.

Recently, active packaging or intelligent packaging systems have been actively developed using the functional properties of curcumin (23).

Highly porous curcumin-loaded polymer mats can be used for the rapid detection of volatile amines (35).

4.7.3.3 Photoluminescence and Photocatalysis

Smart packaging materials were developed to monitor the quality, maturity, and safety of foods in real time. Photoactivated materials capable of photoluminescence or photocatalysis can be used to fabricate this smart and active food packaging (36).

Photoluminescence-based sensors can be used to monitor environmental changes in food packaging, such as gas leakage, freshness, and the presence of specific microorganisms. Photocatalysis-based devices can be used as color indicators, as well as to inhibit a broad spectrum of microbes and for gas removal.

The latest research on the application of photoactivated materials in smart and active food packaging has been reviewed (36).

The preparation, operating principles, applications of photoactivated materials in food packaging materials are discussed, as well as their potential advantages and disadvantages. The here presented knowledge may stimulate more research on the development and application of innovative smart and active food packaging materials in the food industry (36).

4.8 Packages for Pourable Food

Many pourable food products, such as fruit juice, ultra-high-temperature processed milk, wine, tomato sauce, etc., are sold in packages made of a sterilized packaging material (37, 38).

A typical example of such a package is the parallelepiped-shaped package for liquid or pourable food products, which is formed by folding and sealing laminated strip packaging material. The packaging material has a multilayer structure consisting of a layer of fibrous material (e.g., paper), covered on both sides with layers of heat-seal plastic material (e.g., PE) and, in the case of aseptic packages for long-storage products, such as ultra-high-temperature processed milk, also contains a layer of an oxygen-barrier (e.g., aluminum film), which is superimposed on a layer of heat-seal plastic material and is in turn covered with another layer of heat-seal plastic material, eventually defining the inner face of the package contacting the food product.

Such packages are normally made on fully automatic packaging machines, on which a continuous tube is formed from the packaging material supplied in strip form. The strip of packaging material is sterilized on the packaging machine itself, e.g., by applying a chemical sterilizing agent, such as a hydrogen peroxide solution, which, after sterilization, is removed and vaporized by heating. The so sterilized strip of packaging material is maintained in a closed sterile environment, and is folded and sealed longitudinally to form a vertical tube.

The tube is then filled with the sterilized or sterile-processed food product, and is sealed and cut at equally spaced cross sections to form pillow packs, which are then folded mechanically to form the finished packages.

To open such packages, various solutions have been proposed, one of which consists of forming a preferential tear line at the corner of a flap on the package. The package is opened by lifting the flap and cutting or tearing along the perforations. Packages of this type, however, cannot, obviously, be closed once opened, and must therefore be handled with care to prevent spillage of the remaining food product inside the package (37).

One problem of such opening devices is that the cap must be removable from the frame or tamper-proof ring with practically no effort when unsealing the package. For which purpose, the opening devices are made of easy-break plastic material, normally PE (38).

However, PE has the drawback of being a poor oxygen barrier. So, when the hole is formed through the full thickness of the packaging material, an additional cover element or patch, defined by a small sheet of heat-seal plastic material, must be applied over the hole on the side of the packaging material, eventually forming the inside of the package; and the opposite side of the packaging material must be fitted with an oxygen-barrier element, e.g., a pull tab, which is heat sealed to the patch and comprises a layer of aluminum.

To solve this above-mentioned problem, packages have been provided with closable opening devices, which substantially comprise a frame defining an opening and fitted over a hole or a pierceable or tear-off portion in a wall of the package; and a cap hinged to the frame.

The cap is normally molded integrally with the frame, and is initially sealed to the frame, along a peripheral edge surrounding the opening, by a thin breakable annular connecting portion. Once unsealed, the cap is movable between a closed position cooperating in a fluidtight manner with the frame, and an open position. Alternatively, threaded caps are also used, which are separate from and initially screwed to the frame.

Figure 4.7 shows a perspective view of a top portion of a sealed package for pourable food products having a closable opening device.

The cap is fitted initially to the frame in a sealed position, wherein the cap is screwed completely onto the collar (37).

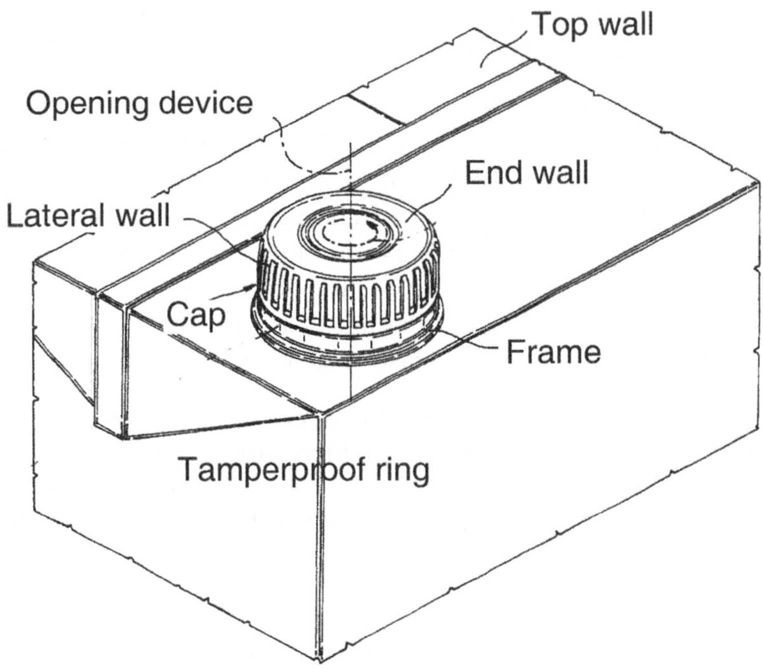

Figure 4.7 Sealed package for pourable food products (37).

4.9 Food Storage

4.9.1 *Trays*

A tray of plastics material for packaging foodstuffs such as meat and fish which may release serum or other liquids has been described, which is able to avoid these foods being in prolonged contact with the released liquids and which can prevent these liquids from being visible within the package (39).

Here, an absorbent layer may simply be fixed to the inner surface of a tray of plastics material or be interposed between two sheets of plastics material, which when in contact with the food has holes through which the liquid released from the food is conveyed to the interposed absorbent layer.

Such a tray has the disadvantage of being made of materials of different types, that is, foamed plastics, generally PS, and paper or hydrophilic materials that are difficult to separate from each other, with the result that there is little or no possibility of recycling them. Moreover, the cost of the production of such trays, because of the presence of the paper, is considerably higher than that of conventional, non-absorbent trays.

These problems can be resolved by a tray of plastics material for foodstuffs which may release liquids, consisting of a base and side walls constituted by an upper layer and a lower layer, the layers forming the side walls which are joined together while those forming the base are spaced apart from each other to define an interspace. The construction of such a tray is shown in Figures 4.8 and 4.9.

A tray has a body 1 having a base 2 and side walls 3. Also, such a tray is composed of an upper layer 4 and a lower layer 5, both of plastic materials, the layers constituting the side walls 3 being joined together while those forming the base 2 are spaced apart to define an interspace 6. The layers 4 and 5 are joined firmly together, for example by heat-sealing or glueing. The upper layer 4 is pierced by holes 7 having a diameter of 0.2–1.5 *mm*. The surface 8 of the layer 4 facing the interspace 6 is rough as a result of either mechanical abrasion or treatment with suitable chemical reagents.

Preferably the lower layer is also of foamed plastics material containing a surfactant and its surface facing the interspace is rough. The foamed plastics material used for the production of the tray is

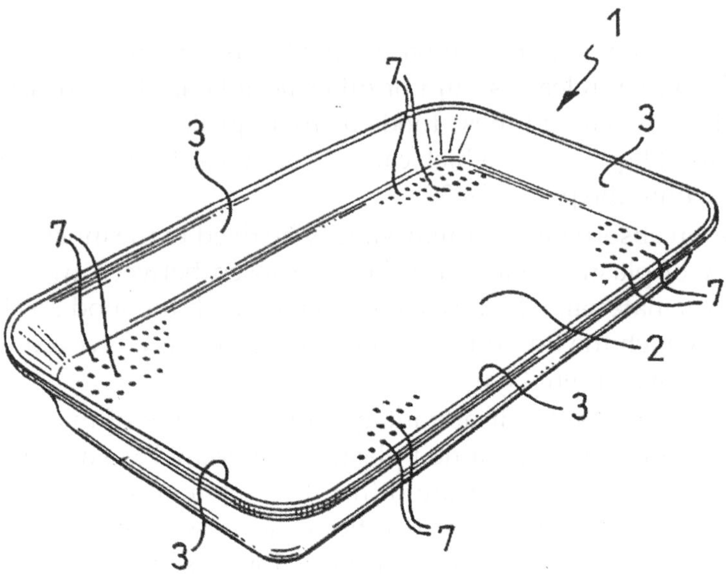

Figure 4.8 Tray (Full view) (39).

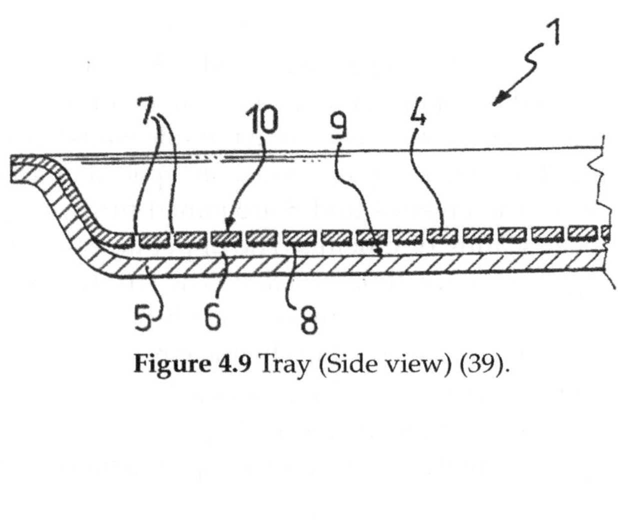

Figure 4.9 Tray (Side view) (39).

preferably selected from the group of PS, PP, PE, PET, PVC, and their copolymers.

The surfactant in the foamed plastics material used in the production of the tray preferably constitutes from 0.2% to 10% by weight of the total weight of the material, which may be selected from surfactants in common use, whether anionic, cationic or nonionic.

A surfactant which has been found to be particularly useful is the aliphatic sulfonate sold by the firm Novacrome under the name Hostastat System E 39045 (39).

4.9.2 *Storage Container*

A storage container for containing food products is composed of (40):

1. A base,
2. A continuous side wall extending upwardly from a periphery of said base, said base and said continuous side wall defining a reservoir, said base having at least one main rib extending into said reservoir spanning substantially a length of said base such that said reservoir is divided into a plurality of cells, said continuous side wall having a perimeter rib extending about said continuous side wall and into said reservoir, said perimeter rib having an upper surface spaced at an approximately uniform distance from the base,
3. An absorbent material arranged in at least one of said plurality of cells, and
4. A cover formed of liquid permeable material, a periphery of said cover engaging said upper surface of said perimeter rib and said at least one main rib, said cover forming a support surface for supporting food products placed within said storage container such that when food products are placed upon said cover, said cover remains spaced from said absorbent material.

The storage container is shown in detail in Figure 4.10.

The tray portion 13 can be formed of PS, PP, or HDPE. It can be thermally formed, or vacuum formed, to produce a lightweight, shaped insert. Preferably, the tray insert 10 substantially conforms to the shape of the bottom portion of the interior of the storage

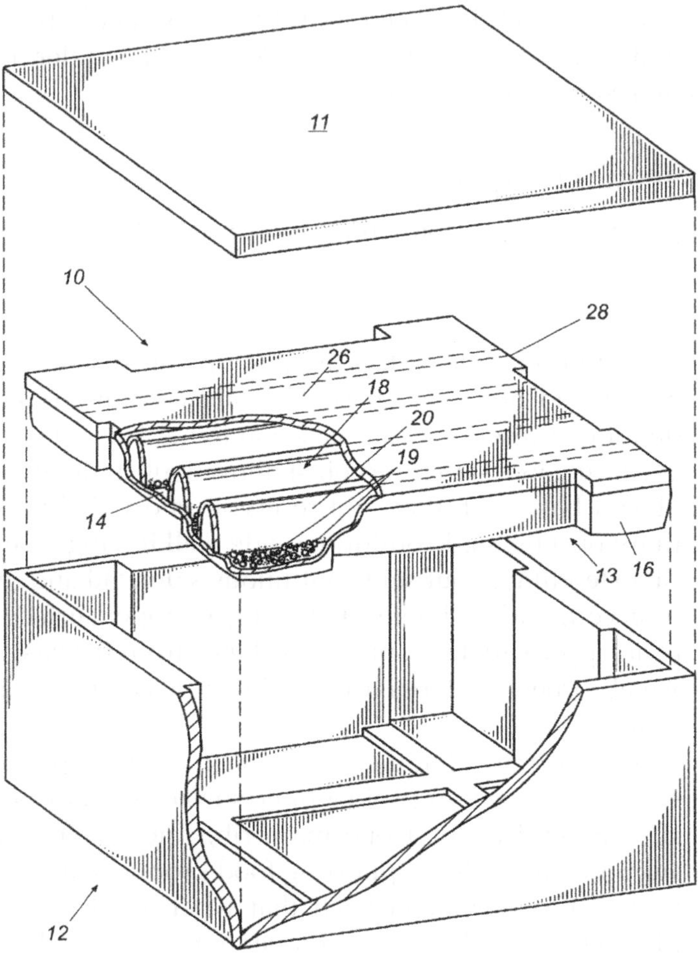

Figure 4.10 Storage container (40).

container 12 in which the insert 10 is to be inserted. So configured, the storage container 12 housing the insert 10 provides strength and support to the insert 10 so that the insert 10 can adequately support the weight of the food products resting on the insert 10 and the weight of the fluid laden absorbent material 19 residing within the reservoir 18. The tray insert 10 incorporates one or more ribs 20 protruding from the base 14, and can be arranged in numerous configurations, including parallel, diverging and interlacing, among others, depending upon the desired support characteristics required for a particular application. The ribs 20 serve as structural supports for food products, which are placed in the storage container 12 above the absorbent tray insert 10 and which rest on the insert. The ribs 20 retain the products above the bottom of the storage container 12 and the base 14 of the tray insert 10, thereby forming a space in which fluids can accumulate and mix with the absorbent material 19 and in which expansion of the absorbent material 19 can occur without restriction. Additionally, the ribs 20 provide structural support to the tray insert 10 such that the insert can be formed to quite a large size without the disadvantages of the prior art devices which are size limited due to a lack of incorporated structural support components. The ribs 20 also provide the benefit of preventing migration of the absorbent material 19 between the ribs 20, thereby substantially retaining the absorbent material 19 in a uniform distribution throughout the tray insert 10.

A cover 26 formed of a liquid permeable sheet, is placed across the tray insert 10 to cover the reservoir 18. The cover 26 is preferably made of material such as perforated film, apertured film, nonwoven fabric, nonwoven laminated to apertured film, multi-composites or other suitable pervious material that can be bonded to the upper edge 28, for example, of insert 10, such as by heat sealing, applying an adhesive or by other suitable means, to create a barrier between the food products 22 placed on the cover 26 and the reservoir 18. The cover 26 also can be bonded to the upper surfaces of the ribs 20 and rib segments for added structural support. As such, the cover 26 forms an additional support surface for the stored products 22 as the cover 26 spans the gaps formed between the raised ribs 20 to form a false bottom support surface in the bottom of the storage container 12. Therefore, the absorbent material 19 within the reservoir 18 and below the cover 26 is not compressed by the weight of the food products 22 placed on the cover 26, thereby allowing the absorbent material 19 to absorb and retain greater quantities of exuded fluid.

4.9.3 *Storage System*

Containers have been described that provide an easy storage and transport of food (41).

Conventional food storage containers are configured to hold a single type of food. This can be disadvantageous when separate foods are to be transported due to the difficulty in packing, loading, and transporting multiple containers for use at a single meal. So, it may be advantageous to provide a food storage system that is preferably adapted to allow different foods to be transported without using multiple separate, detached containers.

The food storage system is shown in Figure 4.11.

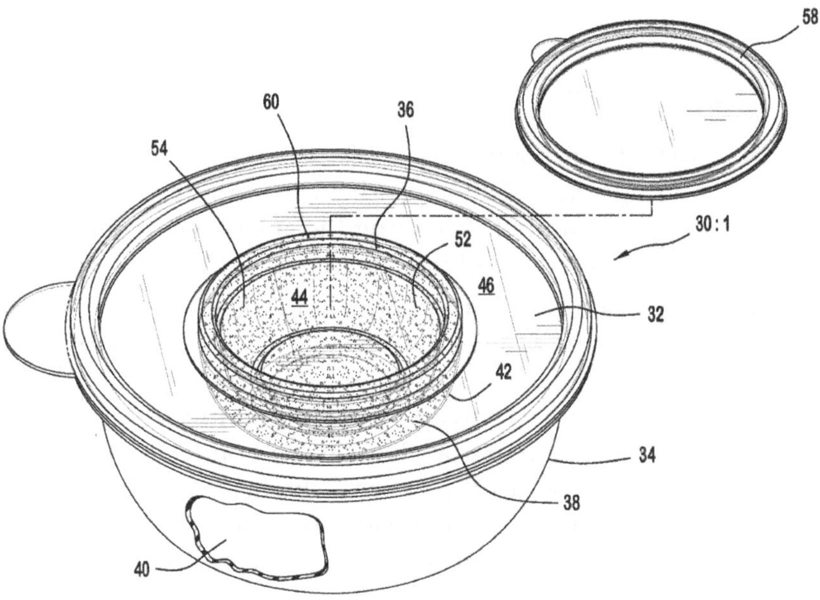

Figure 4.11 Food storage system (41).

Here, the secondary bowl 36 includes a sidewall 42 containing a thermal energy storage material 38 therein (41). The thermal energy storage material 38 is preferably any suitable freezable and/or reheatable material, such as a refreezable gel or the like. It is preferred that the gel can undergo repeated cycles of freezing and thawing in order to provide cooling. Heat-storing materials may also be utilized such that the thermal energy storage material 38 can be placed in a microwave oven to allow the thermal energy storage material 38 to provide warmth to food.

The sidewall 42 defines a second food receiving cavity 44. It is preferred, but not necessary, that a portion of the main lid 32 form a portion of the sidewall 42 of the secondary bowl 36.

The main lid 32 may have a major surface 46 facing in a first direction generally away from the main bowl 34, when the main lid 32 is attached thereto. The portion 48 of the main lid 32 preferably extends outwardly therefrom in a direction generally opposite to the first direction. The portion 48 of the main lid defines an outer wall 50 of the sidewall 42 of the secondary bowl 36.

The main lid 32, the main bowl 34, and the secondary bowl 36, are preferably made of a polymeric material, such as PE, PET, PVC, poly(phthalate carbonate), or any other suitable material.

It is also preferred that a panel 52 is shaped to be generally coextensive with the portion 48 of the main lid 32 forming the outer wall 50 and is attached to the main lid 32 to form an inner wall 54 of the secondary bowl 36. The inner wall 52 defines a food receiving surface for the second food receiving cavity 44 and is generally spaced from the outer wall 50 to define a chamber 56 therebetween

It is preferable that the thermal energy storage material 38 is located within the chamber 56. It is also preferable that the thermal energy storage material 38 is a gel. The gel may be colored or translucent depending on the particular esthetic look desired. It is preferred that the inner wall 54 is welded to the main lid 32. Alternatively, the inner wall 54 may be secured to the main lid via an adhesive, or by any other known method of securing two components of a food storage system.

The food storage system 30:1 can separately hold food items in the first or second food receiving cavities 40, 44. The secondary bowl 36 is configured so that the thermal energy storage material 38 can heat/or cool food contained in the second food receiving cavity 44. It is preferred that the secondary bowl 36 is generally translucent so that the secondary bowl 36 has a color of the thermal energy storage material 38 therein.

When both the main bowl 34 and the secondary bowl 36 are generally translucent, the entire food storage system 30:1-30:3 takes on the general color of the thermal energy storage material 38. This allows for attractive decorative options. It is preferred that the secondary bowl 36 is generally centered in the main bowl 34 (41).

4.10 Special Materials for Food Packaging

The polymers that are most frequently used in food packaging are PE, PP, PS, PVC, and PET (42, 43).

High-density poly(ethylene) (HDPE) is used in applications such as milk bottles and bags. Low-density poly(ethylene) is used for trays and general-purpose containers.

Poly(propylene) (PP) has an excellent chemical resistance, it is strong, and has the lowest density of the plastics used in packaging. It has a high melting point, making it ideal for hot-fill liquids. It is employed in film and microwavable containers (44).

Poly(ethylene terephthalate) (PET) is clear, tough, and has good gas and moisture barrier properties. Soft drink bottles are generally made of PET. It has good resistance to heat, mineral oils, solvents, and acids. Hence, it is becoming the packaging material of choice for many food products, especially beverages and mineral waters. The use of PET to make plastic bottles for carbonated drinks is increasing (45)

However, despite their enormous versatility, a limiting property of polymeric materials in food packaging is their inherent permeability to gases and vapors, including oxygen, carbon dioxide, and organic vapors. Biopolymers are notorious for their high water vapor permeability. This has boosted interest in developing new strategies to enhance barrier properties and to carry out research aimed at understanding the structure–barrier properties relationship. This review emphasizes the impact of nanostructures on barrier properties of polymers (46).

4.10.1 Film-Forming Biodegradable Classes

The principal film-forming biodegradable classes of substances used in packaging are collected in Table 4.3.

Table 4.3 Principal film-forming biodegradable classes of substances used in packaging (47).

Material	References
Poly(saccharide)s	
Cellulose	(48, 49)
Chitosan	(50, 51)
Starch	(52, 53)
Hemicellulose	(54, 55)
Alginate	(56, 57)
Agarose	(58, 59)
Glucomannan	(60, 61)
Pullulan	(62, 63)
Proteins	
Soy protein	(64, 65)
Whey protein	(66, 67)
Zein	(64, 67)
Collagen	(68, 69)
Gelatine	(70, 71)
Casein	(72, 73)
Lipids and waxes	
Plant oil and animal fat	(74, 75)
Waxes	(76, 77)
Other materials	
PLA	(78, 79)
PVA	(80, 81)

4.10.2 *Cytotoxicity*

In recent years, the problem of plastic pollution in the marine environment has attracted an increased public awareness and scientific research. Poly(styrene) (PS) nanoplastics alter the cytotoxicity of human pharmaceuticals on marine fish cell lines.

In a study, PS nanoplastics were prepared by a mini-emulsion polymerization using sodium dodecyl sulfate (SDS) as a stabilizer. As biological models, marine fish cell lines SAF-1 and DLB-1 were used. The SAF-1 cell line is a fibroblast-like line derived from gilthead seabream (*Sparus aurata* L.) caudal fin. The DLB-1 cell line is a neuronal line derived from European sea bass (*Dicen trarchus labrax* L.) brain. As chemicals for testing, gemfibrozil, propranolol, furosemide, fluoxetine, metformin, nicotine, paracetamol (acetaminophen), and acetylsalicylic acid were used. These materials are shown in Figure 4.12.

The estimated lethal doses (LD 50, LD25; and LD10) to SAF-1 cell line after 24 h exposure to pharmaceuticals, alone and combined with 10 $mg\,l^{-1}$ of 100 nm PS particles, are shown in Table 4.4. In addition, the estimated lethal doses of DLB-1 are shown in Table 4.5.

Table 4.4 Estimated lethal doses for SAF-1 (82).

Chemical	Pharmaceutical			Pharmaceutical + PS		
	LD 50	LD 25	LD 10	LD 50	LD 25	LD 10
Acetylsalicylic Acid	–	0.215	0.048	–	0.6760	.104
Acetaminophen	0.822	0.425	0.299	–	—	
Gemfibrozil	0.351	0.173	0.075	0.242	0.1940	.082
Fluoxetine	0.013	0.009	0.008	0.014	0.0110	.009
Furosemide	0.103	0.024	0.008	0.850	0.4530	.206
Metformin	–	0.336	0.112	–	0.5010	.145
Nicotine	0.426	0.124	0.032	0.599	0.2020	.092
Propranolol	0.198	0.184	0.174	0.062	0.0350	.025

The lethal doses were calculated through interpolation of a nonlinear regression with a four-parameter dose-response curve. Values missing indicate that lethal doses were out of the curve range (82).

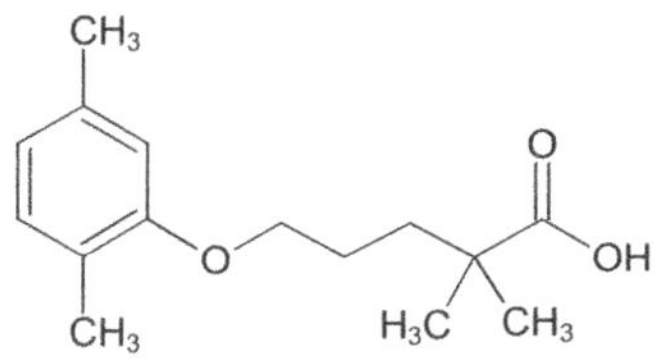

Gemfibrozil

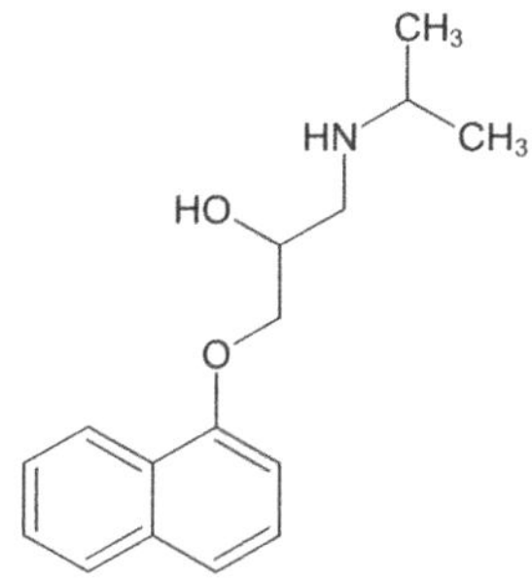

Propranolol

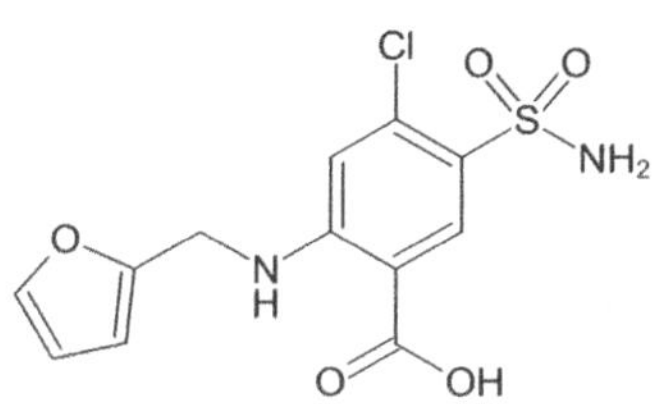

Furosemide

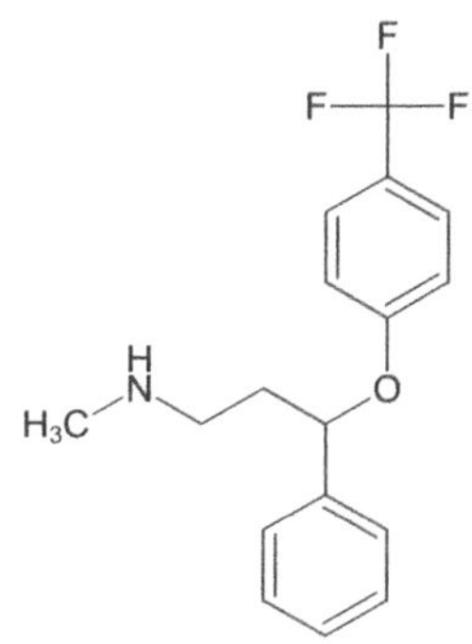

Fluoxetine

Figure 4.12 Chemicals for toxicity measurement.

Table 4.5 Estimated lethal doses for DLB-1 (82).

Chemical	Pharmaceutical			Pharmaceutical + PS		
	LD 50	LD 25	LD 10	LD 50	LD 25	LD 10
Acetylsalicylic Acid	–	0.674	0.486	–	–	0.830
Acetaminophen	–	0.711	0.384	0.081	0.022	0.011
Gemfibrozil	0.262	0.165	0.126	0.249	0.172	0.140
Fluoxetine	0.019	0.013	0.011	0.025	0.017	0.014
Furosemide	–	0.776	0.585	0.727	0.420	0.287
Metformin	–	0.674	0.486	–	–	–
Nicotine	–	0.994	0.718	–	–	–
Propranolol	0.002	–	–	0.005	–	–

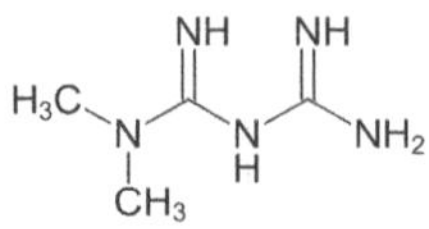

Metformin

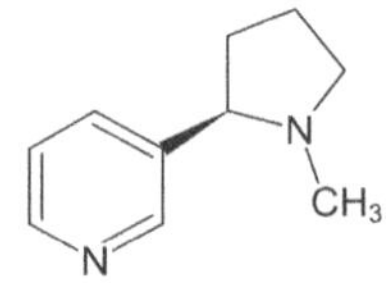

Nicotine

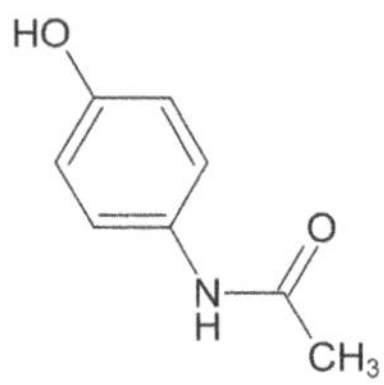

Paracetamol

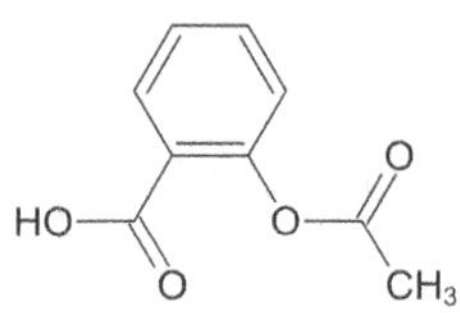

Acetylsalicylic acid

Figure 4.12 (cont) Chemicals for toxicity measurement.

4.10.3 Microwavable Food Packaging Materials

Packaging for microwavable food is typically made from plastic materials (83). The Food and Drug Administration (FDA) establishes regulations and testing standards for food containers and food packaging. Packaging manufacturers must disclose the chemical nature of the container or film, the components, as well as any potential health hazards. In addition, the FDA tests food contact surfaces for migration of polymers and polymer additives into food.

An FDA approved packaging container or food wrapping film ensures it is safe for use as a food contact surface and complies with all industry standards. Containers and wrapping films must also be pleasing in appearance and color to consumers (83).

Plastics used to make containers for microwavable food must have mechanical strength, i.e., the packaging needs to be sturdy, rigid in shape, and capable of supporting the food within before and after microwave heating in conventional home (500 *W*) and restaurant/convenience store microwave ovens (1600 *W*). Packaging also needs to be heat resistant and capable of withstanding temperatures of up to 110°C for microwave ovens and up to 210°C for dual oven applications.

The electromagnetic radiation used in microwave ovens does not directly interact with the packaging. However, food products that contain oils and fats can create localized temperatures of as much as 150°C on the container surface. This localized heating may result in polymer degradation or conformation changes. Thus, the packaging should also maintain its visual appearance after heating in a microwave oven and should resist migration into food substances.

For example, a conventional lid for a microwave food container must be sufficiently transparent so that the food product inside the package is visible. Typically, the lid may be fabricated from general purpose PS, which has a heat distortion temperature of 90°C. Such a lid is prone to deformation in the presence of oily food products, for example, because oily food products may generate heat upon warming in a microwave oven (83).

4.10.4 Film-Forming Materials

The bio-nanocomposite materials represent an important alternative to the traditional fossil-based plastics used as packaging (47). These are biocompatible, biodegradable, and can exhibit superior performances, both mechanical and chemical. Biopolymers like chitosan, carboxymethyl cellulose, starch, alginate, casein, carrageenan, or cellophane can solve the ecological problem of packages due to their biodegradability and nontoxic nature.

However, poly(saccharide)s have some drawbacks besides their clear advantages. Some mechanical properties are inferior, and they have a low water resistance or high permeability. Consequently, by embedding various nanomaterials into the natural polymeric film, better thermal and mechanical properties can be obtained, but the water vapor and oxygen barrier properties can also be enhanced without losing the biodegradability or the nontoxic character of the biopolymer.

Some types of antimicrobial packaging materials, their main polymer, antimicrobial agent, and the types of food tested are presented in Table 4.6 (47). Some of the antimicrobial agents are shown in Figure 4.13.

The natural antimicrobial compounds can be obtained from vegetal sources like cloves, cinnamon, thyme, ginger, oregano, rosemary,

Table 4.6 Types of antimicrobial packaging materials (47).

Used for	Material	Antimicrobial Agent	Reference
Iceberg lettuce	Cellulose	Clove and oregano oils	(84)
Cucumber	Chitosan	Limonene	(85)
Tomato	Chitosan	TiO_2 nanoparticles	(86)
Strawberries	Chitosan/CMC	Chitosan/citric acid	(87)
Strawberries	PLA	AgNPs	(88)
Strawberries	Chitosan/CMC	*Mentha spicata* oil	(89)
Strawberries	Gelatin	Butylated hydroxyanisole	(90)
Fish	PLA	Thymol	(91)
Carp fillets	Alginate/CMC	*Ziziphora clino-podioides* oil / ZnO	(92)
Rainbow trout fillet	Chitosan	Grape seed extract	(93)
Salmon	PLA	Glycerol monolaurate	(94)
Shrimps	Chitosan	Carvacrol	(95)
Shrimps	Gelatin	ZnO /clove oil	(96)
Chicken	Chitosan	Acerola residue extract	(97)
Poultry	Chitosan	Ginger oil	(98)
Chicken	Gelatin	Thyme oil	(99)
Chicken	Pullulan	Nisin	(63)
Chicken	Chitosan/PET	Plantaricin	(100)
Ostrich meat	Kefiran/PU	*Zataria multiflora* oil	(101)
Lamb meat	Chitosan	*Satureja* plant oil	(102)
Ground beef	PLA/NC	*Mentha piperita, Bunium percicum*	(103)
Ham	Chitosan/starch	Gallic acid	(104)
Salami	Whey protein	*Cinnamomum cassia, Rosmarinus officinalis* oils	(105)
Cheese	Chitosan/PVA	TiO_2	(106)
Cheese	Cellulose/Chitosan	Monolaurin	(107)
Cheese	Starch	Clove leaf oil	(108)
Cheese	Agar	Enterocin	(109)
Cheese	Zein	Pomegranate peel extract	(110)
Peanuts (roasted)	Banana flour	Garlic essential oil	(111)

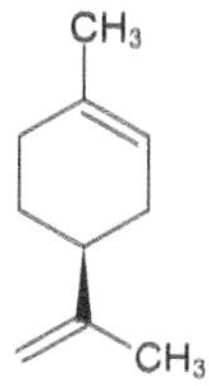

Limonene

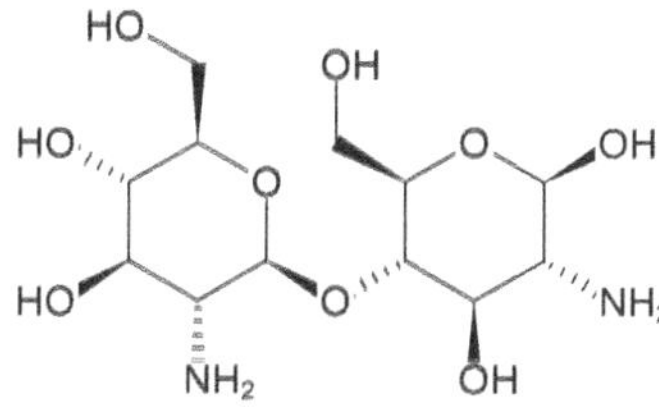

Chitosan

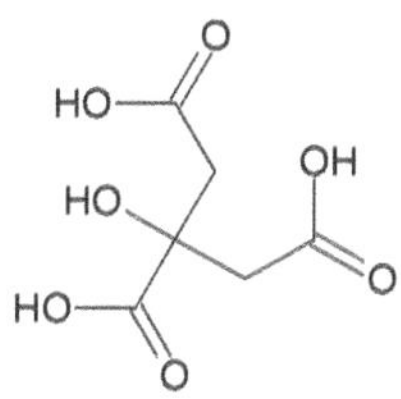

Citric acid

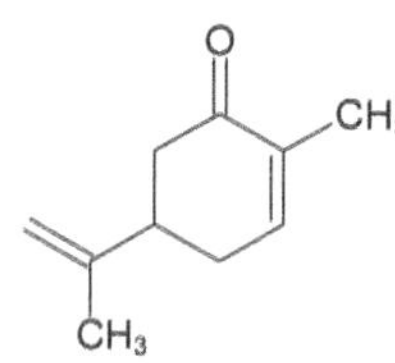

Mentha spicata oil

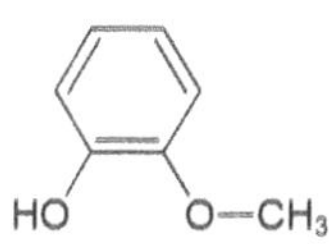

Hydroxyanisole

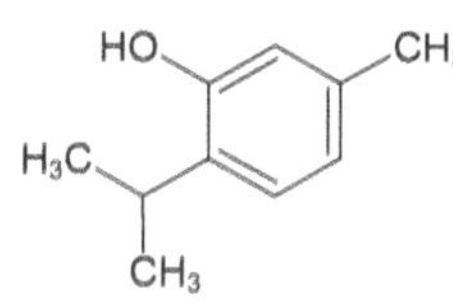

Thymol

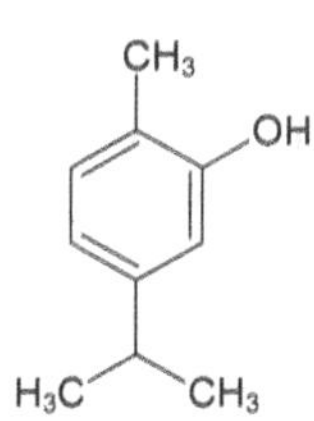

Carvacrol

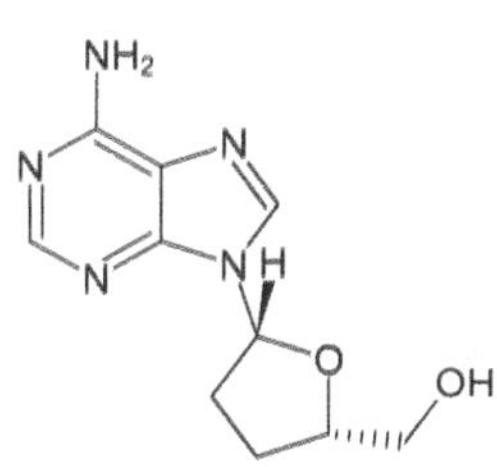

Acerola

Figure 4.13 Antimicrobial agents.

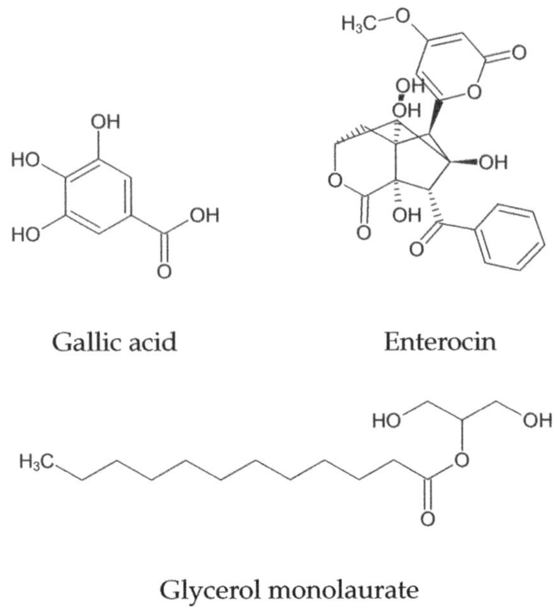

Figure 4.13 (cont) Antimicrobial agents.

garlic, etc. All these compounds have good potential for the meat industry (112, 113).

In a similar way, other natural antimicrobial agents can be isolated from substances produced by bacterial or fungal activity such as pediocin, nisin, and various bacteriocins. Currently, the meat industry is using some technologies based on blocking microbial activity that can enhance the organoleptic properties and packaging performance (114–118). Various synthetic antimicrobial agents have been reported for bread and pastry packaging (119–121) or for fruits and vegetables (122–124).

4.10.5 *Antimicrobial Biodegradable Food Packaging*

The issues of antimicrobial packaging materials have been reviewed (47). The antibacterial, antifungal, and antioxidant activities can be induced by the main polymer used for packaging or by addition of various components from natural agents (bacteriocins, essential

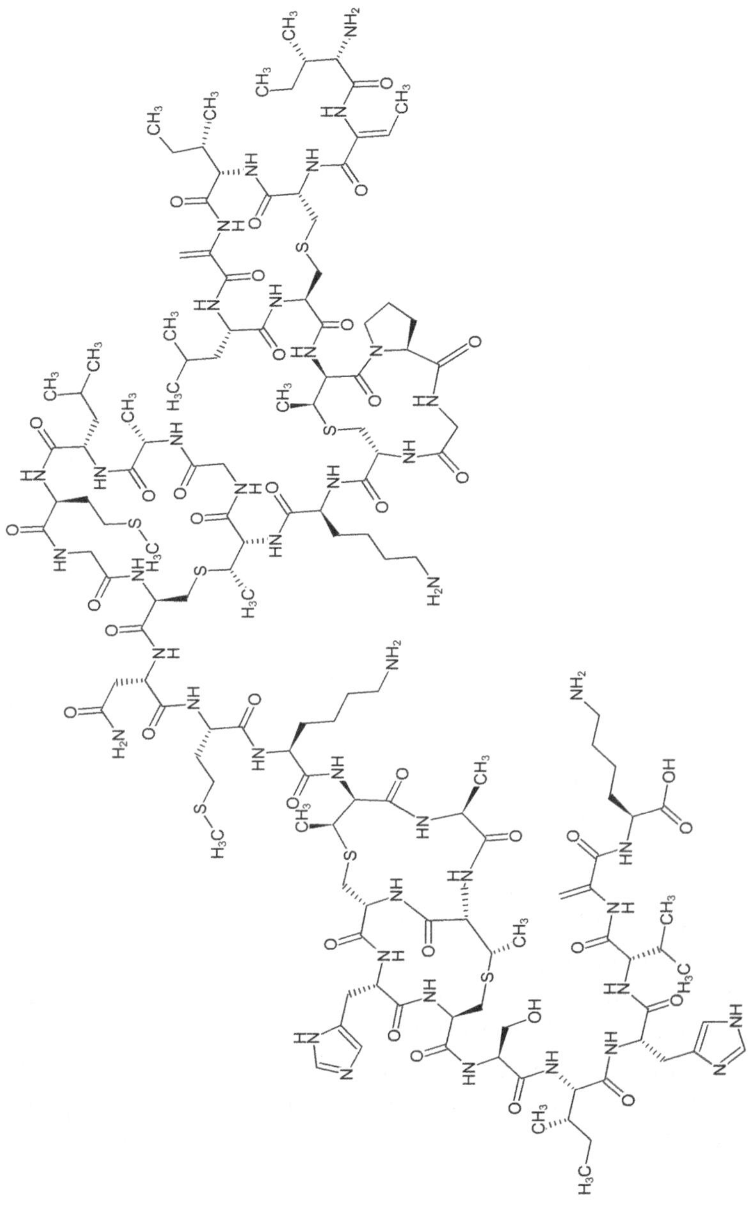

Nisin

Figure 4.13 (cont) Antimicrobial agents.

oils, natural extracts, etc.) to synthetic agents, both organic and inorganic (Ag, ZnO, TiO_2 nanoparticles, synthetic antibiotics).

The antimicrobial packaging strategies can be classified into two types (47):

1. Packaging materials with a direct contact between the antimicrobial surface and the preserved food, in which the active agents can migrate into the food. Such packaging is used for food that is wrapped in foils or under vacuum.
2. Having the antimicrobial agent inside the package, but not in direct contact with the food, such as in modified atmosphere packaging.

The general trend in the evolution of packaging has been to move away from inert and polluting plastic waste towards the antimicrobial active, biodegradable or edible biopolymer film packaging. Like in many domains, this transition is an evolution rather than a revolution, and changes are coming in small steps.

Changing the public perception and industry focus on the antimicrobial packaging solutions will enhance the shelf life and provide healthier food, thus diminishing the waste of agricultural resources, but will also reduce the plastic pollution generated by humankind as most new polymers used for packaging are from renewable sources and are biodegradable (47). Poly(saccharide)s (like chitosan, cellulose and derivatives, starch, etc.), lipids and proteins (from vegetal or animal origin), and some other specific biopolymers, like poly(lactic acid) or poly(vinyl alcohol), have been used as a single component or in blends to obtain antimicrobial packaging materials.

Where the package's antimicrobial and antioxidant activities need a larger spectrum or a boost, certain active substances are embedded, encapsulated, coated, grafted into or onto the polymeric film (47).

Ways to decrease the use of conventional polymers in the plastics industry has become an acute concern since industrially produced plastic waste, mainly conventional food packaging, has become an environmental crisis worldwide (125).

4.10.5.1 Chitosan

Chitosan-based packaging materials have become a recent focus for the biodegradable food packaging sector due to their biodegradability, nontoxic nature, and antimicrobial properties. Chitosan is obtained from chitin, which is the most abundant biopolymer in nature after cellulose. Chitosan is an ideal biomaterial for active packaging as it can be fabricated alone or combined with other polymers as well as metallic antimicrobial particles, either as layers or as coacervates for examination as functional components of active packaging systems.

Chitosan-metal/metal oxide bio-nanocomposites have seen growing interest as antimicrobial packaging materials, with several different mechanisms of inhibition speculated to include direct physical interactions or chemical reactions (i.e., the production of reactive oxygen species as well as the increased dissolution of toxic metal cations).

The use of chitosan and its metal/metal oxide (i.e., titanium dioxide (TNP), zinc oxide, and silver nanoparticles) bio-nanocomposites in packaging applications have been presented in a review (125).

The antimicrobial activity is shown in Table 4.7.

4.10.5.2 Bio-nanocomposites

Eco-friendly, high-performance packaging materials, i.e., bio-nanocomposites, have attracted great attention in recent years. The recent innovation in bio-nanocomposites for food packaging applications have been detailed in a review (126). The current trends and research over the last three years of the various bio-nanocomposites, including inorganic, organic nanomaterials, and nanohybrids, which are suitable as food packaging materials due to their advanced high mechanical, thermal, barrier, antimicrobial, and antioxidant properties among others are described in detail.

In addition, the legislation, migration studies, and SWOT analysis (127) on bio-nanocomposite film have been discussed. It has been observed that the multifunctional properties of the bio-nanocomposite materials have the potential to improve the quality and safety of the food together with no/or fewer negative impacts on the environment. However, more studies need to be performed on bio-nanocomposite materials to determine the migration levels and formulate relevant legislation (126).

Bio-nanocomposites are bio-based polymers made up of two major components in which one acts as a matrix called biopolymer

Table 4.7 Antimicrobial activity of chitosan TNPs (125, 128).

Amount/[%]	Microbes	Results
0	*B. cereus*	70.14 ± 0.02
0	*S. aureus*	81.49 ± 0.20
0	*C. albicans*	72.46 ± 0.08
0	*A. niger*	54.25 ± 0.02
0	*E. coli*	55.11 ± 0.03
2	*B. cereus*	75.50 ± 0.15
2	*S. aureus*	82.45 ± 0.04
2	*C. albicans*	76.36 ± 0.31
2	*A. niger*	86.19 ± 0.25
2	*E. coli*	60.00 ± 0.04
5	*B. cereus*	78.87 ± 0.05
5	*S. aureus*	82.99 ± 0.18
5	*C. albicans*	62.55 ± 0.07
5	*A. niger*	88.49 ± 0.27
5	*E. coli*	57.45 ± 0.11
10	*B. cereus*	79.47 ± 0.01
10	*S. aureus*	82.25 ± 0.14
10	*C. albicans*	51.64 ± 0.06
10	*A. niger*	90.99 ± 0.21
10	*E. coli*	72.77 ± 0.15
15	*B. cereus*	85.85 ± 0.21
15	*S. aureus*	84.62 ± 0.18
15	*C. albicans*	44.82 ± 0.09
15	*A. niger*	93.99 ± 0.29
15	*B. cereus*	60.00 ± 0.25

(continuous phase) and the second is a reinforcement agent (dispersed phase) with dimensions in the range of 1–100 *nm*.

These materials exhibit typical characteristics, such as flexibility, biocompatibility, biodegradability, eco-friendliness, and cost-effectiveness, which can be further enhanced by the reinforcement components. Bio-nanocomposites can be modified by fusion of nanofillers such as montmorillonite, AgO, TiO_2, SiO_2, ZnO and biodegradable polymers, including poly(3-hydroxybutyrate), poly(butylene succinate), PLA and poly(caprolactone), along with natural biopolymers such as starch and chitosan.

The incorporation of antimicrobial agents into the food packaging materials inhibits the growth of microorganisms on food surfaces and thus increases their shelf life. The various experimentally designed antimicrobial nanocomposites are PLA/halloysite, PLA/silver nanoparticles, hydroxyapatite/titania, layered silicate, and cellulose nanofibers, which could inhibit the growth of different pathogens such as *Listeria monocytogenes*, *Escherichia coli*, and *Bacillus subtilis*. Halloysite and hydroxyapatite are shown in Figure 4.14. These advantageous properties of antimicrobial bio-nanocomposites suggest their practical use in the food packaging industry (129).

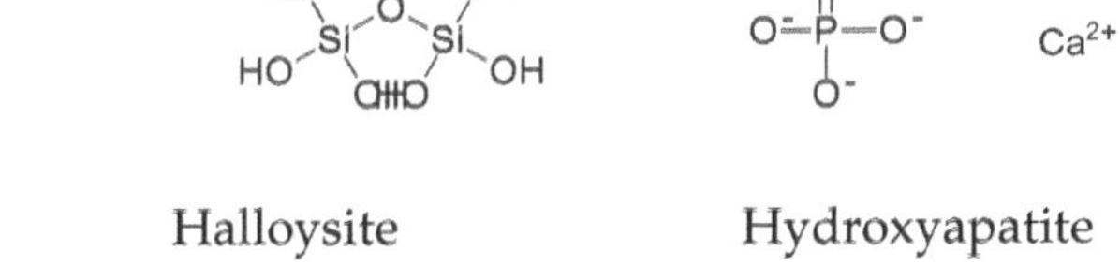

Figure 4.14 Halloysite and Hydroxyapatite.

The food industry has been trying to introduce natural bioactive compounds as food preservatives (130). The following materials are of interest:

1. Natural bioactives can potentially replace synthetic additives in food preservation,

2. Essential oils, plant extracts and food byproducts can provide bioactives for further active packaging development,
3. Cyclodextrins can successfully form inclusion complexes with hydrophobic natural substances,
4. Thermal and oxidative stabilization provided by cyclodextrins enables industrial application of natural bioactives in active packaging,
5. Cyclodextrin:Natural bioactives inclusion complexes have great prospects in active food packaging.

4.10.6 Multilayer Films for Food Packaging

Synthetic plastic packaging is by far the most commonly used for packaging; however, it leads to considerable environmental problems. The use of biodegradable plastics derived from natural polymers can address this issue. Nevertheless, natural polymers have poor barrier properties. An overview of the various polymers used in multilayer packaging has been given in a review (131). Also, their physical, mechanical and barrier properties are explained in detail.

The application of a multilayer film combines the unique features of different polymers that may produce packaging materials with improved performance, such as mechanical and barrier properties. The properties of multilayer films with different types of polymers (biomass, synthesis-derived and microbial-derived) are discussed.

Multilayer films possess superior properties in comparison to single-layer films in terms of physical, mechanical and gas barrier properties (131). Natural polymers for multilayer films are shown in Table 4.8.

4.10.7 Bilayer Film with Dihydromyricetin

A novel bilayer film composed of chitosan (CS) and sodium alginate (SA) incorporated with inclusion complex of dihydromyricetin with hydroxypropyl-β-cyclodextrin by layer-by-layer casting was prepared (132). Dihydromyricetin and hydroxypropyl-β-cyclodextrin are shown in Figure 4.15.

The results of structural and physical characterization revealed that CS/SA/hydroxypropyl-β-cyclodextrin films were hydrophilic

Table 4.8 Natural polymers for multilayer films (131).

Type of polymers	References
Cassava starch + Chitosan	(133)
SA + CMC	(134)
EC + HPMC	(135)
Chitosan + Alginate	(136)
CMC + Chitosan	(137, 138)
Bovine + Bovine + Bovine gelatins	(139)
SPI + Zein	(140)
Corn wheat starch + Zein	(141, 142)
Chitosan + Gelatin	(143–145)
Kefiran + Gelatin	(146)
BC + Zein	(147)
HPS + Zein	(148)
PHA substrate + Electrospun PHA coating	(149)
PHBV + TPS	(3)
WG + PHB	(150)
Zein + PHBV	(151)
Zein + PHB	(151)
SPI + PHB	(152)
Compound	**Shortcut**
Sodium alginate	SA
Carboxymethyl cellulose	CMC
Ethyl cellulose	EC
Hydroxypropyl methyl cellulose	HPMC
Soy protein isolate	SPI

Table 4.8 (cont) Natural polymers for multilayer films (131).

Type of polymers	References
PLA + PCL	(153)
PLA + PVA	(154)
SPS + PLA	(155)
Cassava starch + PLA	(156)
Pea starch + PLA	(157)
MFC + PLA	(158)
HPMC + PLA	(159)
Chitosan + PLA	(160)
TPS + PLA	(161)
Chitosan + Alginate + PVA	(162)
ALG + PLA	(163)
Chitosan + PVA	(164)
Starch + PCL	(165)
TPS + PCL	(166)
Fish gelatin + PLA	(167–170)
Bovine gelatin + PLA	(171, 172)
Compound	**Shortcut**
Bacterial cellulose	BC
Hydroxypropyl starch	HPS
Poly(hydroxyalkanoate)	PHA
Poly(3-hydroxybutyrate-*co*-3-hydroxyvalerate)	PHBV
Wheat gluten	WG
Poly(hydroxy butyrate)	PHB
Poly(hydroxy butyrate-*co*-valerate)	PHBV
Soy protein isolate	SPI
Poly(lactic acid)	PLA
Poly(e-caprolactone)	PCL
Poly(vinyl alcohol)	PVA
Sugar palm starch	SPS
Microfibrillated cellulose	MFC
Thermoplastic corn starch	TPS
Sodium alginate	ALG

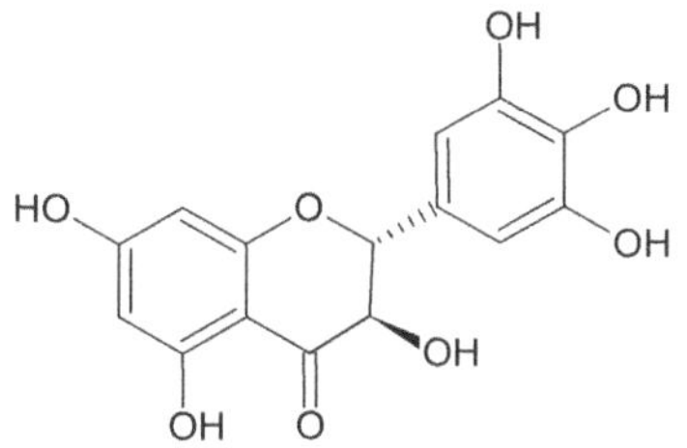

Dihydromyricetin

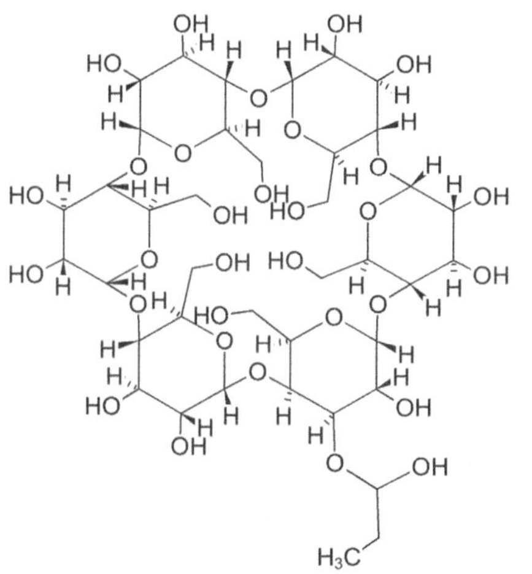

Hydroxypropyl-β-cyclodextrin

Figure 4.15 Compounds for inclusion complex.

and had good compatibility, thermal stability, and UV blocking capability, with a smooth and uniform appearance.

Compared with a chitosan/sodium alginate film, the chitosan/-sodium alginate/hydroxypropyl-β-cyclodextrin-10% film exhibited an improved tensile strength of 45.57 MPa and an elongation at break of 22.46%. The incorporation of hydroxypropyl-β-cyclodextrin was shown to enhance the antioxidant activity of the bilayer film, as well as the antibacterial activity against *Staphylococcus aureus*. The chitosan/sodium alginate/hydroxypropyl-β-cyclodextrin-10% film had a superior performance as a food packaging material on preservation of fried meatballs by effectively reducing lipid oxidation and microbial growth. After storage for 7 d, the thiobarbituric acid reactive substance and total plate count of fried meatballs packaged with chitosan/sodium alginate/hydroxypropyl-β-cyclodextrin-10% film were 31.72 $\mu g\, g^{-1}$ (132).

So, this study suggests that chitosan/sodium alginate/hydroxypropyl-β-cyclodextrin film could be used as a promising active packaging material for application in the food industry (132).

References

1. G.A. Sus, R. Dorsten, J. Hong, G. Schackmuth, A. Courier, K. Keough, R. Terrazzano, M.G. Ceste, C.C. Pinnow, C. Juneau, S. Konold, R. Montafi, B. Nunley, and C.E. Rose, Automated method for packaging food, US Patent 7 356 980, assigned to Restaurant Technology Inc., April 15, 2008.
2. B. Hellmann, Multilayer packaging: Innovative and sustainable, Sustainability News, BASF, 2019.
3. L.S. Dilkes-Hoffman, S. Pratt, P.A. Lant, I. Levett, and B. Laycock, Polyhydroxyalkanoate coatings restrict moisture uptake and associated loss of barrier properties of thermoplastic starch films, *Journal of Applied Polymer Science*, Vol. 135, p. 46379, 2018.
4. G. Faraca and T. Astrup, Plastic waste from recycling centres: Characterisation and evaluation of plastic recyclability, *Waste Management*, Vol. 95, p. 388, 2019.
5. I. Rangwalla and B. Sullivan, Electron beam (EB) curing of inks and in-situ crosslinking of substrates to provide sustainable and recyclable flexible packaging solutions, US Patent Application 20 220 001 660, assigned to Energy Sciences Inc, January 06, 2022.

6. M.A. Edlein and D.R. Kyle, Method of providing a printed thermoplastic film having a radiation-cured overprint coating, US Patent 6 528 127, assigned to Cryovac LLC, March 04, 2003.
7. M. Laksin, S. Chatterjee, and V. Linzer, Method and apparatus for wet trapping with energy-curable flexographic liquid inks, US Patent 6 772 683, assigned to Sun Chemical Corp., August 10, 2004.
8. M.J. Mossbrook, D.R. Kyle, and M.A. Edlein, Printed thermoplastic film with radiation-cured overprint varnish, US Patent 7 063 882, assigned to Cryovac LLC, June 20, 2006.
9. K.M. Birchmeier, P.A. Cogley, N. Sakamoto, K. Otsuka, and T. Kado, Multi-component food packaging for microwave oven, US Patent 9 737 085, assigned to Dai Nippon Printing Co. Ltd., Kraft Foods Global Brands LLC, August 22, 2017.
10. N.D.J. Aubee, D. Checknita, P.S. Chisholm, P. Lam, S. Marshall, D.P. Sauvageau, T. Tikuisis, and T.Y. Chuang, Barrier film for food packaging, US Patent 9 644 087, assigned to Nova Chemicals International SA, May 09, 2017.
11. M. McLeod and N. van Nguyen, High density polyethylene barrier grade resins and films, methods for making same, US Patent 6 777 520, assigned to Fina Technology Inc, August 17, 2004.
12. S.E. Amos, K.E. Nielsen, C. McRoberts, and M.A. Wicki, Modifying agents for polyolefins, US Patent 5 981 636, assigned to 3M Innovative Properties Co., November 09, 1999.
13. X.E. Zhao and D.L. Dotson, Bicyclo[2.2.1]heptane dicarboxylate salts as polyolefin nucleators, US Patent 6 465 551, assigned to Milliken and Co, October 15, 2002.
14. D.L. Dotson and X.E. Zhao, Metals salts of hexahydrophthalic acid as nucleating additives for crystalline thermoplastics, US Patent 6 599 971, assigned to Milliken and Co., July 29, 2003.
15. K.L. Yam, H. Zhao, and C.C. Lai, Frozen food packaging in Y.H. Hui, L.M. Cunha, I. Guerrero-Legarreta, H.H. Liang, Y.M. Lo, D.L. Marshall, W. Nip, F. Shahidi, F. Sherkat, R. Einger, K.L. Yam, and J.D. Culbertson, (Eds.), *Handbook of Food Science, Technology, and Engineering*, pp. 2388–2394. Taylor & Francis, Boca Raton, FL, USA, 2004.
16. M.P. Noriega, O. Estrada, and I. Lopez, Computational model to design plastic multi-layer films for food packaging to assure a shelf life at the best cost, *Journal of Plastic Film & Sheeting*, Vol. 30, p. 48, 2014.
17. P. Tyagi, K.S. Salem, M.A. Hubbe, and L. Pal, Advances in barrier coatings and film technologies for achieving sustainable packaging of food products – A review, *Trends in Food Science & Technology*, Vol. 115, p. 461, 2021.

18. F.-J. Wang, L.-Q. Wang, X.-C. Zhang, S.-F. Ma, and Z.-C. Zhao, Study on the barrier properties and antibacterial properties of cellulose-based multilayer coated paperboard used for fast food packaging, *Food Bioscience*, Vol. 46, p. 101398, 2022.
19. B.A. Annous, C. Saade, and K.L. Yam, Chlorine dioxide gas releasing package label insert for enhancing microbial safety of food, US Patent 10 537 125, assigned to US Department of Agriculture, January 21, 2020.
20. A. Athanassiou, (Ed.), *Sustainable Food Packaging Technology*, Wiley Online Library, 2021.
21. J.S. Thomas and J.S. Beer, Microwavable self-venting package, US Patent Application 20 100 266 732, assigned to Fres-co System USA Inc., October 21, 2010.
22. K. Keller, Ventable, microwave-safe food package, US Patent 6 607 764, assigned to Karl Keller, August 19, 2003.
23. S. Roy, R. Priyadarshi, P. Ezati, and J.-W. Rhim, Curcumin and its uses in active and smart food packaging applications - A comprehensive review, *Food Chemistry*, Vol. 375, p. 131885, 2022.
24. H. Ahari and S.P. Soufiani, Smart and active food packaging: Insights in novel food packaging, *Frontiers in Microbiology*, Vol. 12, 2021.
25. M. Stoica, C. Dima, and P. Alexe, Eco-friendly nanocomposites from bacterial cellulose and biopolyesters as a sustainable alternative for plastic food packaging in A.D. Galaz and D.S. Bailey, (Eds.), *Food Packaging and Preservation Techniques Applications and Technology*, pp. 113–127. Nova Science Publishers, New York, 2018.
26. P.A. Pawar and A.H. Purwar, Biodegradable polymers in food packaging, *American Journal of Engineering Research*, Vol. 2, p. 151, 2013.
27. C.J. Moore, Synthetic polymers in the marine environment: A rapidly increasing, long-term threat, *Environmental Research*, Vol. 108, p. 131, 2008.
28. H. Williams and F. Wikström, Environmental impact of packaging and food losses in a life cycle perspective: A comparative analysis of five food items, *Journal of Cleaner Production*, Vol. 19, p. 43, 2011.
29. A. Valdés, A.C. Mellinas, M. Ramos, N. Burgos, A. Jiménez, and M.C. Garrigós, Use of herbs, spices and their bioactive compounds in active food packaging, *RSC Advances*, Vol. 5, p. 40324, 2015.
30. M. Ydollahi, H. Ahari, and A.A. Anvar, Antibacterial activity of silver-nanoparticles against Staphylococcus aureus, *African Journal of Microbiology Research*, Vol. 10, p. 850, 2016.
31. C.V. Garcia, G.H. Shin, and J.T. Kim, Metal oxide-based nanocomposites in food packaging: Applications, migration, and regulations, *Trends in Food Science & Technology*, Vol. 82, p. 21, 2018.

32. A.M. Pisoschi, A. Pop, C. Georgescu, V. Turcuş, N.K. Olah, and E. Mathe, An overview of natural antimicrobials role in food, *European Journal of Medicinal Chemistry*, Vol. 143, p. 922, 2018.
33. J. Ju, X. Chen, Y. Xie, H. Yu, Y. Guo, Y. Cheng, H. Qian, and W. Yao, Application of essential oil as a sustained release preparation in food packaging, *Trends in Food Science & Technology*, Vol. 92, p. 22, 2019.
34. A. Bahrami, R. Delshadi, E. Assadpour, S.M. Jafari, and L. Williams, Antimicrobial-loaded nanocarriers for food packaging applications, *Advances in Colloid and Interface Science*, Vol. 278, p. 102140, 2020.
35. D. Kossyvaki, A. Barbetta, M. Contardi, M. Bustreo, K. Dziza, S. Lauciello, A. Athanassiou, and D. Fragouli, Highly porous curcumin-loaded polymer mats for rapid detection of volatile amines, *ACS Applied Polymer Materials*, p. xx, 05 2022.
36. H. Xu, L. Chen, D.J. McClements, Y. Hu, H. Cheng, C. Qiu, H. Ji, C. Sun, Y. Tian, M. Miao, and Z. Jin, Progress in the development of photoactivated materials for smart and active food packaging: Photoluminescence and photocatalysis approaches, *Chemical Engineering Journal*, Vol. 432, p. 134301, 2022.
37. M. Laciacera and C. Miani, Closable opening device for sealed packages of pourable food products, US Patent 6 279 779, assigned to Tetra Laval Holdings and Finance SA, August 28, 2001.
38. C. Casale and A. Sorbara, Reclosable opening device for packages of pourable food products, US Patent 8 714 380, assigned to Tetra Laval Holdings and Finance SA, May 06, 2014.
39. F. Lanzani and R. Mauri, Tray of plastics material for food substances which tend to release liquids, US Patent 5 720 999, assigned to Sirap Gema SpA, February 24, 1998.
40. W.M. Brander and R.E. d'Ablaing, Storage container for containing food products, US Patent 6 152 295, assigned to Maxwell Chase Technologies LLC, November 28, 2000.
41. D. Roth and H. Roth, Food storage system, US Patent 20 120 061 404, assigned to Cool Gear International LLC, March 15, 2012.
42. G. Bureau, *Food Packaging Technology*, 1995.
43. K. Marsh and B. Bugusu, Food packaging - roles, materials, and environmental issues, *Journal of Food Science*, Vol. 72, p. R39, 2007.
44. F. Michaels, *Modern Plastics Encyclopedia*, 1995.
45. R.W.G. van Willige, J.P.H. Linssen, M.B.J. Meinders, H.J. van der Stege, and A.G.J. Voragen, Influence of flavour absorption on oxygen permeation through LDPE, PP, PC and PET plastics food packaging, *Food Additives & Contaminants*, Vol. 19, p. 303, 2002.
46. A. Arora and G.W. Padua, Review: Nanocomposites in food packaging, *Journal of Food Science*, Vol. 75, p. R43, 2010.

47. L. Motelica, D. Ficai, A. Ficai, O.C. Oprea, D.A. Kaya, and E. Andronescu, Biodegradable antimicrobial food packaging: Trends and perspectives, *Foods*, Vol. 9, 2020.
48. A. Khezrian and Y. Shahbazi, Application of nanocompostie chitosan and carboxymethyl cellulose films containing natural preservative compounds in minced camel's meat, *International Journal of Biological Macromolecules*, Vol. 106, p. 1146, 2018.
49. S.-W. Zhao, C.-R. Guo, Y.-Z. Hu, Y.-R. Guo, and Q.-J. Pan, The preparation and antibacterial activity of cellulose/ZnO composite: A review, *Open Chemistry*, Vol. 16, p. 9, 2018.
50. M. Kaya, S. Khadem, Y.S. Cakmak, M. Mujtaba, S. Ilk, L. Akyuz, A.M. Salaberria, J. Labidi, A.H. Abdulqadir, and E. Deligöz, Antioxidative and antimicrobial edible chitosan films blended with stem, leaf and seed extracts of Pistacia terebinthus for active food packaging, *RSC Advances*, Vol. 8, p. 3941, 2018.
51. L. Sun, J. Sun, D. Liu, M. Fu, X. Yang, and Y. Guo, The preservative effects of chitosan film incorporated with thinned young apple polyphenols on the quality of grass carp (Ctenopharyngodon idellus) fillets during cold storage: Correlation between the preservative effects and the active properties of the film, *Food Packaging and Shelf Life*, Vol. 17, p. 1, 2018.
52. C.A. Ubaque Beltran and I. Fernando Florez, Preparation of biopackages from native starches and essential oils prolonging shelf life in strawberries, *Revista Colombiana de Investigaciones Agroindustriales*, Vol. 5, p. 71, 2018.
53. A.T. Issa, K.A. Schimmel, M. Worku, A. Shahbazi, S.A. Ibrahim, and R. Tahergorabi, Sweet potato starch-based nanocomposites: Development, characterization, and biodegradability, *Starch-Stärke*, Vol. 70, p. 1700273, 2018.
54. G. Tedeschi, S. Guzman-Puyol, L. Ceseracciu, U.C. Paul, P. Picone, M. Di Carlo, A. Athanassiou, and J.A. Heredia-Guerrero, Multifunctional bioplastics inspired by wood composition: Effect of hydrolyzed lignin addition to xylan–cellulose matrices, *Biomacromolecules*, Vol. 21, p. 910, 2020.
55. P. Nechita and M. Roman, Review on polysaccharides used in coatings for food packaging papers, *Coatings*, Vol. 10, p. 566, 2020.
56. M.J. Fabra, I. Falcó, W. Randazzo, G. Sánchez, and A. López-Rubio, Antiviral and antioxidant properties of active alginate edible films containing phenolic extracts, *Food Hydrocolloids*, Vol. 81, p. 96, 2018.
57. M.S.A. Aziz, H.E. Salama, and M.W. Sabaa, Biobased alginate/castor oil edible films for active food packaging, *LWT - Food Science and Technology*, Vol. 96, p. 455, 2018.

58. Z. Hu, P. Hong, M. Liao, S. Kong, N. Huang, C. Ou, and S. Li, Preparation and characterization of chitosan–agarose composite films, *Materials*, Vol. 9, p. 816, 2016.
59. B. Kusznierewicz, H. Staroszczyk, E. Malinowska-Pańczyk, K. Parchem, and A. Bartoszek, Novel ABTS-dot-blot method for the assessment of antioxidant properties of food packaging, *Food Packaging and Shelf Life*, Vol. 24, p. 100478, 2020.
60. J. Lei, L. Zhou, Y. Tang, Y. Luo, T. Duan, and W. Zhu, High-strength konjac glucomannan/silver nanowires composite films with antibacterial properties, *Materials*, Vol. 10, p. 524, 2017.
61. C. Wu, Y. Li, J. Sun, Y. Lu, C. Tong, L. Wang, Z. Yan, and J. Pang, Novel konjac glucomannan films with oxidized chitin nanocrystals immobilized red cabbage anthocyanins for intelligent food packaging, *Food Hydrocolloids*, Vol. 98, p. 105245, 2020.
62. S. Li, J. Yi, X. Yu, Z. Wang, and L. Wang, Preparation and characterization of pullulan derivative/chitosan composite film for potential antimicrobial applications, *International Journal of Biological Macromolecules*, Vol. 148, p. 258, 2020.
63. A.H.A. Hassan and C.N. Cutter, Development and evaluation of pullulan-based composite antimicrobial films (CAF) incorporated with nisin, thymol and lauric arginate to reduce foodborne pathogens associated with muscle foods, *International Journal of Food Microbiology*, Vol. 320, p. 108519, 2020.
64. F.T. da Silva, K.F. da Cunha, L.M. Fonseca, M.D. Antunes, S.L.M. El Halal, Â.M. Fiorentini, E. da Rosa Zavareze, and A.R.G. Dias, Action of ginger essential oil (Zingiber officinale) encapsulated in proteins ultrafine fibers on the antimicrobial control in situ, *International Journal of Biological Macromolecules*, Vol. 118, p. 107, 2018.
65. I. Echeverría, M.E. López-Caballero, M.C. Gómez-Guillén, A.N. Mauri, and M.P. Montero, Active nanocomposite films based on soy proteins-montmorillonite-clove essential oil for the preservation of refrigerated bluefin tuna (Thunnus thynnus) fillets, *International Journal of Food Microbiology*, Vol. 266, p. 142, 2018.
66. F. Kouravand, H. Jooyandeh, H. Barzegar, and M. Hojjati, Characterization of cross-linked whey protein isolate-based films containing Satureja khuzistanica Jamzad essential oil, *Journal of Food Processing and Preservation*, Vol. 42, p. e13557, 2018.
67. A. Altan, Z. Aytac, and T. Uyar, Carvacrol loaded electrospun fibrous films from zein and poly(lactic acid) for active food packaging, *Food Hydrocolloids*, Vol. 81, p. 48, 2018.
68. M. Ahmad, N.P. Nirmal, M. Danish, J. Chuprom, and S. Jafarzedeh, Characterisation of composite films fabricated from collagen/chi-

tosan and collagen/soy protein isolate for food packaging applications, *RSC Advances*, Vol. 6, p. 82191, 2016.
69. B. Ocak, Film-forming ability of collagen hydrolysate extracted from leather solid wastes with chitosan, *Environmental Science and Pollution Research*, Vol. 25, p. 4643, 2018.
70. K.J. Figueroa-Lopez, J.L. Castro-Mayorga, M.M. Andrade-Mahecha, L. Cabedo, and J.M. Lagaron, Antibacterial and barrier properties of gelatin coated by electrospun polycaprolactone ultrathin fibers containing black pepper oleoresin of interest in active food biopackaging applications, *Nanomaterials*, Vol. 8, p. 199, 2018.
71. H. Kim, S.-E. Beak, and K.B. Song, Development of a hagfish skin gelatin film containing cinnamon bark essential oil, *LWT - Food Science and Technology*, Vol. 96, p. 583, 2018.
72. E. Abdollahzadeh, S.M. Ojagh, A.A.I. Fooladi, B. Shabanpour, and M. Gharahei, Effects of probiotic cells on the mechanical and antibacterial properties of sodium-caseinate films, *Applied Food Biotechnology*, Vol. 5, p. 155, 2018.
73. E. Chevalier, G. Assezat, F. Prochazka, and N. Oulahal, Development and characterization of a novel edible extruded sheet based on different casein sources and influence of the glycerol concentration, *Food Hydrocolloids*, Vol. 75, p. 182, 2018.
74. L. Akyuz, M. Kaya, S. Ilk, Y.S. Cakmak, A.M. Salaberria, J. Labidi, B.A. Yılmaz, and I. Sargin, Effect of different animal fat and plant oil additives on physicochemical, mechanical, antimicrobial and antioxidant properties of chitosan films, *International Journal of Biological Macromolecules*, Vol. 111, p. 475, 2018.
75. B. Tonyali, A. McDaniel, J. Amamcharla, V. Trinetta, and U. Yucel, Release kinetics of cinnamaldehyde, eugenol, and thymol from sustainable and biodegradable active packaging films, *Food Packaging and Shelf Life*, Vol. 24, p. 100484, 2020.
76. C.K. Saurabh, S. Gupta, and P.S. Variyar, Development of guar gum based active packaging films using grape pomace, *Journal of Food Science and Technology*, Vol. 55, p. 1982, 2018.
77. D. Zhang, G. Yu, Z. Long, H. Xiao, and L. Qian, Bio-wax latex-modified paper as antimicrobial and water-vapor-resistant packaging material, *Journal of Wood Chemistry and Technology*, Vol. 36, p. 182, 2016.
78. C. Swaroop and M. Shukla, Nano-magnesium oxide reinforced polylactic acid biofilms for food packaging applications, *International Journal of Biological Macromolecules*, Vol. 113, p. 729, 2018.
79. Ł. Łopusiewicz, F. Jędra, and M. Mizielińska, New poly (lactic acid) active packaging composite films incorporated with fungal melanin, *Polymers*, Vol. 10, p. 386, 2018.
80. P. Shao, Z. Yan, H. Chen, and J. Xiao, Electrospun poly(vinyl alcohol)/permutite fibrous film loaded with cinnamaldehyde for active

food packaging, *Journal of Applied Polymer Science*, Vol. 135, p. 46117, 2018.

81. M.P. Arrieta, C.L. de Dicastillo, L. Garrido, K. Roa, and M.J. Galotto, Electrospun PVA fibers loaded with antioxidant fillers extracted from Durvillaea antarctica algae and their effect on plasticized PLA bionanocomposites, *European Polymer Journal*, Vol. 103, p. 145, 2018.
82. M. Almeida, M.A. Martins, A.M.V. Soares, A. Cuesta, and M. Oliveira, Polystyrene nanoplastics alter the cytotoxicity of human pharmaceuticals on marine fish cell lines, *Environmental Toxicology and Pharmacology*, Vol. 69, p. 57, 2019.
83. N. Takamura, Transparent poly(phenylene ether) compositions, their methods of manufacture, and food packaging films and containers derived therefrom, US Patent 9 845 388, assigned to SABIC Global Technologies BV, December 19, 2017.
84. M. Llana-Ruiz-Cabello, S. Pichardo, J.M. Bermudez, A. Baños, J.J. Ariza, E. Guillamón, S. Aucejo, and A.M. Cameán, Characterisation and antimicrobial activity of active polypropylene films containing oregano essential oil and allium extract to be used in packaging for meat products, *Food Additives & Contaminants: Part A*, Vol. 35, p. 783, 2018.
85. G. Maleki, N. Sedaghat, E.J. Woltering, M. Farhoodi, and M. Mohebbi, Chitosan-limonene coating in combination with modified atmosphere packaging preserve postharvest quality of cucumber during storage, *Journal of Food Measurement and Characterization*, Vol. 12, p. 1610, 2018.
86. P. Kaewklin, U. Siripatrawan, A. Suwanagul, and Y.S. Lee, Active packaging from chitosan-titanium dioxide nanocomposite film for prolonging storage life of tomato fruit, *International Journal of Biological Macromolecules*, Vol. 112, p. 523, 2018.
87. J. Yan, Z. Luo, Z. Ban, H. Lu, D. Li, D. Yang, M.S. Aghdam, and L. Li, The effect of the layer-by-layer (LBL) edible coating on strawberry quality and metabolites during storage, *Postharvest Biology and Technology*, Vol. 147, p. 29, 2019.
88. C. Zhang, W. Li, B. Zhu, H. Chen, H. Chi, L. Li, Y. Qin, and J. Xue, The quality evaluation of postharvest strawberries stored in nano-ag packages at refrigeration temperature, *Polymers*, Vol. 10, p. 894, 2018.
89. Y. Shahbazi, Application of carboxymethyl cellulose and chitosan coatings containing Mentha spicata essential oil in fresh strawberries, *International Journal of Biological Macromolecules*, Vol. 112, p. 264, 2018.
90. L. Li, H. Wang, M. Chen, S. Jiang, S. Jiang, X. Li, and Q. Wang, Butylated hydroxyanisole encapsulated in gelatin fiber mats: Volatile release kinetics, functional effectiveness and application to strawberry preservation, *Food Chemistry*, Vol. 269, p. 142, 2018.

91. J.-Y. Zhu, C.-H. Tang, S.-W. Yin, and X.-Q. Yang, Development and characterisation of polylactic acid–gliadin bilayer/trilayer films as carriers of thymol, *International Journal of Food Science & Technology*, Vol. 53, p. 608, 2018.
92. F. Rezaei and Y. Shahbazi, Shelf-life extension and quality attributes of sauced silver carp fillet: A comparison among direct addition, edible coating and biodegradable film, *LWT - Food Science and Technology*, Vol. 87, p. 122, 2018.
93. P. Hassanzadeh, M. Moradi, N. Vaezi, M.-H. Moosavy, and R. Mahmoudi, Effects of chitosan edible coating containing grape seed extract on the shelf-life of refrigerated rainbow trout fillet, in *Veterinary Research Forum*, Vol. 9, p. 73. Faculty of Veterinary Medicine, Urmia University, Urmia, Iran, 2018.
94. Y. Ma, L. Li, and Y. Wang, Development of PLA-PHB-based biodegradable active packaging and its application to salmon, *Packaging Technology and Science*, Vol. 31, p. 739, 2018.
95. Q. Wang, J. Lei, J. Ma, G. Yuan, and H. Sun, Effect of chitosan-carvacrol coating on the quality of Pacific white shrimp during iced storage as affected by caprylic acid, *International Journal of Biological Macromolecules*, Vol. 106, p. 123, 2018.
96. M. Ejaz, Y.A. Arfat, M. Mulla, and J. Ahmed, Zinc oxide nanorods/-clove essential oil incorporated Type B gelatin composite films and its applicability for shrimp packaging, *Food Packaging and Shelf Life*, Vol. 15, p. 113, 2018.
97. M.D.C.P. Zegarra, A.M.P. Santos, A.M.A.D. Silva, and E.D. Melo, Chitosan films incorporated with antioxidant extract of acerola agroindustrial residue applied in chicken thigh., *Joural of Food Processing and Perservation*, Vol. 42, 2018.
98. V.G.L. Souza, J.R.A. Pires, É.T. Vieira, I.M. Coelhoso, M.P. Duarte, and A.L. Fernando, Shelf life assessment of fresh poultry meat packaged in novel bionanocomposite of chitosan/montmorillonite incorporated with ginger essential oil, *Coatings*, Vol. 8, p. 177, 2018.
99. L. Lin, Y. Zhu, and H. Cui, Electrospun thyme essential oil/gelatin nanofibers for active packaging against Campylobacter jejuni in chicken, *LWT - Food Science and Technology*, Vol. 97, p. 711, 2018.
100. W. Yang, Y. Xie, J. Jin, H. Liu, and H. Zhang, Development and application of an active plastic multilayer film by coating a plantaricin BM-1 for chilled meat preservation, *Journal of Food Science*, Vol. 84, p. 1864, 2019.
101. F.H. Rad, A. Sharifan, and G. Asadi, Physicochemical and antimicrobial properties of kefiran/waterborne polyurethane film incorporated with essential oils on refrigerated ostrich meat, *LWT - Food Science and Technology*, Vol. 97, p. 794, 2018.

102. M. Pabast, N. Shariatifar, S. Beikzadeh, and G. Jahed, Effects of chitosan coatings incorporating with free or nano-encapsulated Satureja plant essential oil on quality characteristics of lamb meat, *Food Control*, Vol. 91, p. 185, 2018.
103. F. Talebi, A. Misaghi, A. Khanjari, A. Kamkar, H. Gandomi, and M. Rezaeigolestani, Incorporation of spice essential oils into poly-lactic acid film matrix with the aim of extending microbiological and sensorial shelf life of ground beef, *LWT - Food Science and Technology*, Vol. 96, p. 482, 2018.
104. Y. Zhao, J.S. Teixeira, M.M. Gänzle, and M.D.A. Saldaña, Development of antimicrobial films based on cassava starch, chitosan and gallic acid using subcritical water technology, *The Journal of Supercritical Fluids*, Vol. 137, p. 101, 2018.
105. R. Ribeiro-Santos, N.R. de Melo, M. Andrade, G. Azevedo, A.V. Machado, D. Carvalho-Costa, and A. Sanches-Silva, Whey protein active films incorporated with a blend of essential oils: Characterization and effectiveness, *Packaging Technology and Science*, Vol. 31, p. 27, 2018.
106. A.M. Youssef, S.M. El-Sayed, H.S. El-Sayed, H.H. Salama, F.M. Assem, and M.H. Abd El-Salam, Novel bionanocomposite materials used for packaging skimmed milk acid coagulated cheese (Karish), *International Journal of Biological Macromolecules*, Vol. 115, p. 1002, 2018.
107. M. Lotfi, H. Tajik, M. Moradi, M. Forough, E. Divsalar, and B. Kuswandi, Nanostructured chitosan/monolaurin film: Preparation, characterization and antimicrobial activity against listeria monocytogenes on ultrafiltered white cheese, *LWT - Food Science and Technology*, Vol. 92, p. 576, 2018.
108. S.-Y. Yang, L. Cao, H. Kim, S.-E. Beak, and K.B. Song, Utilization of foxtail millet starch film incorporated with clove leaf oil for the packaging of queso blanco cheese as a model food, *Starch - Stärke*, Vol. 70, p. 1700171, 2018.
109. M.V. Guitián, C. Ibarguren, M.C. Soria, P. Hovanyecz, C. Banchio, and M.C. Audisio, Anti-Listeria monocytogenes effect of bacteriocin-incorporated agar edible coatings applied on cheese, *International Dairy Journal*, Vol. 97, p. 92, 2019.
110. M. Mushtaq, A. Gani, A. Gani, H.A. Punoo, and F.A. Masoodi, Use of pomegranate peel extract incorporated zein film with improved properties for prolonged shelf life of fresh Himalayan cheese (Kalari/kradi), *Innovative Food Science & Emerging Technologies*, Vol. 48, p. 25, 2018.
111. A. Orsuwan and R. Sothornvit, Active banana flour nanocomposite films incorporated with garlic essential oil as multifunctional pack-

aging material for food application, *Food and Bioprocess Technology*, Vol. 11, p. 1199, 2018.

112. S. He, Q. Yang, X. Ren, J. Zi, S. Lu, S. Wang, Y. Zhang, and Y. Wang, Antimicrobial efficiency of chitosan solutions and coatings incorporated with clove oil and/or ethylenediaminetetraacetate, *Journal of Food Safety*, Vol. 34, p. 345, 2014.
113. K. Feng, P. Wen, H. Yang, N. Li, W.Y. Lou, M.H. Zong, and H. Wu, Enhancement of the antimicrobial activity of cinnamon essential oil-loaded electrospun nanofilm by the incorporation of lysozyme, *RSC Advances*, Vol. 7, p. 1572, 2017.
114. F. Fratianni, L. De Martino, A. Melone, V. De Feo, R. Coppola, and F. Nazzaro, Preservation of chicken breast meat treated with thyme and balm essential oils, *Journal of Food Science*, Vol. 75, p. M528, 2010.
115. I.S. Arvanitoyannis and A.C. Stratakos, Application of modified atmosphere packaging and active/smart technologies to red meat and poultry: A review, *Food and Bioprocess Technology*, Vol. 5, p. 1423, 2012.
116. M. Mulla, J. Ahmed, H. Al-Attar, E. Castro-Aguirre, Y.A. Arfat, and R. Auras, Antimicrobial efficacy of clove essential oil infused into chemically modified LLDPE film for chicken meat packaging, *Food Control*, Vol. 73, p. 663, 2017.
117. M. Radulescu, S. Popescu, D. Ficai, M. Sonmez, O. Oprea, A. Spoiala, A. Ficai, and E. Andronescu, Advances in drug delivery systems, from 0 to 3D superstructures, *Current Drug Targets*, Vol. 19, p. 393, 2018.
118. K. Szabo, B.-E. Teleky, L. Mitrea, L.-F. Călinoiu, G.-A. Martău, E. Simon, R.-A. Varvara, and D.C. Vodnar, Active packaging-poly(vinyl alcohol) films enriched with tomato by-products extract, *Coatings*, Vol. 10, p. 141, 2020.
119. S. Rivero, L. Giannuzzi, M.A. Garcia, and A. Pinotti, Controlled delivery of propionic acid from chitosan films for pastry dough conservation, *Journal of Food Engineering*, Vol. 116, p. 524, 2013.
120. F.A. Lopes, N. de Fátima Ferreira Soares, C. de Cássia Pires Lopes, W.A. da Silva, J.C.B. Júnior, and E.A.A. Medeiros, Conservation of bakery products through cinnamaldehyde antimicrobial films, *Packaging Technology and Science*, Vol. 27, p. 293, 2014.
121. A. Mihaly-Cozmuta, A. Peter, G. Craciun, A. Falup, L. Mihaly-Cozmuta, C. Nicula, A. Vulpoi, and M. Baia, Preparation and characterization of active cellulose-based papers modified with TiO_2, Ag and zeolite nanocomposites for bread packaging application, *Cellulose*, Vol. 24, p. 3911, 2017.
122. L. Al-Naamani, J. Dutta, and S. Dobretsov, Nanocomposite zinc oxide-chitosan coatings on polyethylene films for extending storage life of okra (Abelmoschus esculentus), *Nanomaterials*, Vol. 8, p. 479, 2018.

123. Y. Xing, W. Li, Q. Wang, X. Li, Q. Xu, X. Guo, X. Bi, X. Liu, Y. Shui, and H. Lin, Antimicrobial nanoparticles incorporated in edible coatings and films for the preservation of fruits and vegetables, *Molecules*, Vol. 24, p. 1695, 2019.
124. Y. Xing, X. Li, X. Guo, W. Li, J. Chen, Q. Liu, Q. Xu, Q. Wang, H. Yang, and Y. Shui, Effects of different TiO_2 nanoparticles concentrations on the physical and antibacterial activities of chitosan-based coating film, *Nanomaterials*, Vol. 10, p. 1365, 2020.
125. A. Babaei-Ghazvini, B. Acharya, and D.R. Korber, Antimicrobial biodegradable food packaging based on chitosan and metal/metal-oxide bio-nanocomposites: A review, *Polymers*, Vol. 13, p. 2790, Aug 2021.
126. K.Y. Perera, S. Jaiswal, and A.K. Jaiswal, A review on nanomaterials and nanohybrids based bio-nanocomposites for food packaging, *Food Chemistry*, Vol. 376, p. 131912, 2022.
127. D. Leigh, *SWOT Analysis*, chapter 5, pp. 115–140. John Wiley & Sons, Ltd., 2009.
128. U. Siripatrawan and P. Kaewklin, Fabrication and characterization of chitosan-titanium dioxide nanocomposite film as ethylene scavenging and antimicrobial active food packaging, *Food Hydrocolloids*, Vol. 84, p. 125, 2018.
129. R. Sharma, S.M. Jafari, and S. Sharma, Antimicrobial bio-nanocomposites and their potential applications in food packaging, *Food Control*, Vol. 112, p. 107086, 2020.
130. T. Rodrigues Arruda, P. Campos Bernardes, A. Robledo Fialho e Moraes, and N. de Fátima Ferreira Soares, Natural bioactives in perspective: The future of active packaging based on essential oils and plant extracts themselves and those complexed by cyclodextrins, *Food Research International*, Vol. 156, p. 111160, 2022.
131. A.R. Alias, M.K. Wan, and N.M. Sarbon, Emerging materials and technologies of multi-layer film for food packaging application: A review, *Food Control*, p. 108875, 2022.
132. M. Li, X. Luo, R. Zhu, K. Zhong, W. Ran, Y. Wu, and H. Gao, Development and characterization of active bilayer film incorporated with dihydromyricetin encapsulated in hydroxypropyl-β-cyclodextrin for food packaging application, *Food Hydrocolloids*, p. 107834, 2022.
133. C. Valencia-Sullca, M. Vargas, L. Atarés, and A. Chiralt, Thermoplastic cassava starch-chitosan bilayer films containing essential oils, *Food Hydrocolloids*, Vol. 75, p. 107, 2018.
134. Y. Han, M. Yu, and L. Wang, Physical and antimicrobial properties of sodium alginate/carboxymethyl cellulose films incorporated with cinnamon essential oil, *Food Packaging and Shelf Life*, Vol. 15, p. 35, 2018.

135. W. Guiga, Y. Swesi, S. Galland, E. Peyrol, P. Degraeve, and I. Sebti, Innovative multilayer antimicrobial films made with nisaplin or nisin and cellulosic ethers: Physico-chemical characterization, bioactivity and nisin desorption kinetics, *Innovative Food Science & Emerging Technologies*, Vol. 11, p. 352, 2010.
136. J. Zhu, H. Wu, and Q. Sun, Preparation of crosslinked active bilayer film based on chitosan and alginate for regulating ascorbate-glutathione cycle of postharvest cherry tomato (Lycopersicon esculentum), *International Journal of Biological Macromolecules*, Vol. 130, p. 584, 2019.
137. H. Arnon, Y. Zaitsev, R. Porat, and E. Poverenov, Effects of carboxymethyl cellulose and chitosan bilayer edible coating on postharvest quality of citrus fruit, *Postharvest Biology and Technology*, Vol. 87, p. 21, 2014.
138. A.H. Basta, K. Khwaldia, H. Aloui, and H. El-Saied, Enhancing the performance of carboxymethyl cellulose by chitosan in producing barrier coated paper sheets, *Nordic Pulp & Paper Research Journal*, Vol. 30, p. 617, 2015.
139. J.F. Martucci and R.A. Ruseckaite, Biodegradable three-layer film derived from bovine gelatin, *Journal of Food Engineering*, Vol. 99, p. 377, 2010.
140. S.Y. Cho, S.Y. Lee, and C. Rhee, Edible oxygen barrier bilayer film pouches from corn zein and soy protein isolate for olive oil packaging, *LWT - Food Science and Technology*, Vol. 43, p. 1234, 2010.
141. G. Zuo, X. Song, F. Chen, and Z. Shen, Physical and structural characterization of edible bilayer films made with zein and corn-wheat starch, *Journal of the Saudi Society of Agricultural Sciences*, Vol. 18, p. 324, 2019.
142. C. Chen, Y. Du, G. Zuo, F. Chen, K. Liu, and L. Zhang, Effect of storage condition on the physico-chemical properties of corn–wheat starch/zein edible bilayer films, *Royal Society Open Science*, Vol. 7, p. 191777, 2020.
143. E. Poverenov, R. Rutenberg, S. Danino, B. Horev, and V. Rodov, Gelatin-chitosan composite films and edible coatings to enhance the quality of food products: Layer-by-layer vs. blended formulations, *Food and Bioprocess Technology*, Vol. 7, p. 3319, 2014.
144. G.P. Cardoso, M.P.D. Andrade, L.M. Rodrigues, A.A. Massingue, P.R. Fontes, A.d.L.S. Ramos, and E.M. Ramos, Retail display of beef steaks coated with monolayer and bilayer chitosan-gelatin composites, *Meat Science*, Vol. 152, p. 20, 2019.
145. H. Haghighi, R. De Leo, E. Bedin, F. Pfeifer, H.W. Siesler, and A. Pulvirenti, Comparative analysis of blend and bilayer films based on chitosan and gelatin enriched with LAE (lauroyl arginate ethyl)

with antimicrobial activity for food packaging applications, *Food Packaging and Shelf Life*, Vol. 19, p. 31, 2019.
146. W. Wang, J. Xiao, X. Chen, M. Luo, H. Liu, and P. Shao, Fabrication and characterization of multilayered kafirin/gelatin film with one-way water barrier property, *Food Hydrocolloids*, Vol. 81, p. 159, 2018.
147. Z. Wan, L. Wang, X. Yang, J. Guo, and S. Yin, Enhanced water resistance properties of bacterial cellulose multilayer films by incorporating interlayers of electrospun zein fibers, *Food Hydrocolloids*, Vol. 61, p. 269, 2016.
148. X. Chen, F. Cui, H. Zi, Y. Zhou, H. Liu, and J. Xiao, Development and characterization of a hydroxypropyl starch/zein bilayer edible film, *International Journal of Biological Macromolecules*, Vol. 141, p. 1175, 2019.
149. J.L. Castro-Mayorga, M.J. Fabra, L. Cabedo, and J.M. Lagaron, On the use of the electrospinning coating technique to produce antimicrobial polyhydroxyalkanoate materials containing in situ-stabilized silver nanoparticles, *Nanomaterials*, Vol. 7, p. 4, 2016.
150. M.J. Fabra, A. Lopez-Rubio, and J.M. Lagaron, Effect of the film-processing conditions, relative humidity and ageing on wheat gluten films coated with electrospun polyhydryalkanoate, *Food Hydrocolloids*, Vol. 44, p. 292, 2015.
151. M.J. Fabra, A. Lopez-Rubio, and J.M. Lagaron, High barrier polyhydroxyalcanoate food packaging film by means of nanostructured electrospun interlayers of zein, *Food Hydrocolloids*, Vol. 32, p. 106, 2013.
152. P.R. Salgado, D.A. D'Amico, I.T. Seoane, M. Iglesias Montes, A.N. Mauri, and V.P. Cyras, Improvement of water barrier properties of soybean protein isolate films by poly (3-hydroxybutyrate) thin coating, *Journal of Applied Polymer Science*, Vol. 138, p. 49758, 2021.
153. K.M. Van de Voorde, J.K. Pokorski, and L.T.J. Korley, Exploring morphological effects on the mechanics of blended poly (lactic acid)/poly (ε-caprolactone) extruded fibers fabricated using multilayer coextrusion, *Macromolecules*, Vol. 53, p. 5047, 2020.
154. A. Tampau, C. González-Martínez, A.A. Vicente, and A. Chiralt, Enhancement of PLA-PVA surface adhesion in bilayer assemblies by PLA aminolisation, *Food and Bioprocess Technology*, Vol. 13, p. 1215, 2020.
155. M.L. Sanyang, S.M. Sapuan, M. Jawaid, M.R. Ishak, and J. Sahari, Development and characterization of sugar palm starch and poly(lactic acid) bilayer films, *Carbohydrate Polymers*, Vol. 146, p. 36, 2016.

156. J. Muller, C. González-Martínez, and A. Chiralt, Combination of poly(lactic) acid and starch for biodegradable food packaging, *Materials*, Vol. 10, p. 952, 2017.
157. X. Zhou, R. Yang, B. Wang, and K. Chen, Development and characterization of bilayer films based on pea starch/polylactic acid and use in the cherry tomatoes packaging, *Carbohydrate Polymers*, Vol. 222, p. 114912, 2019.
158. Ç. Meriçer, M. Minelli, M.G. De Angelis, M.G. Baschetti, A. Stancampiano, R. Laurita, M. Gherardi, V. Colombo, J. Trifol, and P. Szabo, Atmospheric plasma assisted PLA/microfibrillated cellulose (MFC) multilayer biocomposite for sustainable barrier application, *Industrial Crops and Products*, Vol. 93, p. 235, 2016.
159. A. Aydogdu, E. Yildiz, Z. Ayhan, Y. Aydogdu, G. Sumnu, and S. Sahin, Nanostructured poly(lactic acid)/soy protein/HPMC films by electrospinning for potential applications in food industry, *European Polymer Journal*, Vol. 112, p. 477, 2019.
160. H. Gartner, Y. Li, and E. Almenar, Improved wettability and adhesion of polylactic acid/chitosan coating for bio-based multilayer film development, *Applied Surface Science*, Vol. 332, p. 488, 2015.
161. E. Benetto, C. Jury, E. Igos, J. Carton, P. Hild, C. Vergne, and J. Di Martino, Using atmospheric plasma to design multilayer film from polylactic acid and thermoplastic starch: A screening life cycle assessment, *Journal of Cleaner Production*, Vol. 87, p. 953, 2015.
162. C. Zhuang, Y. Jiang, Y. Zhong, Y. Zhao, Y. Deng, J. Yue, D. Wang, S. Jiao, H. Gao, H. Chen, et al., Development and characterization of nano-bilayer films composed of polyvinyl alcohol, chitosan and alginate, *Food Control*, Vol. 86, p. 191, 2018.
163. C.-H. Gu, J.-J. Wang, Y. Yu, H. Sun, N. Shuai, and B. Wei, Biodegradable multilayer barrier films based on alginate/polyethyleneimine and biaxially oriented poly (lactic acid), *Carbohydrate Polymers*, Vol. 92, p. 1579, 2013.
164. J. Ding, R. Zhang, S. Ahmed, Y. Liu, and W. Qin, Effect of sonication duration in the performance of polyvinyl alcohol/chitosan bilayer films and their effect on strawberry preservation, *Molecules*, Vol. 24, p. 1408, 2019.
165. H.M. Heidemann, M.E.R. Dotto, J.B. Laurindo, B.A.M. Carciofi, and C. Costa, Cold plasma treatment to improve the adhesion of cassava starch films onto PCL and PLA surface, *Colloids and Surfaces A: Physicochemical and Engineering Aspects*, Vol. 580, p. 123739, 2019.
166. A. Tampau, C. González-Martínez, and A. Chiralt, Biodegradability and disintegration of multilayer starch films with electrospun PCL fibres encapsulating carvacrol, *Polymer Degradation and Stability*, Vol. 173, p. 109100, 2020.

167. S.F. Hosseini, Z. Javidi, and M. Rezaei, Efficient gas barrier properties of multi-layer films based on poly(lactic acid) and fish gelatin, *International Journal of Biological Macromolecules*, Vol. 92, p. 1205, 2016.
168. K.-Y. Lee and K.B. Song, Preparation and characterization of an olive flounder (paralichthys olivaceus) skin gelatin and polylactic acid bilayer film, *Journal of Food Science*, Vol. 82, p. 706, 2017.
169. M. Nagarajan, T. Prodpran, S. Benjakul, and P. Songtipya, Properties and characteristics of multi-layered films from tilapia skin gelatin and poly (lactic acid), *Food Biophysics*, Vol. 12, p. 222, 2017.
170. K. Nilsuwan, S. Benjakul, and T. Prodpran, Physical/thermal properties and heat seal ability of bilayer films based on fish gelatin and poly (lactic acid), *Food Hydrocolloids*, Vol. 77, p. 248, 2018.
171. J.F. Martucci and R.A. Ruseckaite, Three-layer sheets based on gelatin and poly(lactic acid), Part 1: Preparation and properties, *Journal of Applied Polymer Science*, Vol. 118, p. 3102, 2010.
172. J.F. Martucci and R.A. Ruseckaite, Biodegradation behavior of three-layer sheets based on gelatin and poly(lactic acid) buried under indoor soil conditions, *Polymer Degradation and Stability*, Vol. 116, p. 36, 2015.

5

Materials

Polymers for packaging applications have been described in a monograph (1).

5.1 Nanocomposites

The issues related to nanocomposites for food packaging applications have been reviewed (2). The development of nanocomposites is a strategy to improve the physical properties of polymers, such as mechanical strength, thermal stability, and gas barrier properties (3–5).

When used in food packaging, nanocomposites are better able to withstand the stress of thermal food processing, transportation, and storage (3, 6).

Because of their improved mechanical properties, nanocomposites may allow downgauging, thus reducing materials usage. Particle fillers used and proposed in the literature include the nanoclays montmorillonite (MMT) and kaolinite, carbon nanotubes, and graphene nanosheets.

In food packaging, a major emphasis is on the development of high barrier properties against the migration of oxygen, carbon dioxide, flavor compounds, and water vapor. Decreasing the permeability of water vapor is a critical issue in the development of biopolymers as sustainable packaging materials.

Challenges in the development of suitable materials are to increase the compatibility between clays and polymers and to reach a complete dispersion of nanoplates.

Nanocomposites may advance the utilization of biopolymers in food packaging.

5.1.1 PVC/Ag Nanocomposites

Silver nanoparticles have been synthesized and characterized. Their antimicrobial activity against *Escherichia coli, Staphylococcus aureus, Bacillus cereus, Pseudomonas fluorescens* and microflorae derived from raw chicken, beef or cooked ham was determined (7).

Poly(vinyl chloride) (PVC) films or antimicrobial Ag/PVC nanocomposite films were manufactured using a solvent casting method and the mechanical and thermal properties of these materials were determined.

These antimicrobial nanocomposite films were used to wrap chicken breast fillets, followed by modified atmosphere packaging. Here, conventional laminates were used and a gas mix of 60% N_2 and 40% CO_2 was employed.

In general, Gram-negative bacteria were more sensitive to silver nanoparticles than Gram-positive bacteria, and microflora isolated from meat products were more resistant than pure culture bacteria.

However, the most sensitive bacteria to silver nanoparticles were *Pseudomonas fluorescens.* No significant differences in tensile strength and elongation at break were observed, but glass transition temperatures of Ag/PVC nanocomposite films were lower then PVC control films.

Antimicrobial Ag/PVC nanocomposite films significantly extended the shelf life of chicken breast fillets and reduced lipid oxidation of chicken breast fillets compared to PVC-wrapped equivalents (7).

5.1.1.1 Poly(phenylene ether)

It has been discovered that blending general purpose poly(styrene) (PS) with a master batch that contains poly(phenylene ether) (PPE) and conventional PS provides a sufficiently transparent material with desirable thermal stability that can be used in fabricating microwavable food containers.

The master batch contains (8):

1. 40% to 75% of a PPE,
2. 25% to 60% of a styrenic polymer,

3. 0.01% to 2.5% of an optical enhancing additive,
4. 0.003% to 0.012% of a colorant, and
5. 0% to 3% of an unsaturated di- or trifunctional carboxylic acid.

The master batch is characterized by a heat deflection temperature that is 92°C to 146°C as measured according to ASTM D648 (9) on 6.4 *mm* thick bars at 1.82 *MPa* and a Haze value of less than 5.

The polymer composition can be used to fabricate microwavable food packages and food wrapping films because it has improved thermal stability compared to conventional general purpose PS. The improved thermal stability is due to the blending of the PPE resin as a master batch with conventional general purpose PS. Surprisingly, the PPE in the master batch has unique characteristics in that it is miscible and forms a homogeneous phase with PS. At certain ratios an increase in the glass transition temperature of PS is observed, leading to the increased thermal stability of the packaging material.

The PPE may have 2,6-dimethyl-1,4-phenylene ether units, 2,3,6-trimethyl-1,4-phenylene ether units, or a combination thereof.

Optical enhancing agents are typically selected from the group consisting of α-hydroxyketones or unsubstituted or substituted trihydrocarbyl phosphites.

α-Hydroxyketones include, for example, 1-hydroxy-2-propanone, acetoin (3-hydroxy-2-butanone), 2-hydroxyacetophenone, benzoin (2-hydroxy-2-phenylacetophenone), 2-hydroxy-1-phenyl-2-*p*-tolyl-ethanone, and the like.

Suitable trihydrocarbyl phosphites include trioctyl phosphite, tridecyl phosphite, tridodecyl phosphite, phenyl didecyl phosphite, decyl diphenyl phosphite, triphenyl phosphite, and tritolyl phosphites.

Suitable trihydrocarbyl phosphites may further include Spiro diphosphites such as, for example, 3,9-bis[2,4-bis(1,1-dimethylethyl)phenoxy]-2,4,8,10-tetraoxa-3,9-diphosphaspiro[5.5]undecane.

Some of these compounds are shown in Figure 5.1.

5.1.2 Montmorillonite Clays

Montmorillonite (MMT) clays consist of nanometer scale platelets of magnesium aluminum silicate. Their dimensions, 1 *nm* thick and

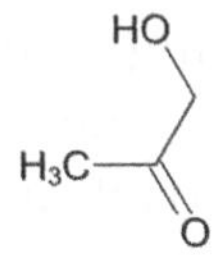

1-Hydroxy-2-propanone

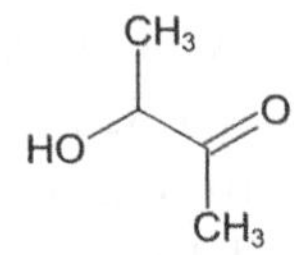

Acetoin

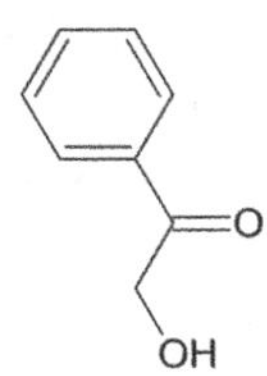

2-Hydroxyacetophenone

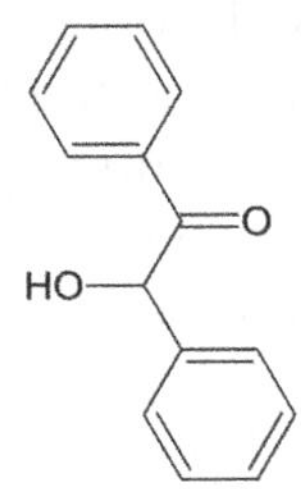

Benzoin

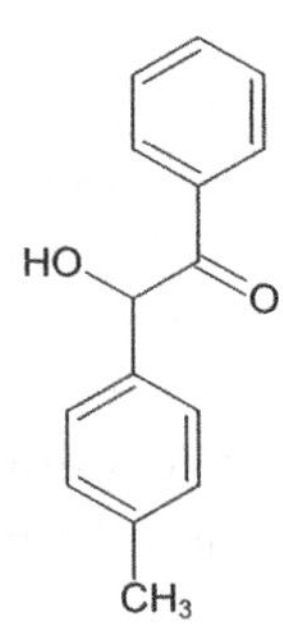

2-Hydroxy-1-phenyl-2-*p*-tolyl-ethanone

Figure 5.1 Optical enhancing agents (8).

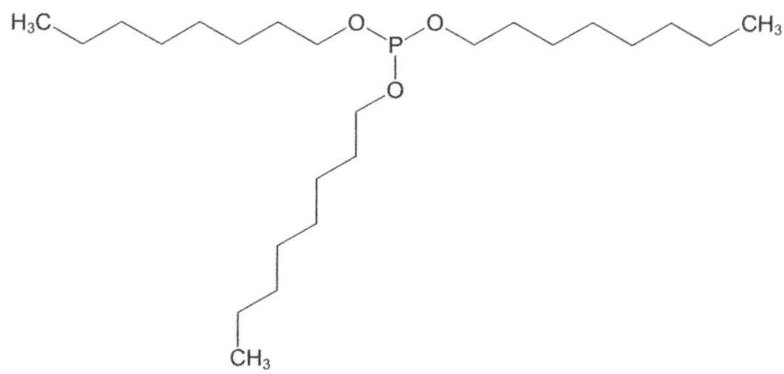

Trioctyl phosphite

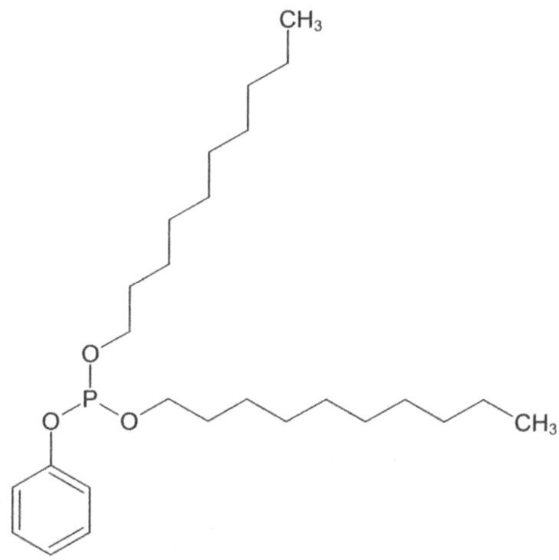

Phenyl didecyl phosphite

Figure 5.1 (cont) Optical enhancing agents (8).

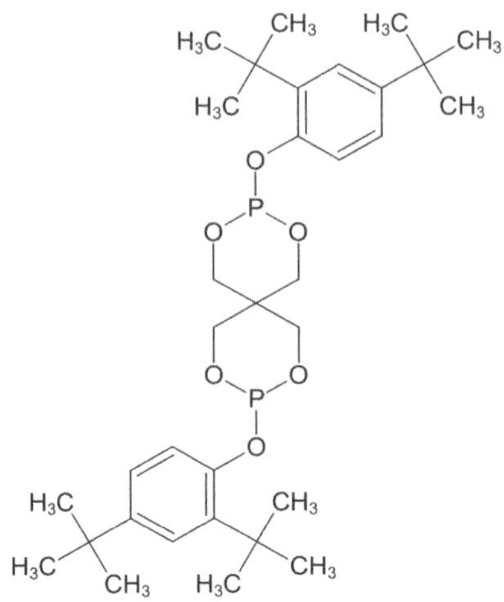

3,9-Bis[2,4-bis(1,1-dimethylethyl)phenoxy]-
2,4,8,10-tetraoxa-3,9-diphosphaspiro[5.5]undecane

Figure 5.1 (cont) Optical enhancing agents (8).

100 to 500 *nm* in diameter, result in platelets with a high aspect ratio (10).

The clay structure is formed by hundreds of layered platelets stacked into particles or tactoids 8 to 10 μm in diameter. The effect of nanoclays on polymer properties stems mainly from their high surface to volume ratio, since interactions between polymer and filler are governed by interfacial forces. Clay particles should be exfoliated as individual platelets and uniformly dispersed within the polymer matrix in order to take full advantage of the potential high surface area (3).

Exfoliated nanoclays are effective at improving gas barrier properties of polymeric materials. When dispersed into polymers, they create a maze structure that presents a tortuous path to moving gases, greatly slowing their permeation rate (11).

Traditional composite structures contain large quantities of filler (approximately 60% vol), but in nanocomposites dramatic changes in properties are possible at very low loads (<2% vol).

5.1.3 *Graphene Nanoplates*

Carbon-based graphene nanoplates (GNPs) can form heat-resistant, high barrier nanocomposites that are promising in food packaging applications. GNPs, 20 to 60 *nm* in thickness and 0.5 to 25 μm in diameter, dispersed in poly(methyl methacrylate) (PMMA) were recently reported to increase the glass transition temperature for PMMA by 30°C at 1 to 5% loading (12). GNPs were dispersed by high-speed shearing methods.

5.1.4 *Carbon Nanotubes*

Carbon nanotubes (CNTs), another carbon-based nanofiller, have attracted considerable attention due to their intrinsic mechanical and electrical properties. Improvements in modulus and strength of 30% and 15%, respectively, have been reported for 1% loading of functionalized single-walled carbon nanotubes in epoxy. However, the use of CNTs in nanocomposites to date has been limited by challenges in processing and dispersion, and their prohibitively high cost (13).

5.1.5 *Micro- and Nanoplastics*

Microplastics are ubiquitous and their presence and accumulation in the environment threatens the ecological equilibrium, aquatic environment, human health, food sustainability and security (14–16).

The environmental and metabolic threats to marine organisms are attributable to the stable chemical and physical properties of the parent plastics (17).

The plastics can absorb and adsorb toxic inorganic and organic compounds which can leach into the tissues of the marine organisms. Polychlorinated biphenyl compounds and polycyclic aromatic hydrocarbon compounds are the most plastic/microplastic adsorbent (18–21).

The release rate of these chemical compounds is favored by the pH of the medium, temperature, heat, UV light, mechanical and chemical abrasion (22).

The manufacturing additives, such as bisphenol A, phthalates, nonylphenol, urethane foam and brominated flame retardant, used in production of plastics can become toxic residues (20, 23).

The ubiquitous nature and regular occurrence of microplastics and nanoplastics in the food chain and the environment are important issues of research (24).

5.2 Biopolymers

The bioplastics industry is a young, innovative sector with great economic and ecological potential for a low-carbon, circular bioeconomy that utilizes resources more efficiently (25, 26). The market for biopolymers is expected to show some growth and new demands are envisioned for applications which have clear benefits to the consumer and the environment (27). The global production of biopolymers has grown from just under 300,000 tonnes in 2009 to 2.11 million tonnes in 2019, and is predicted to grow to 2.42 million tonnes by the year 2024 (28).

Bioplastics produced in 2015 globally had a share of biobased non-degradable and biodegradable plastics of about 64% and 36%, respectively (29, 30).

The envisioned growth in the share of biodegradable polymers to meet the environmental demand for reduced pollution was analyzed in 2019 when the share of biodegradable plastics grew to about 56% (31).

The production of biodegradable and non-biodegradable bioplastics production capacities is shown in Table 5.1.

A chapter in a monograph describes the development of biodegradable packaging materials as part of the solution for controlling waste disposal (32). The biopolymers used in developing biodegradable packaging materials include poly(*d,l*-lactic acid), thermoplastic starch, poly(hydroxyalkanoate)s, synthetic polyesters from glycols and dicarboxylic acids, poly(caprolactone) (PCL), and poly(vinyl alcohol). The technology of applying the nanocomposite concept for developing novel biodegradable materials, legislative issues concerned with biodegradability, and current applications of biodegradable materials in food packaging are also discussed.

Table 5.1 Production of biodegradable and non-biodegradable bioplastics (33, 34).

Year	Biodegradable amount/[Mt]	Non-biodegradable amount/[Mt]
2015	957	2995
2016	964	3192
2017	880	1174
2018	1077	934
2019	1174	941
2020	1239	848
2021	1553	864
	Forecast	
2022	3694	1025
2023	4116	1101
2024	4360	1150
2025	4583	2140
2026	5297	2297

5.2.1 *Proteins*

The recent developments in biodegradable packaging, such as edible films/coatings and innovative packaging materials based on plant protein sources, have been reviewed (35). The impact of different additives on the functionality of plant protein-based biodegradable materials was also investigated. In addition, plant protein-based packaging characteristics and their application in food were also introduced.

In recent years, plant protein-based films as a replacement for synthetic plastics have attracted the attention of the global food packaging industry due to their biodegradability and unique properties (36).

Biopolymer-based films need a filler to show improved packaging properties. One of the latest strategies introduced to food packaging technology is the production of nanocomposite films which are multiphase materials containing a filler with at least one dimension less than 100 *nm*.

The functional performance, barrier, mechanical, optical, thermal and antimicrobial properties of plant protein-based materials can be extended by incorporating nanomaterials (36).

5.2.1.1 *Protein Nanocomposites*

The film-forming ability of various proteins has been utilized in industrial applications for a long time (37)

Animal-derived proteins used in commercial applications are mainly casein, whey protein, collagen, egg white, and fish myofibrillar protein (38).

Plant-based proteins (39) under consideration include soybean protein (40), zein (corn protein), and wheat gluten (41).

Compared with nonionic poly(saccharide) films, protein films have better oxygen barrier properties and lower water vapor permeability due to their more polar nature and more linear (nonring) structure, and lower free volume (42).

However, serious concerns remain regarding their performance in food packaging, including their high modulus, high water adsorption, and high gas permeability. Significant efforts have been made to improve the properties of various proteins applying nanocomposites technology, mainly using nanoclays.

5.2.1.2 *Whey Protein*

To apply a whey protein isolate as a food packaging material industrially, a three-step study was performed (43). First, the whey protein isolate coating solution was screened for plasticizer type, plasticizer concentration, and pH to minimize its oxygen permeability. The production batch size was scaled-up from 1 to 320 *L*. Then, the poly(ethylene terephthalate) (PET) film was sequentially coated with the whey protein isolate and laminated with a linear low-density poly(ethylene) (LLDPE) film using an industrial manufacturing facility. The developed multilayer (PET/whey protein isolate/linear low-density poly(ethylene) (LLDPE)) film was compared to the aluminum oxide (AlOx)-deposited multilayer (PET/AlOx/LLDPE) film. The PET/whey protein isolate/LLDPE films show ed high oxygen and water vapor barrier properties (43).

Finally, the PET/whey protein isolate/LLDPE films were applied as frozen shrimp fried rice (SFR) packaging. For 6 months, they delayed peroxide formation in the SFR samples, as much as the PET/AlOx/LLDPE films did. These results indicated that the developed film has high potential as a packaging material for frozen foods in practical applications (43).

Whey protein has received significant attention as edible film and coating material. The formation of whey protein transparent films was reported (44), which also acted as oxygen barrier. TiO_2 was added to form a nanocomposite with improved antimicrobial properties.

The potential of whey TiO_2 nanocomposites was tested for use as food-grade, biodegradable packaging materials (45). Addition of small amounts of less than 1% of TiO_2 nanoparticles significantly increased the tensile properties of a WPI film (1.69 to 2.38 *MPa*). Similar studies regarding interactions on ZnO–whey protein nanocomposite have been reported (46).

For applications of food packaging materials, high oxygen and water vapor barriers are the prerequisite conditions for preserving the quality of the products throughout their whole lifecycle (47).

Currently available polymers and/or biopolymer films are mostly used in combination with barrier materials derived from oil-based plastics or aluminum to enhance their low barrier properties. In order to replace these nonrenewable materials, several research efforts have been focused on the development of sustainable coatings, while maintaining the functional properties of the resulting packaging materials.

An introduction to food packaging requirements has been presented that highlights the prior art on the use of whey-based coatings for their barrier properties, and describes the key properties of an innovative packaging multilayer material that includes a whey-based layer (47).

Whey protein coatings were already tested as edible films on, among others, peanuts, salmon, fruits, or cereals, whereby whey coating offered good aroma, fat, humidity, and oxygen barriers. It could be shown that they helped to improve the shelf life of, for example, peanuts, by retarding the lipid oxidation causing rancidity (48). In addition, those edible films were reported not to modify the sensory attributes of the coated goods or their aspects, while providing some health benefits for the consumer (49).

The here developed whey protein formulations had excellent barrier properties almost comparable to the ethylene vinyl alcohol copolymers barrier layer conventionally used in food packaging composites, with a small oxygen barrier when normalized to a thickness of 100 μm (47).

Further requirements of the barrier layer are good adhesion to the substrate and sufficient flexibility to withstand mechanical load while preventing delamination and/or brittle fracture. Whey-protein-based coatings have successfully met these functional and mechanical requirements (47).

5.2.1.3 Soy Protein

Soy protein has been of great interest to researchers for its thermoplastic properties and its potential as a biodegradable plastic. However, because of its poor response against moisture and high rigidity, its biodegradability has not been exploited effectively (50).

Similarly to starch, soy protein is also blended with plasticizers to overcome brittleness. However, the use of plasticizers further decreases barrier properties. Soy protein nanocomposite films showed reduced water vapor permeability and improved elastic modulous and tensile strength compared to counterparts without fillers (51–53).

An increase in the tensile strength by 23% and 47%, respectively, was found without and with ultrasonic treatment in soy protein nanocomposite. The mechanism of interaction between soy protein and MMT clay was investigated by correlating structure and properties (52). The surface electrostatic interaction between the soy protein (+ charged) and the MMT layers (- charge) as well as the hydrogen bonding between the –NH and Si–O groups was understood to be the interacting mechanism for protein/MMT system. Such mechanism resulted in improved mechanical strength of nanocomposite. The Young's modulus (E) increased from 180.2 to 587.6 *MPa* with an increase of the MMT content from 0 to 20 wt%. The tensile strength of the sheets improves from 8.77 to 15.43 *MPa* when the MMT content increased from 0% to 16%.

5.2.1.4 *Proteins for Fish Product Packaging*

Protein-based polymers are helpful because they possess valuable characteristics for the production of food packaging materials (54). Proteins have been intensively studied as bio-based packaging materials due to their film-forming ability, UV-barrier properties, mechanical properties, transparency, and barrier properties against oxygen and carbon dioxide diffusion (55). Gelatin, collagen, fish proteins, corn zein, wheat gluten, whey, and soy protein are widely accepted for film formation due to their nature to form intermolecular bonds (56). Different proteins and other biopolymers-based fish packaging materials incorporated with different bioactive molecules are presented in Table 5.2 (54). Some of these compounds are shown in Figure 5.2.

5.2.2 *Gelatin*

Gelatin is obtained by the denaturation of the triple helix of collagen and can be obtained from the skin, tendon and connective tissues (57–59).

Gelatins are widely used due to their availability, low cost and excellent film-forming capacity (60).

Gelatin-based films are also good oxygen barriers and offer good mechanical resistance and high elasticity, but have low water barrier properties and are prone to environmental conditions such as relative humidity (61).

Many researchers have produced biodegradable film packaging from different gelatins, such as pig skin (62), bovine skin (63), and also cattle bones (64).

It was shown that the properties of gelatin films depend on the gelatin type and the concentrations used. The physical properties of gelatin films can be determined by the characterization of raw materials and an extraction method (65).

5.2.2.1 *Antibacterial Gelatin-Bacterial Cellulose Nanocomposite*

The fabrication of an antibacterial gelatin-bacterial cellulose nanocomposite film was achieved using cinnamaldehyde, c.f. Figure 5.3, as a crosslinking agent and an antibacterial additive in order to improve the structural stability of gelatin in water and

Table 5.2 Bioactive molecules.

Proteins	Biomolecules	Uses	Ref.
Gelatin/ κ-Carrageenan	Anthocyanin and TiO_2 nanoparticles	Monitors fish freshness during storage	(66)
Fish gelatin	Anthocyanins and phenolic acids	Monitors the freshness of shrimp	(67)
Gelatin/ Oxidized chitin nanocrystals	Black rice bran anthocyanins	Antioxidant activity and have great potential in monitoring freshness of shrimp and hairtail (smart packaging), high protein foods	(68)
Gelatin/Chitosan	Squid maillard peptides	Retards lipid oxidation	(69)
Gelatin/Chitosan	0.4% TiO_2-Ag	Improves antibacterial and UV barrier capability	(70)
Gelatin/Chitosan	Polylysine	Controls foodborne pathogens	(71)
Gelatin	Curcumin, betanin and anthocyanin	Color retention	(72)
Gelatin	Dialdehyde κ-Carrageenan (DAK-car) Thymol-loaded zein nanoparticles (ZNP)	Films containing a 1:2 ratio of gelatin films (GEL)/DAK-car and thymol-loaded zein 0.25 and 0.5 mg/mL of thymol have antioxidant and antimicrobial activity in active food packaging	(73)

Table 5.2 (cont) Bioactive molecules.

Proteins	Biomolecules	Uses	Ref.
Gelatin	Eugenol nanofiber and PLA	Antimicrobial and antioxidant films	(74)
Soy Protein Isolate/Gelatin, Essential oils of *Z. multiflora* and *Cinnamon zeylanicum*	The population of *S. aureus, B. cereus, S. typhimurium, L. monocytogens*, and *E. coli* was diminished via the inclusion of 20% *Zataria multiflora* active food packaging, extended shelf life		
Whey protein isolate (WPI)	Emulsions and nanoemulsions of *Citrus sinensis* essential oil peel extract (CSEO)	The film preserves food against oxidation and microbial spoilage, improving shelf life and quality of foodstuffs	(75)
Gelatin	Oxidized guar gum green tea extracts (0.5 and 1.0% w/v).	Antimicrobial against *S. aureus*, higher antioxidant activity due to the presence of green tea extract	(76)

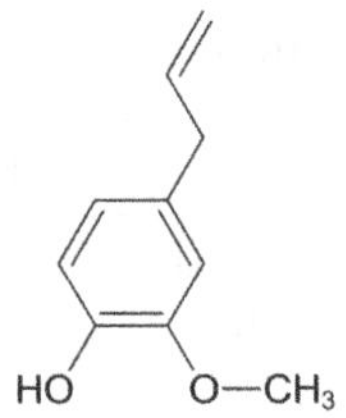

Eugenol

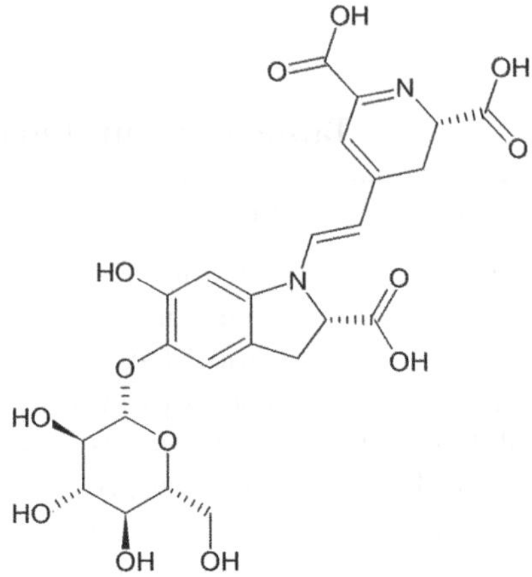

Betanin

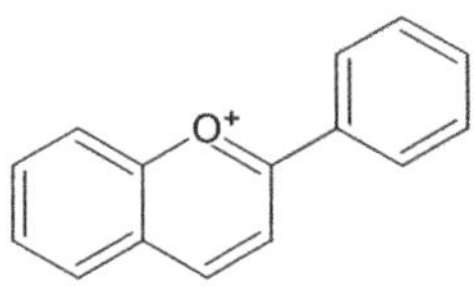

Anthocyanin

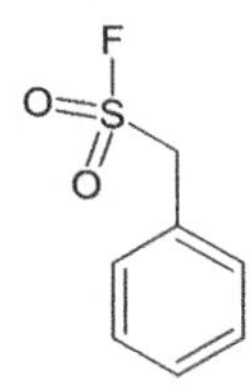

Zein

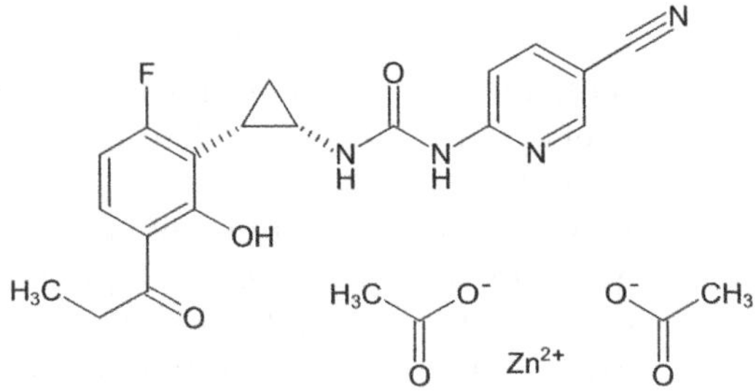

Carrageenan

Figure 5.2 Bioactive molecules.

to promote the antibacterial activity of both gelatin and bacterial cellulose (77).

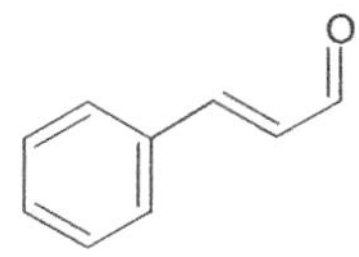

Figure 5.3 Cinnamaldehyde.

The reaction of amine groups of gelatin with the aldehyde group of cinnamaldehyde through Schiff's base and Michael's addition reactions could be accelerated by the heat treatment at 120°C for 3 *h*, resulting in the reduction of the water solubility of the gelatin-bacterial cellulose nanocomposite films.

The supplementation of bacterial cellulose into gelatin enhanced the tensile strength of the composite films and also decreased the water vapor permeability of the composite films. The gelatin-bacterial cellulose nanocomposite films showed nontoxicity against L929 cells and strong antibacterial activity against both *Staphylococcus aureus* and *Escherichia coli* bacteria, suggesting the potential application as an antibacterial food packaging (77).

5.2.3 *Storage Polymers in Bacterial Cells*

Most of the excess sludge from a wastewater treatment plant, such as 60%, is disposed of in landfills. Excess sludge has been proposed as a resource for use in the production of biodegradable plastics.

Storage polymers in bacterial cells can be extracted and used as biodegradable plastics (78). However, widespread applications have been limited by high production cost. In a study, activated sludge bacteria in a conventional wastewater treatment system were induced by controlling the carbon to nitrogen ratio to accumulate storage polymers.

The yield of polymers increased to a maximum 33% of biomass when the C/N ratio was increased from 24 to 144, whereas specific growth yield decreased with increasing C/N ratio (78).

5.2.4 *Bio-Based Nanocomposites*

Biopolymers have attracted much attention as potential replacements for conventional plastic packaging materials due to an increased interest in sustainable development. Biopolymers include plant-derived materials, e.g., starch, cellulose, and other poly(saccharide)s, animal products (proteins, poly(saccharide)s), microbial products (polyhydroxybutyrate), and polymers that are synthesized chemically from naturally derived monomers, e.g., poly(lactic acid).

Most reports on the formation and properties of biopolymer films are focused on their application as edible films (79–82).

For packaging applications, biopolymers present relatively poor mechanical and barrier properties, which limit their industrial use. Especially challenging are the development of moisture barrier properties due to the hydrophilic nature of biopolymers. However, it has been suggested that inherent shortcomings of biopolymer-based packaging materials may be overcome by nanocomposite technology (13).

Efforts for developing protein films as biodegradable packaging materials have increased in recent years. One of the drawbacks of such films is that their functional properties are generally inferior to those of synthetic films (82). Although protein films typically have excellent gas barrier properties, they tend to have higher water vapor permeability, are mechanically weaker, and have lower elongation as compared to most synthetic polymers.

The various methods that have been reported to enhance the physical and barrier properties of protein films have been reviewed. The effects of film formulation and preparation conditions, including plasticization, pH alteration, and lipid addition, on the functional properties of protein films are discussed. The article also offers a comprehensive review of the various methods used to modify protein film functional properties, including crosslinking techniques induced by heat, chemicals, enzymes, and irradiation, and use of nanocomposites to enhance their barrier properties. A critical comparison of the different technical solutions is offered, and the challenges and potential applications of these films are discussed (82).

5.2.5 *Starch*

A review of starch-based materials has been presented, with emphasis on methods to improve their properties and widen their application in food packaging (83). The preparation methods, common additives, characterization techniques, and applications of starch-based biodegradable materials are summarized. In addition, advanced technological approaches to prepare innovative starch-based biodegradable materials are also shown.

Starch-based plastics can be applied for agricultural, medical and pharmaceutical purposes, and food packaging (84).

The techniques for starch film production can be classified into wet and dry processes. The wet process is mainly applied at lab scale mainly because it is considerably time-consuming. The dry process can be useful in an industrial setting.

The major processes for starch-based film production are casting (wet process), molding and extrusion (dry processes) (84).

Different types of starch, e.g., thermoplastic starch, cornstarch and commercial starch are usually utilized in the food packaging applications (85).

It was reported that the merging of PCL as an external agent with thermoplastic starch increased the stabilization and mechanical characteristics of the thermoplastic starch matrix (86). Among the different added concentrations of PCL, 15% addition of PCL in the thermoplastic starch produced the most pronounced effect in incrementing the stabilization properties of the blends.

Cassava starch films were modified by using natural plant extracts, including basil and green tea extract, as smart packaging biodegradable plastic material (87). The presence of high polyphenolic components in the extract provided additional coating edges to the films and helped in the preservation of the food by retarding the oxidative reactions. Moreover, the presence of pH-dependent components, i.e., chlorophyll and carotenoids, in the extracts also allowed these films to be pH sensitive. Biodegradability tests showed that the engineered material was eco-friendly and could be completely degraded in less than two weeks. Furthermore, the thermal analysis of the blend films demonstrated that the packaging material was able to withstand temperatures up to 240°C without losing any textural property (85).

5.2.6 *Dry Heating Treatment*

The modification of starch was studied by dry heating treatment and the production of cassava starch films or sheets, by casting or extrusion processes, respectively (88).

The molecular changes promoted by the dry heating treatment process resulted in a new matrix. Furthermore, the slower drying in casting, and the faster drying in extrusion, promoted different molecular arrangements, which affected the material performance.

The main results showed that a dry heating treatment for 4 h in cassava starch produced bio-based plastics with increased mechanical properties and opacity while decreasing water permeability, moisture content and light transmittance. So, a dry heating treatment seems to be an interesting alternative to produce biodegradable plastics (88).

5.2.7 *Thermoplastic Starch*

The environmental and health risks associated with the emission of greenhouse gases and plastic pollution, combined with the increasing price and future shortage of petroleum, have been the main driving forces for the development of greener materials from renewable resources (89). Although such increased emphasis on sustainability has transversally crossed all the technological sectors and end markets of mass produced goods, it is especially the packaging industry, traditionally dominated more than others by disposable fossil-based plastics, that has been facing growing pressure to provide and implement commercially attractive and competitive alternatives, with an optimal balance in terms of functionality, cost, eco-efficiency, product availability, and production capacity.

In this regard, public organizations, industrial manufacturers, and enterprises from different industries, ranging from food packaging to agriculture to medical care, are joining efforts aimed at the large-scale production and commercialization of biodegradable packaging solutions that can meet the legislative requirements imposed by the restrictions in the use of disposable plastics and the demand for more responsible waste management arising from an increased environmental awareness. So, thermoplastic starch and its composites exhibit great commercial and technological potential as sustainable packaging materials. An overview of their processing methods, properties and suitability for packaging applications is presented in a chapter of a book (89).

5.2.8 *Modified Starch*

Biodegradable films based on starch modified with ascorbic acid or caffeine and dialdehyde starch were evaluated for their applications in the food packaging industry (90). These compounds are shown in Figure 5.4.

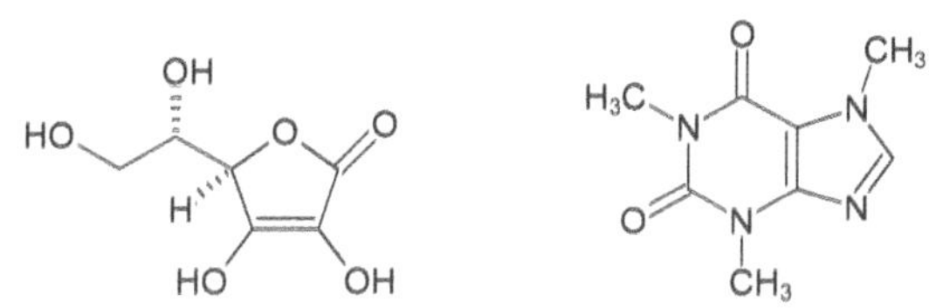

Ascorbic acid Caffeine

Figure 5.4 Compounds for modified starch.

The morphology of the film surface was observed with scanning electron microscopy (SEM) and atomic force microscopy (AFM). The chemical structure was examined with Fourier transform infrared (FTIR) spectroscopy and X-ray diffraction. The mechanical, thermal and barrier properties of the films were examined. The crosslinking and hydrophobic effect of dialdehyde starch on starch-based films was proven.

Vitamin C addition prompts the crystalline structure formation and improves the hydrophilicity and antioxidant activity of the starch blends. Caffeine addition results in higher starch film hydrophobicity, whereas caffeine combined with dialdehyde starch reduces crystalline order in starch. Both caffeine and ascorbic acid incorporation influence the film stiffness, and dialdehyde starch causes an increase in brittleness. The high susceptibility to biodegradation and good antioxidant activity confirms the potential application of examined systems in the packaging industry (90).

5.2.9 *Starch-Based Foams*

Starch has been used as a raw material to develop foams for packaging (91). The different strategies for improving the mechanical properties, processibility, and antimicrobial activity of starch-based foams have been reviewed. Also reviewed are methods for decreasing their sensitivity to humidity, aimed at employing them as a competitive replacement for expanded poly(styrene) in packaging (91).

The synthesis methods of starch-based polymer foams have been detailed in a chapter of a monograph (92).

Starch-based foams are one of the possible alternative materials for making biodegradable foams. The starch is abundant and economical, which makes it a substance of choice to produce foams. These are produced mainly by the extrusion method.

Cellulose nanofibrils are also used along with starch to produce biodegradable foams. This involved freeze-drying with cellulose nanofibrils/starch water suspensions within a range of 1% to 7.5% solids content. The mechanical and thermal properties of the composite foams were found to be higher than that of the foams produced by a common starch.

A co-rotating twin-screw extruder machine was used to develop cushion foam sheets by blending wheat and starch (93). The extruded materials were suitable for cushioning since they had comparable physical properties with the commercial plastic foams. Their dynamic cushion curve test indicated that the starch-based foam sheets provided good shock absorption properties.

Biodegradable foam trays have been developed by blending cassava starch with fibers and chitosan. This was achieved by hot-mold baking in an oven with a controlled temperature at 250°C for 5 *min* (94). The water solubility and water absorption indexes of these foams were better than those of PS foams.

Starch-based foams combined with natural resources, such as grape waste, thistle waste, barley and straw fibers, have been developed using a microwave foaming process (95). The product formed could be used for proving structural support, due to its high stiffness and strength. It can also be used for packing, since it is biodegradable.

5.2.10 *Residual Starch from Potato*

During the industrial processing of potato materials, a large amount of residual starch is produced from the washing of the raw material (96). Starch-like biopolymers have been proposed for the formulation of biodegradable materials.

The application of starch in the packaging production depends on chemical, physical and functional properties to form gels and films.

Packaging films have been developed from potato processing of residual starch combined with some polar organic solvents, i.e., glycerol and acetic acid that can be fully obtained from renewable sources.

The characterization of the so obtained material was performed by evaluating the microbiological stability, mechanical properties, formed bonds and barrier characteristics.

In order to demonstrate the real application in fruit preservation, the material was directly applied onto papaya fruit (Carica papaya) by the immersion method.

The films achieved the expected results for microbiological analysis, water vapor permeability and Fourier transform infrared spectroscopy. The qualitative analysis of the shelf stability test showed the potential of the films in fruit preservation envisioned for application in the food industry for packaging (96).

5.2.10.1 *Starch Nanocomposites*

Starch has been extensively investigated as a material for food packaging applications due to its environmental compatibility, wide availability, and low cost (97, 98).

The addition of inorganic materials and synthetic polymers (3, 99, 100) has been proposed to improve water resistance of starch. Also, blends of starch with nanoclays have been investigated.

5.2.11 *Cellulose*

5.2.11.1 *Cellulose Nanocomposites*

Nanocomposites from fruit and vegetable purees and cellulose nanofibers (CNFs) have been recently studied as film-forming edible materials. Cellulose nanofibers were added to improve tensile properties, water vapor permeability, and glass transition temperature of mango puree films (101).

The tensile strength increased (4.09 to 8.76 *MPa*) with an increase in the concentration of the CNF from 0% to 36%. It was proposed that due to the formation of fibril network within the matrix, there was an effective increment in tensile strength and Young's modulus, especially at higher concentrations of CNFs. The addition of CNFs was also effective to improve the water vapor barrier of the films (2.66 to 1.67 $g\,mm\,kPa^{-1}h^{-1}m^{-2}$). The water vapor permeability was significantly decreased when a CNF was incorporated at loadings of at least 10% (10 *g* per 100 *g*).

Cellulose films as such are poor water vapor barriers because of the inherent hydrophilic nature of poly(saccharide)s.

A cellulose derivative was found (102), i.e., hydroxypropyl methyl cellulose (HPMC), to be a promising material for edible coatings or films for packaging.

Furthermore, nanocomposites were proposed using chitosan (CS) as nanofiller in HPMC to improve mechanical and film barrier properties (103). Here, different concentrations of CS nanoparticles were incorporated into HPMC to evaluate changes in mechanical properties, water vapor permeability, and oxygen permeability. Incorporation of chitosan nanoparticles in the films improved tensile strength (30.7 to 66.9 *MPa*) and film barrier properties. SEM analysis revealed that chitosan nanoparticles tended to fill porous spaces in the HPMC matrix, improving film tensile properties and water vapor permeability. The oxygen permeability was reduced significantly by incorporating CS-poly(methacrylic acid) in HPMC matrix. A further reduction in permeability was observed with smaller nanoparticle sizes.

Thus, HPMC-CS nanocomposite could be a potential material for food packaging applications to extend the shelf life of foods.

5.2.12 *Biodegradable Polymer Reinforced with Cellulose Nanocrystals*

Paper is an inherently hydrophilic material due to the hydroxyl groups contained in cellulose, which limits its applications in packaging materials (104). In a study, a simple one-step method for preparing hydrophobic paper by introduction of chitin nanocrystals and hexadecyltrimethoxysilane was proposed. Hexadecyltrimethoxysilane is shown in Figure 5.5.

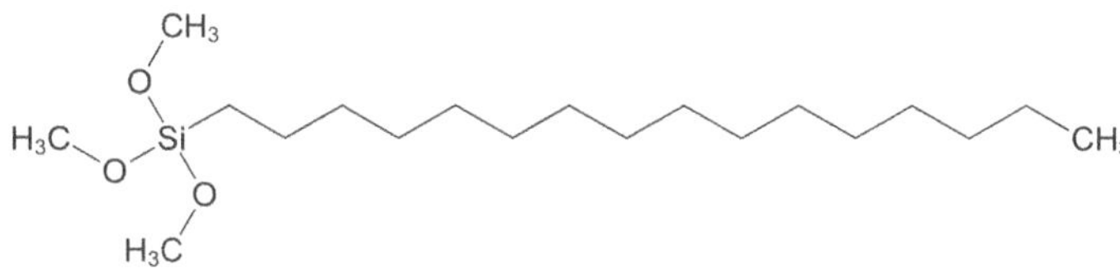

Figure 5.5 Hexadecyltrimethoxysilane.

The surface chemistry and the morphology of coated paper were studied in detail. A 2% chitin nanocrystals/hexadecyltrimethoxysilane coated paper was highly hydrophobic with a water contact angle greater than 130°, mainly due to the low surface energy provided by hexadecyltrimethoxysilane. Compared with uncoated paper, the tensile strength of the hydrophobic paper and the water vapor barrier were significantly improved due to the addition of chitin nanocrystals.

The hydrophobic paper showed an excellent environmental tolerance and mechanical stability. The survival rate of L929 cells on the modified paper was more than 80% by CCK-8 assay, which proved that the nanocoating was safe.

In addition, the obtained hydrophobic coated paper can hold water, cola, green tea, milk, coffee, and orange juice for a long time and can be used as a straw to absorb liquid. This work prepared high-performance and green hydrophobic paper, which exhibits wide application in food packaging (104).

The massive consumption of conventional food packaging has increased disposal of non-eco-friendly packaging waste, which is severely damaging the environment.

Therefore, the replacement of convention packaging by sustainable packaging is an important alternative to reduce the enormous volume of plastic waste (105). Bionanocomposite films composed of a blend of poly(butylene adipate-*co*-terephthalate, poly(lactic acid) and cellulose nanocrystals extracted from agro-waste were investigated. The characterization of the cellulose nanocrystals confirmed that the bionanocomposite films presented better hydrophobic characteristics and thermal stability than the blend film.

In addition, the tensile strength, elongation at break, and Young's modulus were around 52%, 29%, and 118%, respectively, i.e., higher than the blend films. These mechanical values were comparable to values of commercial plastic materials that are extensively used in food packaging. Thus, these bionanocomposite films might be an interesting alternative to produce sustainable food packaging materials (105).

5.2.12.1 Cellulose-Based Hydrogel

Cellulose has smart features for hydrogel production, such as cost-effectiveness, water absorption, swelling ability, biodegradability, flexibility, and stimuli responsiveness. Hydrogels are tailored foods with three-dimensional crosslinked networks in the modern world (106). Cellulose-based hydrogels have been used in the food industry as an important means for various aspects due to their functional properties.

Hydrogels are produced from biopolymers such as poly(saccharide)s, proteins, or lipids. In the last few years, cellulose and its derivatives were investigated for the synthesis of hydrogels, which is a basic, simple, attractive, and environmentally friendly alternative innovation for the food industry.

Due to their structural interchangeable/stimuli-responsive property, cellulose-based hydrogels have been widely studied in the areas of functional food, food packaging, drug delivery, and food safety.

The recent progress and application of cellulose-based hydrogels in the food industry has been reviewed (106). Also, information about the development of edible functional cellulose-based hydrogels, their prospects, and opportunities for future research has been detailed.

Hydrogels are produced by physical, chemical, and radiation crosslinking, which is the key concept and mechanism that improves their properties (107).

Hydrogen bonding, covalent bonds, van der Waals interactions, and other chemical interactions form the crosslinked network in hydrogels (108).

5.2.12.2 *Cellulose Sulfate Hydrogel*

Cellulose sulfate is an ester of cellulose that can be synthesized by homogeneous, quasi-homogeneous, and heterogeneous sulfation procedures (106). Cellulose sulfate contains high amounts of water solubility, a high degree of substitution, and antimicrobial characteristics. It is easily soluble in water when the sulfonic acid groups are increased from 528 $\mu molg^{-1}$ to 689 $\mu molg^{-1}$, resulting in a transparent solution (109).

The transitional reaction of cellulose sulfate produces cellulose nitrate. The sulfation was achieved by the dissolution of cellulose in *N,N*-dimethylformamide combined with a mixture of sulfating agent (chlorosulfuric acid) and acetylating agent (acetic anhydride) followed by acetyl group cleavage, and upon precipitation converting cellulose acetate into cellulose sulfate. This process is called acetosulfation and is a quasi-homogeneous process (106).

5.2.12.3 *Pyranoflavylium and Cellulose Acetate*

Pyranoflavylium cations are synthetic analogs containing the same basic chromophore as the pyranoanthocyanins that form in red wine during maturation and are responsible for its final color (110).

A pyranoflavylium salt was incorporated into cellulose acetate-based films to build up a colorimetric pH indicator as a smart label for monitoring the freshness of food. The pigment in an amount of 0.1% was immobilized in a cellulose acetate matrix in the absence of glycerol and the presence of different amounts of 10% to 40% glycerol (111).

The films were obtained by a casting method and characterized by their thickness, morphology and barrier properties, thermogravimetric analysis, and color.

The chromatic responsive properties of the films were tested in solutions at different pH values (pH 4 to 8) and in the headspace of biogenic amines and ammonia-containing solutions at different concentrations. The results showed that the films without glycerol are not color-responsive to different pH values.

On the other hand, the incorporation of glycerol above 20% demonstrated an effective pH responsiveness after 20 *min* of immersion and after exposition to an amine-rich environment. The glycerol-containing films showed relevant and remarkable color change at the pH range of food spoilage, indicating great potential for application as a food freshness indicator (111).

5.2.12.4 Cellulose-Based Multilayer Coated Paperboards

Some petroleum-based packaging materials have a series of safety issues. So, the use of biodegradable and nontoxic packaging materials for foods is preferred. However, biodegradable alternatives often need to be modified to overcome their own shortcomings (112).

Carboxymethyl chitosan (CMCS)/carboxymethyl cellulose (CMC) and poly(lactic acid) (PLA) nano zinc oxide ZnO nanoparticles multilayer coated paperboard was prepared to explore its potential use in food package applications. The multilayer coating reduced the oxygen transmission rate by 99% and doubled the barrier properties for soybean oil penetration and water vapor compared with a single PLA coating layer in an environment of 23°C and 50% relative humidity.

The oil resistance time of the multilayer coating increased 28 times, and the water vapor barrier properties increased twofold compared to a single CMCS/CMC coating layer.

The water vapor barrier properties increased by a factor of 3 even at 38°C and 90% relative humidity. Compared with the CMCS/CMC with PLA coating without nanoparticles, the barrier performance of the ZnO nanoparticle coating was further improved. The oxygen, water vapor, and heptane vapor barrier properties all doubled, and the oil resistance time reached 235 *h* (112).

5.2.12.5 Cellulose Fiber-Based Food Packaging Papers

The application of paper for food packaging is limited by a relatively strong affinity for water and oil (113). Two types of coated paper were prepared and the influence of the air permeability and polarity on the oil resistance of the coated papers was investigated.

Reducing the air permeability improved the grease resistance of the coated papers. High-surface-energy coatings exhibited an enhanced oil resistance because of a high content of polar components that offered high resistance to grease. The mechanical properties of the paper were also improved by applying the coating. These naturally derived materials are an alternative to the fluoride-containing materials that are currently available on the market for improving the wettability of paper and the quality and safety of grease-rich foods (113).

Cellulose fiber-based food packaging papers with enhanced barrier and antibacterial properties could be successfully fabricated by a sequential deposition of multilayers consisting of chitosan and carboxymethyl cellulose on the surface of a paper (114).

The multilayers effectively improved not only the barrier properties of papers against grease, oil, water, air and water vapor, but also the mechanical properties of the papers. Moreover, the multilayer-modified papers showed improved antibacterial activities against *E. coli* and *S. aureus*. A multilayer-modified paper exhibited the strongest antibacterial activity with growth inhibition rate of 95.8% against *E. coli* and 98.9% against *S. aureus*.

No obvious cytotoxicity was detected for the so obtained modified paper. Therefore, in consideration of its mechanical, barrier, antibacterial and safe performances, it seems that this cellulose fiber-based paper can be used as a promising material for food packaging (114).

5.2.12.6 *Papers with Cellulose Nanofibers*

Waterborne barrier coatings have the potential to generate good barrier layers, but their performance is often less than expected. An improved performance of these coatings has been shown when they are applied on paper that has a cellulose nanofiber layer (115).

Papers with a cellulose nanofiber layer were coated with barrier coatings at different coat weights applied as single layers, as double layers, and single layers pressed together in a hot press in order to generate a packaging system that has good barrier properties.

The performance of double-layer samples resulted in moisture transmission rates that were 40% to 70% of the value of the single-layer systems.

Hot-pressing of two dry layers showed no advantage compared to the single-layer system. A barrier pigment added to one formulation improved the performance further and followed the same trends. Three-dimensional models of diffusion through layers that have defects help explain the results. This study shows a potential path to produce paper-based packaging materials that have both good oxygen and water vapor barrier properties (115).

5.2.12.7 Cold Plasma Treated Cellulose

The strong hydrogen bonding system makes cellulose hard to dissolve, and so limits its applications (116). In a study, an innovative cold plasma technology was used to modify cellulose from sugarcane (*Saccharum officinarum*) bagasse pulp.

Dissolution, structure, and surface chemistry of cellulose before and after the cold plasma treatment were investigated. Results showed that the dissolution rate of cellulose after different cold plasma treatment time (3–12 *min*) and operating voltage (40–70 *kV*) was significantly improved. Roughness, even holes (cold plasma treatment 9 *min* with 50 *kV*) and breakage (cold plasma treatment 9 *min* with 70 *kV*) were observed on the surface. The crystallinity index decreased from 62.31% (control) to 60.88% (cold plasma treatment 3 *min* with 50 *kV*). The hydrogen bonding force was weakened and the peak intensity of double-bond and single-bond stretching vibration groups were enhanced.

Therefore, a cold plasma-modified cellulose may be applied more successfully as biological films for future food packaging (116).

5.2.12.8 Cellulose-Based Antimicrobial Films

Biodegradable and antimicrobial films without antibiotics are of great significance for applications associated with food packaging, meanwhile minimizing the negative impact on the environment (117).

In this work, cellulose-based films with the surface tailor-constructed with ZnO nanopillars were prepared via chemical crosslinking in conjunction with a hydrothermal process for *in-situ* growth of ZnO nanopillars.

As a packaging material, these films possess excellent mechanical properties, oxygen and water vapor barrier, food preservation, biodegradability and a low ZnO^{2+} migration. Moreover, these films show remarkable antimicrobial activity, especially for *Staphylococcus aureus* (Gram-positive bacteria) and *Escherichia coli* (Gram-negative bacteria).

The antimicrobial mechanism of these films was studied using the controlled variable method. The study results showed that the film without UV pretreatment killed bacterial cells mainly by mechanical rupture, while the film with UV pretreatment killed bacterial cells mainly via the synergistic effect of photocatalytic oxidation and mechanical rupture (117).

5.2.12.9 Carboxymethyl Cellulose Capped Zinc Oxide Nanoparticles

Zinc oxide (ZnO) is one of the promising food additives, which adds nutrients and provides antimicrobial properties when incorporated into various food matrices. In a study, carboxymethyl cellulose capped ZnO (ZnO-CMC) was developed via a low-energy and cost-effective technique without calcination or grinding (118).

The fabrication involved two steps:

1. Crosslinking Zn^{2+} ions with carboxymethyl cellulose through electrostatic interactions, and
2. Generation of ZnO nanoparticles with carboxymethyl cellulose as capping agent.

After a mild heating process, the crystalline structure of ZnO-CMC was confirmed by wide-angle X-ray scattering. Both FTIR and AFM studies showed that ZnO was physically trapped by CMC molecules, resulting in a physical barrier to prevent aggregation. The SEM analysis verified that the ZnO-CMC had a size of 50 *nm* to 80 *nm* with comparable morphology to commercial ZnO.

Overall, CMC played a key role in controlling growth and inhibiting agglomeration of ZnO. Given the small and uniform particle size, the obtained ZnO-CMC is ready to be incorporated into different food matrices (118).

5.2.13 *Cellulose Nanofibrils*

A cellulose nanofibril is a natural biodegradable biopolymer with excellent mechanical and barrier properties. However, it is susceptible to moisture-induced deterioration of its properties (119).

The attachment of phenolic acids can improve its hydrophobicity and provide additional active functionalities such as antioxidant properties.

In a study, cellulose nanofibril films were esterified to vanillic and syringic acid through two different reaction mechanisms (119). The films were investigated for evidence of modification, hydrophobicity, mechanical properties, crystallinity, thermal stability, and antioxidant properties.

The results showed that the esterification with vanillic and syringic acids imparted an antioxidant activity to the cellulose nanofibril films. Similarly, the esterification of phenolic acids significantly improved the hydrophobicity of the films with a water contact angle of 94±3°. So, the covalent attachment of phenolic acids can improve the hydrophobicity while providing additional functionality to cellulose nanofibril important for food packaging applications (119).

5.2.13.1 Poly(lactic acid) Nanocomposites

Poly(lactic acid) (PLA) is the most utilized biopolymer for food packaging applications. Its excellent mechanical properties rival and in some cases surpass those of the petro-based plastics like poly(ethylene) (PE). However, noticeable setbacks exist for PLA in the race to replace petro-based plastics with biodegradable biomaterials. Brilliant stiffness, flavor, and aroma barrier is overlapped by high brittleness and poor water vapor barrier among others. The PLA nanocomposites represent the frontier in which these drawbacks could be overcome and improved. Numerous research activities have shown how nanofillers, plasticizers, and copolymers enhance PLA composite performance across a wide range of storage utility applications in food packaging and preservation, from excellent transparency, tensile strength and modulus to superb antioxidant and antimicrobial properties. Therefore, a chapter in a monograph mainly focuses on promising advances in the process and characterization of PLA nanocomposite technology aimed at improving its functionality in the food packaging industry for the storage of different food products (120).

Since PLA has received attention as a sustainable, biocompatible, biodegradable material with good mechanical and optical properties, it has emerged as the most favored bioplastic (31).

Lactic acid, the monomer of PLA, may easily be produced by fermentation of carbohydrate feedstock. Thus, PLA offers more options for disposal and its manufacture is less environmentally burdensome than traditional petroleum-based plastics (121, 122).

However, the large-scale use of PLA as packaging material is still hampered by its high cost and low performance in comparison to commodity polymers. The most important limitation for the application of PLA in food packaging is its low gas barrier properties.

Nanocomposites technology has the potential to improve polymer properties and expand the applications of PLA (121, 123–125).

Nanocomposites of amorphous PLA and chemically modified kaolinite were studied (126). A good interaction was observed between polymer and clay, which led to an increase in oxygen barrier properties of about 50%. The study also included the addition of plasticizers to overcome the inherent brittleness of PLA. In general, plasticizers lower the gas barrier properties of polymers. In that study, the effect of plasticizers on oxygen permeability of PLA was offset by the formation of the kaolinite nanocomposite.

The combination of PLA and montmorillonite layered silicate may result in a nanocomposite with barrier properties suitable for food packaging applications (123, 124).

5.2.13.2 Zein

Zein, a relatively hydrophobic protein found in corn kernels, is known to form films easily (127, 128).

Zein is used in the food industry as a coating agent and has shown potential as a biodegradable polymer (129, 130).

However, zein products, although less water sensitive than other biopolymers, still show high water vapor permeability and low tensile strength when compared with commodity polymers. As was the case for other biopolymers, its inherent brittleness may be ameliorated by the use of plasticizers which, on the other hand, further decrease water vapor and gas barrier properties.

A method for producing a zein protein product from a whole stillage byproduct produced in a corn dry-milling process has been described (131). The method includes separating the whole stillage byproduct into an insoluble solids portion and a centrate (solubles) portion, which includes protein, such as zein protein.

Thereafter, the centrate (solubles) portion can be separated into a water-soluble solids portion and a protein portion, which includes zein protein. The zein protein may be separated out from the protein portion. The remaining protein portion may be further processed to produce a high protein corn meal. The resulting zein protein portion may be further processed to be sold as a zein protein product and/or used as or in, for example, coatings, fibers, adhesives, ceramics, inks, cosmetics, textiles, food products, pharmaceutical, and biodegradable plastics (131).

5.2.14 *Chia Seeds*

Chia seeds are a promising raw material for the development of biodegradable and edible polymers due to their composition and properties (132).

The effects of some drying processes of chia mucilage, i.e., oven and freeze-drying, and the incorporation of chia oil in films for food packaging were evaluated.

The films were formed by casting using chia mucilage and glycerol. The so developed polymers were evaluated by their physicochemical properties, microstructure, thermal properties, and biodegradation. The drying process of mucilage and oil incorporation in films mainly affected the mechanical and color properties.

Freeze-dried mucilage resulted in a superior mechanical performance. Differences were caused by the effect of the drying process on the molecular structure of chia mucilage and the incorporation of oil among the polymer chains. Chia mucilage films were completely soluble in water and biodegraded in a short time in soil.

So, these films are promising biodegradable polymers for the development of eco-friendly food packaging and edible sachets for small pre-measured portions, preventing environmental pollution (132).

In another study, the physical-chemical, optical, water vapor permeability, and mechanical properties of films made with two different types of chia matrixes (defatted flour and whole seeds) were determined (133).

In general, defatted chia flour films exhibited a slightly reddish and yellowish color but still transparent in appearance, were good visible light barriers, and had better mechanical properties than films made with whole seeds. They also have greater water vapor permeability values than synthetic films such as low-density polyethylene. The results of this study demonstrated that defatted chia flour can be used in producing edible films with improved quality characteristics (133).

The use of bio-based plastic packaging materials contribute to creating more sustainable food packaging materials, compared to conventional plastic packaging materials. Although bio-based plastic packaging materials have reduced climate impact, they have other environmental impacts that are less favorable (134). Even though recycling reduces the environmental and climate impacts of bio-based packaging/food systems, their entire life cycle should be assessed to improve sustainability.

The use of conventional nondegradable and noncompostable polymers has a significant negative impact on the environment (135). So, biodegradable polymers have attracted a lot of attention in the field of sustainable development, recycling, or environmental protection.

Morphological and structural transformations resulting from the degradation of the material have a significant impact on the polymer material properties. Thus, evaluation and understanding of the structure, properties, and behavior of advanced materials for perspective food packaging applications is important to avoid degradation during storage.

A review of biodegradable polymer materials with approved or potential applications in food packaging has been presented (135). The major groups of biodegradable natural, microbial, and synthetic polymers are briefly discussed with regard to the properties that are required for packaging materials, like permeability, physical and mechanical properties, antioxidative and antimicrobial additives, plasticization, and biodegradability, based on studies reported in the literature as well as future trends (135).

Non-biodegradable plastic materials include poly(propylene) (PP), poly(ethylene) (PE), poly(styrene) (PS), poly(urethane) (PU), and poly(vinyl chloride) (PVC) (85). These materials are known to be employed for packaging applications owing to their strong mechanical and physical characteristics. Moreover, the properties of such materials are further intensified by adding aluminum foil during processing, which imparts non-recyclability to such materials.

The production of biodegradable plastic materials is believed to be an alternative route for resolving waste management issues of conventional plastics. Biodegradation can be defined as an end-of-life breakdown pathway for the degradation of material enforced via the degradative actions of microorganisms (typically bacteria, algae and fungi). The process of biodegradation utilizes the metabolic activities of microorganisms habituating a particular environment for the complete/incomplete removal of large materials from the environment to release beneficial biodegraded small products in a timely and effectual manner (85).

An ideal biodegradable plastic material is one which is completely degraded into smaller molecules of CO_2 and H_2O at the end of the biodegradation.

Biodegradable plastic materials that are employed for packaging applications are poly(lactic acid), poly(hydroxyalkanoate)s, poly(hydroxybutyrate), cellulose, starch and poly(caprolactone).

Generally, the plastic bags commonly used for food packaging applications composed of PE, are marketed with claims of being degradable, oxo-degradable or biodegradable. However, this does not correlate with the true essence of biodegradability owing to the presence of non-biodegradable synthetic additives in such assemblies.

Numerous studies stating the obvious concerns of whether or not these products provide what they claim can be found in the literature (136).

Hence, the formation of biodegradable plastic to replace such harmful plastics is a current need as they effectively degrade themselves into smaller useful products with time and do not create waste management concerns.

5.2.15 Poly(lactic acid)

Poly(lactic acid) belongs to the class of aliphatic polyesters which is made by copolymerization of poly(D-lactic acid) and poly(L-lactic acid) (85).

It is regarded as one of the most important biodegradable biopolymers generally acquired from the fermentation of renewable food sources such as starch and sugar feedstock. It is a compostable, recyclable, biodegradable and unique polymer having appreciably high transparency and molecular weight (137).

Moreover, by varying the enantiomer ratio in the mixture, the properties of PLA can be varied and controlled as well. Currently, it is employed for short-term food packaging applications. In its pure form, PLA is brittle and rarely ever considered alone for its packaging applications. Generally, blends of PLA are reported where foreign material is introduced into the PLA matrix for improving its properties. Following is a brief discussion of some case studies exploring the biodegradable food packaging potential of PLA.

Composites of PLA/PCL and starch were prepared in twelve different mixing ratios (137). The synthesized composites were then hardened by microwave treatment. This study is of great significance as it validates the formation of chemical linkages among the composite molecules instead of generally formed physical linkages. The authors attributed these findings to the application of uniform and even distribution of heat during synthesis procedure via microwave treatment.

A blending of PLA with a plasticizer poly(1,4-butylene adipate) (PBA) was done to enhance the mechanical properties of the PLA (138). Blends of PLA/PBA were prepared by melt blending.

5.2.16 Poly(hydroxyalkanoate)s

Research into renewable biopolymers as substitutes for full-carbon-backbone plastics from fossil resources is due to the ongoing depletion of fossil resources, the growing piles of plastic waste and plastic pollution of marine environments, and the need to convert waste streams of different industrial origins in a value-added way.

Poly(hydroxyalkanoate)s have the potential to replace established petro-plastics both in bulk applications as packaging material and in niche applications, such as in the medical and electronics fields among others. Moreover, the close relation of poly(hydroxyalkanoate) production and application to the food sector is becoming more and more evident.

Food production provides numerous industrial byproducts which can, on the one hand, be applied to boost growth kinetics of poly(hydroxyalkanoate)-accumulation strains, as evidenced in the case of nitrogenaceous whey retentate, silage residues, or shrimps waste, and, on the other hand, act as feedstocks for poly(hydroxyalkanoate)-biosynthesis under nutritionally unbalanced growth conditions, as demonstrated for carbonaceous surplus materials like whey permeate, lignocellulosics, glycerol, waste lipids, etc.

Poly(hydroxyalkanoate)s have been investigated as materials for safe and smart food storage and packaging, as shown by their beneficial gas barrier properties.

A range of promising poly(hydroxyalkanoate)-based blends of composite materials are accessible to design novel food packaging materials. This encompasses the application of lignocellulosic filler materials from rice, sugar, or wood production, and even the development of more sophisticated formulations resorting to the incorporation of functional nanoparticles into poly(hydroxyalkanoate) matrixes (139).

Poly(hydroxyalkanoate)s are one of the most recent biodegradable thermoplastic materials utilized for food packaging applications. Similar to PLA, it is usually considered that blending poly(hydroxyalkanoate)s with copolymers is an effective method for acquiring diversification in the properties of poly(hydroxyalkanoate)s.

The addition of high barrier property-containing materials, such as whey protein isolate, zein, and pullulan (poly(saccharide) polymer) into poly(hydroxyalkanoate)s, can be used to modify the water and the oxygen barrier properties of the blend (140). Here, improvements of 38% to 48% and 28% to 35% have been observed in the case of WVPC and OPC values acquired with the addition of these materials in the poly(hydroxyalkanoate) matrix.

The added materials were also found to be responsible for manipulating the shape of the blend, i.e., the introduction of pullulan and zein imparted fibrous morphology to the system, while in the case of wey protein isolate, beaded morphology was observed (140).

In another study, nanokeratin was directly extracted from poultry feathers as a nanoadditive for enhancing the adhesive, barrier and mechanical properties of the poly(hydroxyalkanoate)s (141).

This study provided a comparative analysis for the preparation of blends using three different techniques, i.e., solution casting, direct melt molding and electrospun masterbatch melt compounding.

Furthermore, a solution of whey and gelatin at different concentrations in poly(hydroxyalkanoate)s films was fabricated to improve the characteristics of the standard poly(hydroxyalkanoate)s film by making blend film (142). Vapor permeability coefficient values of 0.45 $g\,mm\,m^{-2}d^{-1}kPa^{-1}$ were found. Optical properties of 14.5% opacity and mechanical properties of 2.18% elongation were found to be enhanced via the addition of these biomaterials.

The formation of a blend of poly(hydroxyalkanoate)s and *Posidonia oceania* was reported (143). Here, the effect of *P. oceania* on the characteristics of the poly(hydroxyalkanoate)s was investigated. An improvement of almost 80% was observed in the values of the tensile strength for the blend, i.e., 2.3 *GPa* in comparison to the pure poly(hydroxyalkanoate)s material with 1.24 *GPa*.

Moreover, the impact energy showing mechanical strength and resistivity of the material was also found to be increased from 1.63 $kJ\,m^{-2}$ to 3.8 $kJ\,m^{-2}$.

Degradation studies that were carried out in the marine environment indicated that factors such as seawater temperature and *P. oceania* content directly manipulate the degradation rate. The blend was completely degraded within six months when exposed to seawater under controlled conditions (85).

5.2.17 *Poly(hydroxybutyrate)*

Poly(3-hydroxybutyrate) (PHB) is known for its exceptional properties, including biocompatibility, biodegradability, optical properties and UV-resistivity. Its enhanced properties make it comparable with synthetic polymers such as poly(propylene) (85). Therefore, PHB can be effectively used in food packaging applications, controlled drug delivery carriers, wound dressing and tissue engineering.

However, the poor mechanical strength and thermal characteristics of PHB limit its applicability in these fields to some extent. So, blends of PHB were prepared in order to compensate for these drawbacks.

A grafting technique was used for the preparation of blends of PHB/poly(butylene adipate-*co*-terephthalate) and incorporated antimicrobial additives of 1-allylhydantoin and perfluorooctyl acrylate into the system to achieve active antimicrobial membranes (144). Antimicrobial additives are shown in Figure 5.6.

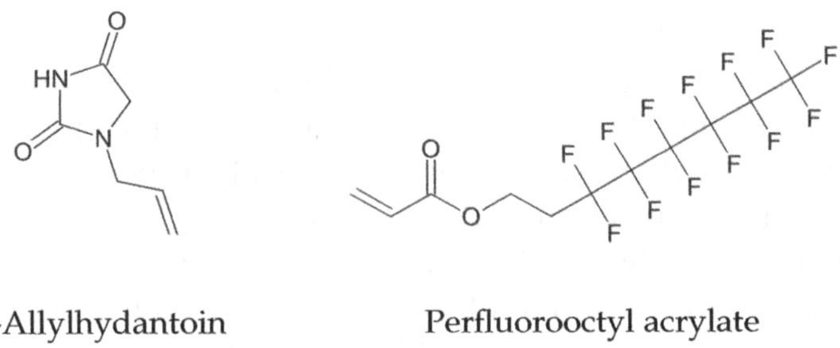

Figure 5.6 Antimicrobial additives.

A UV-light test and storage test showed that the membranes possess enhanced ductility, stability and hydrophobicity for food packaging applications.

Solution casting method was used to design PHB blends containing ZnO nanoparticles in the absence of any surfactant or another dispersing agent for the nanoparticles (145).

The nanomaterial acted as the aggregating/nucleating agent and improved the degree of crystallinity of the matrix, which in turn enhanced the thermal stability by creating diffusional barriers for volatiles generated during the decomposition processes. An increment of 32%, 26% and 43% was observed in PHB/ZnO blends in the case of the properties of tensile strength, impact strength and Young's modulus respectively. This increment was attributed to the strong matrix-nanomaterial interfacial adhesive interactions caused due to hydrogen bonding. The addition of ZnO also improved the antimicrobial potential of the system, which seemed to constantly increase by increasing the amount of ZnO in the blend.

A PLA-PHB blend was formed by the process of melt blending (146). The addition of catechin, c.f. Figure 5.7, in the blend as an antioxidant agent improved the thermal stability of the blend, whereas the introduction of acetyltributyl citrate as plasticizer resulted in the reduction of elastic modulus and rigidity of the blend.

5.2.18 *Poly(caprolactone)*

Poly(caprolactone) (PCL) is a flexible, biodegradable, nontoxic, easily processable and hydrophobic fossil-based polymer material.

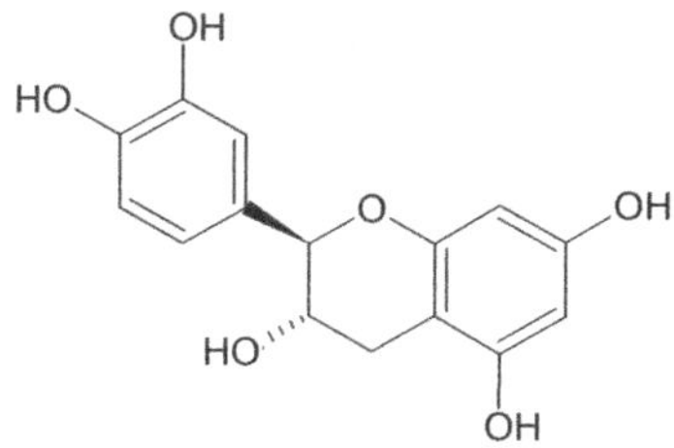

Figure 5.7 Catechin.

The incorporation of PCL as an enhancer in pure chitosan film by the solution casting method generated PCL/chitosan blend films with improved properties (147). The film with composition ratio (PCL:chitosan) 80:20 exhibited improvement in the barrier properties along with the retention in the mechanical properties. The introduction of the hydrophobic polymer (PCL) into the hydrophilic polymer (chitosan) causes an overall increase in the hydrophobicity of the blended films. Consequently, lower water vapor transmission rate values were reported in the case of blended films as compared to the pure chitosan films. Thus, food packaged inside such films will have a moderately longer storage life owing to this feature.

PCL/TPS blends were tested for food packaging applications (86). Pure PCL generally shows good crystalline ability to a certain extent, but to enhance this particular property, blends of PCL were synthesized. The presence of starch enhanced the degree of crystallinity in the blend owing to its reinforcing actions.

A surface modification treatment of layered double hydroxide by utilizing an ionic liquid constituted of positively charged phosphonium ions electrostatically combined with the negative ions of 2-ethylhexanoate acting as a surfactant agent was reported (144).

The ionic liquid is trihexyl(tetradecyl)phosphonium 2-ethylhexanoate.

5.3 Microplastics

The ubiquitous nature and regular occurrence of microplastics and nanoplastics in the food chain and environment have made them a topical subject for investigations (24). The subject has been widely researched but only a few studies have focused on ethical issues.

Microplastics are ubiquitous and their presence and accumulation in the environment threatens the ecological equilibrium, aquatic environment, human health, food sustainability and security (14, 16, 148).

The environmental and metabolic threats to marine organisms are attributable to the stable chemical and physical properties of the parent plastics (17).

5.3.1 *Microplastics in the Environment*

The exposure of plastic waste in the environment causes a deterioration in the mechanical and physicochemical properties and leads to the formation of plastic fragments, which are considered as microplastics when their size is smaller than 5 *mm* (149).

Microplastic pollution has been reported in all kinds of environments worldwide and is considered a potential threat to the health of ecosystems and humans. However, knowledge on the environmental degradation of plastics and the formation of microplastics is still limited.

Potential hotspots for the accumulation of plastic waste have been identified. Because of improper management and disposal practice, a large amount of plastic waste enters the environment through various pathways and causes serious environmental pollution problems (150, 151).

Also, the major mechanisms of plastic degradation are summarized in a review (149). The degradation of plastics is critically important in determining their fate and effects in the environment. Microplastics produced during the degradation of plastics have become an issue of increasing concern in recent years because they are more prevalent and may cause potential health risks to the ecosystem and humankind. Most of the current research on the environmental degradation of plastics has been focused on the early stages of degradation; thus, the formation of microplastics under different environmental conditions is still poorly understood, and the persistence of microplastics in the environment is virtually unknown.

Future research works should further identify the key environmental parameters and properties of plastics affecting the degradation in order to predict the fate of plastics in different environments and facilitate the development of technologies for reducing plastic pollution. Formation and degradation of microplastics, including nanoplastics, should receive more research attention to assess their fate and ecological risks in the environment more comprehensively (149).

The plastics can absorb and adsorb toxic inorganic and organic compounds which can leach into the tissues of the marine organisms. Polychlorinated biphenyls (PCB) and polycyclic aromatic hydrocarbons (PAH) compounds are reportedly the most plastic/microplastic adsorbent (14, 19–21, 152).

The release rate of these chemical compounds is favored by the pH of the medium, temperature, heat, UV light, and mechanical and chemical abrasion (22). The manufacturing additives, such as bisphenol A, phthalates, nonylphenol, urethane foam and brominated flame retardant, used in production of plastics can become toxic residues (20, 23).

5.3.2 *Microplastics in Food*

Several studies have shown the presence of micro- and nano-sized plastic particles in the food chain. A study by Van Cauwenberghe and Janssen (153) discusses microplastics in bivalves cultured for human consumption.

A book chapter describes the issues of micro- and nanoplastics and human health (154).

Microplastics were also found as contaminants in seafood (155), such as in the mussels from UK coastal waters sold in supermarkets (156).

The role of microplastics in seafood destined for human consumption was investigated (16, 19, 157, 158). Here, it was found that the contamination level of seafood sampled from the northern area of the Persian Gulf is about 0.2 to 21 particles per gram (wet weight).

A higher amount of microplastics was found in cooked mussels (156), i.e., about 1.4 particles per gram compared to 0.9 particles per gram in live mussels. This suggested a potential contamination during the processing of the product.

Twenty-three samples of milk from different brands for sale were analyzed for microplastic particles and results found a range of 3 ±

2 to 11 ± 3.54 particles per liter with an overall average of 6.5 ± 2.3 particles per liter.

Similarly, in China, 440 ± 225 particles were found, mainly PE and PE in tap water (159).

An investigation (160) showed a whole-body uptake of nanoplastics especially of 24 *nm* particle size by *Pecten maximus*, a commercially popular mollusk. It was further found that polystyrene nanoplastics (20 *nm*) can bioaccumulate and cause damage to the DNA in the brain tissue of the embryos of zebrafish (*Danio rerio*) (161). These findings demonstrated the ability of nanoplastics to migrate across cellular membranes.

Nanoplastics can penetrate deep tissues and edible parts of fish, cause reproduction and growth defects, metabolic disorder and even death (23, 148, 162).

A study of the Pacific mole crab (*Emerita analoga*) showed that microfiber contamination affects its fecundity and is manifested as a reduction in egg retention and a differential rate of embryonic development (158). There is also an increase in mortality amongst the Pacific mole crab population exposed to microplastics compared to their control group.

Plastic fragments can accumulate in the gastrointestinal tract of marine organisms through ingestion and there is evidence of deep tissue penetration by nanoplastics (24).

5.3.3 *Microplastics in Agroecosystems*

Microplastics enter ecosystems either directly or indirectly through several processes (163) and are 4- to 23-fold more prevalent in the ecosystems than in aquatic ecosystems (164, 165).

Structural and physicochemical changes due to microplastics in soil, their ecotoxicology, and influence on plant growth performance have been investigated (166).

Compost is considered one of the prime sources of microplastics in agricultural environments, where microplastics in compost can influence the carbon cycle in copious ways, such as soil microbial processes, plant growth, or litter decomposition. Compost can be a carrier for MP-bound toxic trace metals into the agroecosystems. Different toxic metals that are associated with the microplastics in compost, i.e., Cr, Pb, Cu, and Ni, have been identified.

The addition of microplastics changes the physical, chemical, and biological properties of soil, which leads to a change in toxic trace metal fractionation and partitioning, as reported in the literature (166).

5.3.4 *Release of Microplastics*

Very little is known about the release of microplastics on land, storage in soils and sediments and transport by run-off and rivers (164).

A theoretical assessment of these processes has been described in an article. A mathematical model of catchment hydrology, soil erosion and sediment budgets was upgraded to enable description of microplastic fate. The Thames River in the UK was used as a case study.

A general lack of data on microplastic emissions to soils and rivers and the mass of microplastics in agricultural soils, limits the study to serve as a purely theoretical, nevertheless rigorous, assessment that can be used to guide future monitoring and impact evaluations. The fundamental assumption on which modeling is based is that the same physical controls on soil erosion and natural sediment transport (for which model calibration and validation are possible), also control microplastic transport and storage. Depending on sub--catchment soil characteristics and precipitation patterns, approximately 16%–38% of the heavier-than-water microplastics hypothetically added to soils (e.g., through routine applications of sewage sludge) are predicted to be stored locally. In the stream, microplastics $<0.2\ mm$ are generally not retained, regardless of their density. Larger microplastics with densities marginally higher than water can instead be retained in the sediment.

It is, however, anticipated that high flow periods can remobilize this pool. Sediments of river sections experiencing low stream power are likely hotspots for deposition of microplastics. Exposure and impact assessments should prioritize these environments (164).

5.4 Single-Use Materials

Single-use plastic is a versatile material (167). It has been adopted as an alternate to traditional materials specifically for use in food packaging due to its inherent characteristics like high durability, inertness, and protective ability. However, single-use plastic has become a curse for living beings today due to its random usage and unplanned rejection by nature.

Today, most of the plastics used in the packaging of beverages, fresh meats, fruits and vegetables are of concern. Single-use packages result in the generation of several billion tons of garbage till date, which pollutes the environment. Recently, it has come to light that microplastics resulting from slow degradation of single-use plastic present in the oceans, are also being consumed by marine organisms such as fishes and shellfish species, which extensively disturbs the marine life.

Hence, a correct strategy to mitigate plastic waste-related issues is becoming inevitable.

In a study, the various strategies undertaken worldwide to mitigate pollution due to generation of plastic waste have been presented (167).

Also, the notable impact of various adopted strategies and recent innovations to replace single-use plastic products is being discussed. A roadmap is suggested which can be used to achieve the milestone of zero plastic waste.

5.5 Edible Films

Edible coatings and films can be used to improve food quality (168). There has been intense research into edible coatings and films for many reasons, including consumer interest in health, food quality, convenience, and safety. It is a fact that edible coatings can conceivably reduce the complexity and improve the recyclability of packaging (168).

5.5.1 *Poly(saccharide)s*

Edible films were prepared using various pectin and pullulan mixing ratios and evaluated for their properties in food packaging applications (169). Pectin and pullulan are shown in Figure 5.8.

Results of ACL-FTIR spectroscopy showed that an intermolecular H-bond was formed between the hydroxyl group of pullulan and the carboxyl group of pectin. When the pullulan content increased,

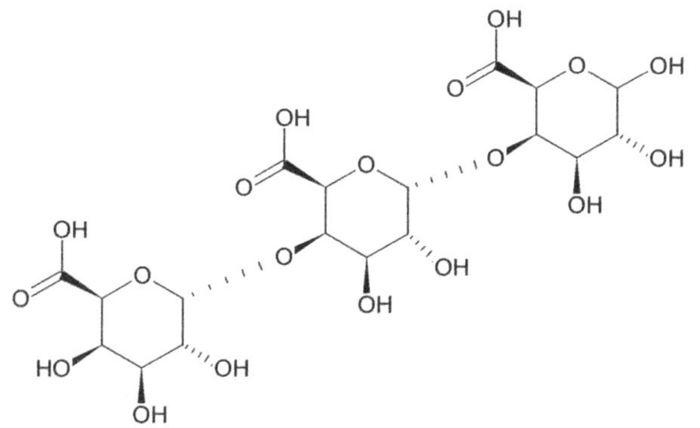

Pectin

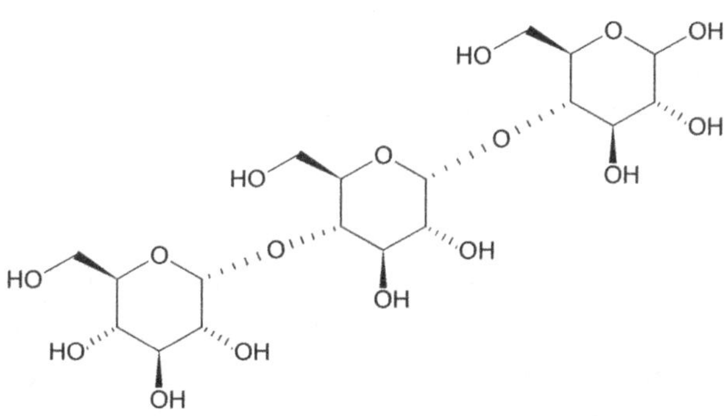

Pullulan

Figure 5.8 Pectin and pullulan.

the surface of the film became smoother and formed a film with a denser structure, leading to an increased water vapor barrier. The blend film with a 50:50 ratio of pullulan and pectin exhibited the highest thermal stability and surface hydrophobicity.

Blending also increased strength while maintaining flexibility and stiffness compared to the individual films. Besides which, the films with ratios above 50:50 displayed the least water and oil absorption values (169).

5.5.2 *Cellulose*

5.5.2.1 *Cellulose with Lysozyme*

Carboxymethyl cellulose (CMC) films containing lysozyme were prepared. Then, the changes in properties of the films were investigated. An enhancement in mechanical properties was observed with increased lysozyme, maximum 0.05 $g\,dl^{-1}$ reached to 39.07 MPa tensile strength (TS) and 25.04% elongation at break (EAB) (170).

Meanwhile, water resistance ability improved, the minimum (0.05 $g\,dl^{-1}$) reached to 0.42 $g\,mm\,(mm^2 h\,kPa^{-1})$, 84.62% of a pure CMC film.

Thermogravimetric tests showed better thermal stability of the films. SEM illustrated a few cracks on the surface of the films and FTIR supported that more intermolecular hydrogen between lysozyme and CMC was formed with increased lysozyme, yet kept increasing to form less intermolecular hydrogen. X-Ray diffraction observed the aggregated lysozyme by crystal structure.

An antibacterial test showed an inhibitory effect on two common foodborne pathogens. Weight loss experiment indicated that films reduced the dry consumption of meat. Overall, the modification of the CMC film by adding lysozyme was effective (170).

5.5.2.2 *Cellulose with Astragalus Honey*

The ability of CMC containing *Astragalus* honey (*Astragalus gossypinus*) to control rancidity and microbial spoilage of pistachio kernel during storage at room temperature without effect on their sensory properties was studied (171).

Pistachio kernels were coated with different amounts of *Astragalus* honey (0, 1 and 2%) and 1% CMC. Samples were stored at room temperature for 120 *d*. Microbial and fungal counts, oxidation parameters, weight loss, and hardness of pistachio samples were increased, but moisture content and sensory scores were decreased during the storage.

Pistachio kernels coated with honey (2%) and CMC (1%) had the lowest microbial and fungal counts, weight loss, peroxide, anisidine, c.f. Figure 5.9 , free fatty acids, and textural hardness during the storage ($P < 0.05$). Pistachio kernels coated with CMC containing honey exhibited the lowest color changes. Sensory score of samples

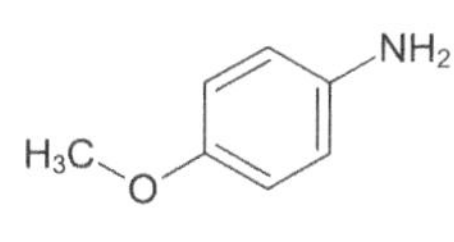

Figure 5.9 Anisidine.

increased after coating with honey and CMC. At the end of 120 *d* storage, the peroxide value and free fatty acid content of the sample coated with CMC and 2% honey were about half of what was considered an acceptable threshold.

So, the CMC coating that contains *Astragalus* honey could extend the shelf life of pistachio kernels (171).

References

1. S. Alavi, S. Thomas, K.P. Sandeep, N. Kalarikkal, J. Varghese, and S. Yaragalla, (Eds.), *Polymers for Packaging Applications*, Apple Academic Press, 2014.
2. J. Sarfraz, T. Gulin-Sarfraz, J. Nilsen-Nygaard, and M.K. Pettersen, Nanocomposites for food packaging applications: An overview, *Nanomaterials*, Vol. 11, p. 10, 2020.
3. S. Sinha Ray and M. Okamoto, Polymer/layered silicate nanocomposites: A review from preparation to processing, *Progress in Polymer Science*, Vol. 28, p. 1539, 2003.
4. S. Ray, S.Y. Quek, A. Easteal, and X.D. Chen, The potential use of polymer-clay nanocomposites in food packaging, *International Journal of Food Engineering*, Vol. 2, 2006.
5. A. Sorrentino, G. Gorrasi, and V. Vittoria, Potential perspectives of bio-nanocomposites for food packaging applications, *Trends in Food Science & Technology*, Vol. 18, p. 84, 2007.

6. E.T. Thostenson, C. Li, and T.-W. Chou, Nanocomposites in context, *Composites Science and Technology*, Vol. 65, p. 491, 2005.
7. S. Azlin-Hasim, M.C. Cruz-Romero, M.A. Morris, S.C. Padmanabhan, E. Cummins, and J.P. Kerry, The potential application of antimicrobial silver polyvinyl chloride nanocomposite films to extend the shelf-life of chicken breast fillets, *Food and Bioprocess Technology*, Vol. 9, 2016.
8. N. Takamura, Transparent poly(phenylene ether) compositions, their methods of manufacture, and food packaging films and containers derived therefrom, US Patent 9 845 388, assigned to SABIC Global Technologies BV, December 19, 2017.
9. Standard test method for deflection temperature of plastics under flexural load in the edgewise position, ASTM Standard D648-18, ASTM International, West Conshohocken, PA, 2018. https://www.astm.org/d0648-18.html.
10. H. Uyama, M. Kuwabara, T. Tsujimoto, M. Nakano, A. Usuki, and S. Kobayashi, Green nanocomposites from renewable resources: Plant oil-clay hybrid materials, *Chemistry of Materials*, Vol. 15, p. 2492, 2003.
11. P. Demetrakakes, Nanocomposites raise barriers, but also face them: Clay-based additives increase the barrier qualities of plastics, but obstacles to commercialization must be overcome, *Nanocomposite Materials*, 2002.
12. T. Ramanathan, A.A. Abdala, S. Stankovich, D.A. Dikin, M. Herrera-Alonso, R.D. Piner, D.H. Adamson, H.C. Schniepp, X.R.R.S. Chen, R.S. Ruoff, S.T. Nguyen, I.A. Aksay, R.K. Prud'Homme, and L.C. Brinson, Functionalized graphene sheets for polymer nanocomposites, *Nature Nanotechnology*, Vol. 3, p. 327, 2008.
13. A. Arora and G.W. Padua, Review: Nanocomposites in food packaging, *Journal of Food Science*, Vol. 75, p. R43, 2010.
14. G.E. De-la-Torre, Microplastics: An emerging threat to food security and human health, *Journal of Food Science and Technology*, Vol. 57, p. 1601, 2020.
15. T. Børresen, Microplastics: A new threat to aquatic food safety?, *Journal of Aquatic Food Product Technology*, Vol. 24, p. 415, 2015.
16. C. Walkinshaw, P.K. Lindeque, R. Thompson, T. Tolhurst, and M. Cole, Microplastics and seafood: Lower trophic organisms at highest risk of contamination, *Ecotoxicology and Environmental Safety*, Vol. 190, p. 110066, 2020.
17. A.L. Andrady, The plastic in microplastics: A review, *Marine Pollution Bulletin*, Vol. 119, p. 12, 2017.

18. Panel on Contaminants in the Food Chain (CONTAM), Presence of microplastics and nanoplastics in food, with particular focus on seafood, *EFSA Journal*, Vol. 14, p. e04501, 2016.
19. A. Naji, M. Nuri, and A.D. Vethaak, Microplastics contamination in molluscs from the northern part of the Persian gulf, *Environmental Pollution*, Vol. 235, p. 113, April 2018.
20. M. Almeida, M.A. Martins, A.M.V. Soares, A. Cuesta, and M. Oliveira, Polystyrene nanoplastics alter the cytotoxicity of human pharmaceuticals on marine fish cell lines, *Environmental Toxicology and Pharmacology*, Vol. 69, p. 57, 2019.
21. C. Barría, I. Brandts, L. Tort, M. Oliveira, and M. Teles, Effect of nanoplastics on fish health and performance: A review, *Marine Pollution Bulletin*, Vol. 151, p. 110791, 2020.
22. T.S. Galloway, Micro- and nano-plastics and human health in *Marine Anthropogenic Litter*, pp. 343–366. Springer, Cham, 2015.
23. M. Shen, Y. Zhang, Y. Zhu, B. Song, G. Zeng, D. Hu, X. Wen, and X. Ren, Recent advances in toxicological research of nanoplastics in the environment: A review, *Environmental Pollution*, Vol. 252, p. 511, 2019.
24. D. Ojinnaka and M. Aw, Micro and nano plastics: A consumer perception study on the environment, food safety threat and control systems, *Biomedical Journal of Scientific & Technical Research*, Vol. 32, 2020.
25. M. Biron, *Thermoplastics and Thermoplastic Composites*, William Andrew, 2018.
26. J. Payne, P. McKeown, and M.D. Jones, A circular economy approach to plastic waste, *Polymer Degradation and Stability*, Vol. 165, p. 170, 2019.
27. M. Murariu and P. Dubois, PLA composites: From production to properties, *Advanced Drug Delivery Reviews*, Vol. 107, p. 17, 2016.
28. C. Vilela, A.F. Sousa, A.C. Fonseca, A.C. Serra, J.F.J. Coelho, C.S.R. Freire, and A.J.D. Silvestre, The quest for sustainable polyesters — insights into the future, *Polym. Chem.*, Vol. 5, p. 3119, 2014.
29. Y. Byun and Y.T. Kim, Bioplastics for food packaging: Chemistry and physics in J.H. Han, (Ed.), *Innovations in Food Packaging*, Food Science and Technology, chapter 14, pp. 353–368. Academic Press, San Diego, 2nd edition, 2014.
30. S. Mangaraj, A. Yadav, L.M. Bal, S.K. Dash, and N.K. Mahanti, Application of biodegradable polymers in food packaging industry: A comprehensive review, *Journal of Packaging Technology and Research*, Vol. 3, p. 77, 2019.

31. L.K. Ncube, A.U. Ude, E.N. Ogunmuyiwa, R. Zulkifli, and I.N. Beas, Environmental impact of food packaging materials: A review of contemporary development from conventional plastics to polylactic acid based materials, *Materials*, Vol. 13, 2020.
32. J.J. De Vlieger, Green plastics for food packaging in R. Ahvenainen, (Ed.), *Novel Food Packaging Techniques*, chapter 24, pp. 519–534. Woodhead Publishing Ltd., Cambridge, 2003.
33. E.B. e.V. (EUBP), European bioplastics, https://www.european-bioplastics.org/news/publications/#MarketData, 2022.
34. E.B. e.V. (EUBP), European bioplastics, https://www.european-bioplastics.org/global-bioplastics-production-will-more-than-triple-within-the-next-five-y 2022.
35. M. Hadidi, S. Jafarzadeh, M. Forough, F. Garavand, S. Alizadeh, A. Salehabadi, A.M. Khaneghah, and S.M. Jafari, Plant protein-based food packaging films; recent advances in fabrication, characterization, and applications, *Trends in Food Science & Technology*, 2022.
36. S. Jafarzadeh, M. Forough, S. Amjadi, V. Javan Kouzegaran, H. Almasi, F. Garavand, and M. Zargar, Plant protein-based nanocomposite films: A review on the used nanomaterials, characteristics, and food packaging applications, *Critical Reviews in Food Science and Nutrition*, pp. 1–27, 2022.
37. B. Cuq, N. Gontard, and S. Guilbert, Proteins as agricultural polymers for packaging production, *Cereal Chemistry*, Vol. 75, p. 1, 1998.
38. R. Zhao, P. Torley, and P.J. Halley, Emerging biodegradable materials: Starch- and protein-based bio-nanocomposites, *Journal of Materials Science*, Vol. 43, p. 3058, 2008.
39. C.-J. Lee, D.A. Scheufele, and B.V. Lewenstein, Public attitudes toward emerging technologies: Examining the interactive effects of cognitions and affect on public attitudes toward nanotechnology, *Science Communication*, Vol. 27, p. 240, 2005.
40. A.H. Brandenburg, C.L. Weller, and R.F. Testin, Edible films and coatings from soy protein, *Journal of Food Science*, Vol. 58, p. 1086, 1993.
41. P. Hernández-Muñoz, A. Kanavouras, P.K.W. Ng, and R. Gavara, Development and characterization of biodegradable films made from wheat gluten protein fractions, *Journal of Agricultural and Food Chemistry*, Vol. 51, p. 7647, 2003.
42. K.S. Miller and J.M. Krochta, Oxygen and aroma barrier properties of edible films: A review, *Trends in Food Science & Technology*, Vol. 8, p. 228, 1997.
43. H.-G. Song, I. Choi, J.-S. Lee, Y. Chang, C.S. Yoon, and J. Han, Whey protein isolate coating material for high oxygen barrier properties: A scale-up study from laboratory to industrial scale and its application

to food packaging, *Food Packaging and Shelf Life*, Vol. 31, p. 100765, 2022.

44. J.H. Han, *Innovations in Food Packaging*, Elsevier, 2005.
45. J.J. Zhou, S.Y. Wang, and S. Gunasekaran, Preparation and characterization of whey protein film incorporated with TiO_2 nanoparticles, *Journal of Food Science*, Vol. 74, p. N50, 2009.
46. L. Shi, J. Zhou, and S. Gunasekaran, Low temperature fabrication of ZnO–whey protein isolate nanocomposite, *Materials Letters*, Vol. 62, p. 4383, 2008.
47. M. Schmid, K. Dallmann, E. Bugnicourt, D. Cordoni, F. Wild, A. Lazzeri, and K. Noller, Properties of whey-protein-coated films and laminates as novel recyclable food packaging materials with excellent barrier properties, *International Journal of Polymer Science*, Vol. 2012, 2012.
48. K. Khwaldia, C. Perez, S. Banon, S. Desobry, and J. Hardy, Milk proteins for edible films and coatings, *Critical Reviews in Food Science and Nutrition*, Vol. 44, p. 239, 2004.
49. C. Onwulata and P. Huth, *Whey Processing, Functionality and Health Benefits*, Vol. 82, John Wiley & Sons, 2009.
50. H. Zheng, F. Ai, P.R. Chang, J. Huang, and A. Dufresne, Structure and properties of starch nanocrystal-reinforced soy protein plastics, *Polymer Composites*, Vol. 30, p. 474, 2009.
51. J.-W. Rhim, J.-H. Lee, and H.-S. Kwak, Mechanical and water barrier properties of soy protein and clay mineral composite films, *Food Science and Biotechnology*, Vol. 14, p. 112, 2005.
52. P. Chen and L. Zhang, Interaction and properties of highly exfoliated soy protein/montmorillonite nanocomposites, *Biomacromolecules*, Vol. 7, p. 1700, 2006.
53. J. Yu, G. Cui, M. Wei, and J. Huang, Facile exfoliation of rectorite nanoplatelets in soy protein matrix and reinforced bionanocomposites thereof, *Journal of Applied Polymer Science*, Vol. 104, p. 3367, 2007.
54. A.L. Abelti, T.A. Teka, S.F. Forsido, M. Tamiru, G. Bultosa, A. Alkhtib, and E. Burton, Bio-based smart materials for fish product packaging: A review, *International Journal of Food Properties*, Vol. 25, p. 857, 2022.
55. S. Roy and J.-W. Rhim, Anthocyanin food colorant and its application in pH-responsive color change indicator films, *Critical Reviews in Food Science and Nutrition*, Vol. 61, p. 2297, 2021.
56. G. Rajeshkumar, S.A. Seshadri, R.R. Richelle, K.M. Mitha, and V. Abinaya, Production of biodegradable films and blends from proteins in S. Mavinkere Rangappa, J. Parameswaranpillai, S. Siengchin, and M. Ramesh, (Eds.), *Biodegradable Polymers, Blends and Composites*, Woodhead Publishing Series in Composites Science and Engineering, chapter 24, pp. 681–692. Woodhead Publishing, 2022.

57. A.R. Alias, M.K. Wan, and N.M. Sarbon, Emerging materials and technologies of multi-layer film for food packaging application: A review, *Food Control*, p. 108875, 2022.
58. N.M. Sarbon, F. Badii, and N.K. Howell, The effect of chicken skin gelatin and whey protein interactions on rheological and thermal properties, *Food Hydrocolloids*, Vol. 45, p. 83, 2015.
59. N.N. Azizun, W.M. Khairul, A.I. Daud, and N.M. Sarbon, Effect of pH on functional, gas sensing and antimicrobial properties of bio-nanocomposite gelatin film for food packaging application, *Journal of Food Science and Technology*, Vol. 58, p. 3338, 2021.
60. A.A. Ahmad and N.M. Sarbon, A comparative study: Physical, mechanical and antibacterial properties of bio-composite gelatin films as influenced by chitosan and zinc oxide nanoparticles incorporation, *Food Bioscience*, Vol. 43, p. 101250, 2021.
61. Z.A.N. Hanani, E. Beatty, Y.H. Roos, M.A. Morris, and J.P. Kerry, Manufacture and characterization of gelatin films derived from beef, pork and fish sources using twin screw extrusion, *Journal of Food Engineering*, Vol. 113, p. 606, 2012.
62. E.M.C. Alexandre, R.V. Lourenço, A.M.Q.B. Bittante, I.C.F. Moraes, and P.J. do Amaral Sobral, Gelatin-based films reinforced with montmorillonite and activated with nanoemulsion of ginger essential oil for food packaging applications, *Food Packaging and Shelf Life*, Vol. 10, p. 87, 2016.
63. I. Chentir, H. Kchaou, M. Hamdi, M. Jridi, S. Li, A. Doumandji, and M. Nasri, Biofunctional gelatin-based films incorporated with food grade phycocyanin extracted from the Saharian cyanobacterium Arthrospira sp., *Food Hydrocolloids*, Vol. 89, p. 715, 2019.
64. M. Ulfa, W. Trisunaryanti, I.I. Falah, and I. Kartini, Characterization of gelatines extracted from cow bone for carbon synthesis, *Journal of Applied Chemistry*, Vol. 8, p. 57, 2015.
65. N.S. Said and N.M. Sarbon, Response surface methodology (RSM) of chicken skin gelatin based composite films with rice starch and curcumin incorporation, *Polymer Testing*, Vol. 81, p. 106161, 2020.
66. M.A. Sani, M. Tavassoli, S.A. Salim, M. Azizi-lalabadi, and D.J. McClements, Development of green halochromic smart and active packaging materials: TiO_2 nanoparticle- and anthocyanin-loaded gelatin/κ-carrageenan films, *Food Hydrocolloids*, Vol. 124, p. 107324, 2022.
67. J. Liu, H. Yong, Y. Liu, Y. Qin, J. Kan, and J. Liu, Preparation and characterization of active and intelligent films based on fish gelatin and haskap berries (Lonicera caerulea L.) extract, *Food Packaging and Shelf Life*, Vol. 22, p. 100417, 2019.
68. Y. Ge, Y. Li, Y. Bai, C. Yuan, C. Wu, and Y. Hu, Intelligent gelatin/oxidized chitin nanocrystals nanocomposite films containing black

rice bran anthocyanins for fish freshness monitorings, *International Journal of Biological Macromolecules*, Vol. 155, p. 1296, 2020.
69. W. Jiang, S. Hu, S. Li, and Y. Liu, Evaluation of the preservation effect of gelatin-water soluble chitosan film incorporated with maillard peptides on bluefin tuna (Thunnus thynnus) slices packaging, *LWT - Food Science and Technology*, Vol. 113, p. 108294, 2019.
70. D. Lin, Y. Yang, J. Wang, W. Yan, Z. Wu, H. Chen, Q. Zhang, D. Wu, W. Qin, and Z. Tu, Preparation and characterization of TiO_2-Ag loaded fish gelatin-chitosan antibacterial composite film for food packaging, *International Journal of Biological Macromolecules*, Vol. 154, p. 123, 2020.
71. F. Liu, Y. Liu, Z. Sun, D. Wang, H. Wu, L. Du, and D. Wang, Preparation and antibacterial properties of ε-polylysine-containing gelatin/chitosan nanofiber films, *International Journal of Biological Macromolecules*, Vol. 164, p. 3376, 2020.
72. A. Etxabide, J.I. Maté, and P.A. Kilmartin, Effect of curcumin, betanin and anthocyanin containing colourants addition on gelatin films properties for intelligent films development, *Food Hydrocolloids*, Vol. 115, p. 106593, 2021.
73. L.Y. Maroufi, M. Ghorbani, M. Tabibiazar, M. Mohammadi, and A. Pezeshki, Advanced properties of gelatin film by incorporating modified kappa-carrageenan and zein nanoparticles for active food packaging, *International Journal of Biological Macromolecules*, Vol. 183, p. 753, 2021.
74. M. Li, H. Yu, Y. Xie, Y. Guo, Y. Cheng, H. Qian, and W. Yao, Fabrication of eugenol loaded gelatin nanofibers by electrospinning technique as active packaging material, *LWT - Food Science and Technology*, Vol. 139, p. 110800, 2021.
75. S. Amjadi, H. Almasi, A. Ghadertaj, and L. Mehryar, Whey protein isolate-based films incorporated with nanoemulsions of orange peel (Citrus sinensis) essential oil: Preparation and characterization, *Journal of Food Processing and Preservation*, Vol. 45, p. e15196, 2021.
76. L. Yavari Maroufi, M. Ghorbani, and M. Tabibiazar, A gelatin-based film reinforced by covalent interaction with oxidized guar gum containing green tea extract as an active food packaging system, *Food and Bioprocess Technology*, Vol. 13, p. 1633, 2020.
77. N. Thongsrikhem, S. Taokaew, M. Sriariyanun, and S. Kirdponpattara, Antibacterial activity in gelatin-bacterial cellulose composite film by thermally crosslinking with cinnamaldehyde towards food packaging application, *Food Packaging and Shelf Life*, Vol. 31, p. 100766, 2022.
78. M. Suresh Kumar, S.N. Mudliar, K.M.K. Reddy, and T. Chakrabarti, Production of biodegradable plastics from activated sludge gener-

ated from a food processing industrial wastewater treatment plant, *Bioresource Technology*, Vol. 95, p. 327, 2004.

79. S. Guilbert, N. Gontard, and L.G.M. Gorris, Prolongation of the shelf-life of perishable food products using biodegradable films and coatings, *LWT-Food Science and Technology*, Vol. 29, p. 10, 1996.
80. F. Debeaufort, J.-A. Quezada-Gallo, and A. Voilley, Edible films and coatings: Tomorrow's packagings: A review, *Critical Reviews in Food Science*, Vol. 38, p. 299, 1998.
81. M.G.A. Vieira, M.A. da Silva, L.O. dos Santos, and M.M. Beppu, Natural-based plasticizers and biopolymer films: A review, *European Polymer Journal*, Vol. 47, p. 254, 2011.
82. M. Wihodo and C.I. Moraru, Physical and chemical methods used to enhance the structure and mechanical properties of protein films: A review, *Journal of Food Engineering*, Vol. 114, p. 292, 2013.
83. H. Cheng, L. Chen, D.J. McClements, T. Yang, Z. Zhang, F. Ren, M. Miao, Y. Tian, and Z. Jin, Starch-based biodegradable packaging materials: A review of their preparation, characterization and diverse applications in the food industry, *Trends in Food Science & Technology*, Vol. 114, p. 70, 2021.
84. L. do Val Siqueira, C.I.L.F. Arias, B.C. Maniglia, and C.C. Tadini, Starch-based biodegradable plastics: Methods of production, challenges and future perspectives, *Current Opinion in Food Science*, Vol. 38, p. 122, 2021.
85. M.I. Din, T. Ghaffar, J. Najeeb, Z. Hussain, R. Khalid, and H. Zahid, Potential perspectives of biodegradable plastics for food packaging application–review of properties and recent developments, *Food Additives & Contaminants: Part A*, Vol. 37, p. 665, 2020.
86. M.P. Guaras, V.A. Alvarez, and L.N. Ludueña, Processing and characterization of thermoplastic starch/polycaprolactone/compatibilizer ternary blends for packaging applications, *Journal of Polymer Research*, Vol. 22, p. 1, 2015.
87. C. Medina-Jaramillo, O. Ochoa-Yepes, C. Bernal, and L. Famá, Active and smart biodegradable packaging based on starch and natural extracts, *Carbohydrate Polymers*, Vol. 176, p. 187, 2017.
88. C.I.A. La Fuente, L. do Val Siqueira, P.E.D. Augusto, and C.C. Tadini, Casting and extrusion processes to produce bio-based plastics using cassava starch modified by the dry heat treatment (DHT), *Innovative Food Science & Emerging Technologies*, Vol. 75, p. 102906, 2022.
89. A.A. Singh and M.E. Genovese, Green and sustainable packaging materials using thermoplastic starch in A. Athanassiou, (Ed.), *Sustainable Food Packaging Technology*, chapter 5, pp. 133–160. Wiley Online Library, 2021.
90. D. Bajer and A. Burkowska-But, Innovative and environmentally safe composites based on starch modified with dialdehyde starch,

caffeine, or ascorbic acid for applications in the food packaging industry, *Food Chemistry*, Vol. 374, p. 131639, 2022.

91. D.R. Tapia-Blácido, G.J. Aguilar, M.T. de Andrade, M.F. Rodrigues-Júnior, and F.C. Guareschi-Martins, Trends and challenges of starch-based foams for use as food packaging and food container, *Trends in Food Science & Technology*, Vol. 119, p. 257, 2022.
92. S. Pachori, A. Sarkar, A. Dutta, J. Palanivelu, and R. Chidambaram, Synthesis methods of starch-based polymer foams and its comparison with conventional polymer foams for food packaging applications in T.J. Gutiérrez, (Ed.), *Polymers for Agri-Food Applications*, chapter 16, pp. 317–338. Springer International Publishing, Cham, 2019.
93. G. Abinader, C. Lacoste, M.L. Baillif, D. Erre, and A. Copinet, Effect of the formulation of starch-based foam cushions on the morphology and mechanical properties, *Journal of Cellular Plastics*, Vol. 51, p. 31, 2015.
94. N. Kaisangsri, O. Kerdchoechuen, and N. Laohakunjit, Biodegradable foam tray from cassava starch blended with natural fiber and chitosan, *Industrial Crops and Products*, Vol. 37, p. 542, 2012.
95. A. López-Gil, F. Silva-Bellucci, D. Velasco, M. Ardanuy, and M.A. Rodriguez-Perez, Cellular structure and mechanical properties of starch-based foamed blocks reinforced with natural fibers and produced by microwave heating, *Industrial Crops and Products*, Vol. 66, p. 194, 2015.
96. K.M. Romeira, G. Abdalla, R.P. Gonçalves, G.S. Pegorin, H.M.C. de Azeredo, C.U. Mussagy, and R.D. Herculano, Residual starch packaging derived from potato washing slurries to preserve fruits, *Food and Bioprocess Technology*, Vol. 14, p. 2248, 2021.
97. A.L. Charles, H.-M. Kao, and T.-C. Huang, Physical investigations of surface membrane–water relationship of intact and gelatinized wheat–starch systems, *Carbohydrate Research*, Vol. 338, p. 2403, 2003.
98. R.P. Singh, J.K. Pandey, D. Rutot, P. Degée, and P. Dubois, Biodegradation of poly (ε-caprolactone)/starch blends and composites in composting and culture environments: The effect of compatibilization on the inherent biodegradability of the host polymer, *Carbohydrate Research*, Vol. 338, p. 1759, 2003.
99. J. Huang, C. He, X. Liu, J. Xu, C.S.S. Tay, and S.Y. Chow, Organic–inorganic nanocomposites from cubic silsesquioxane epoxides: Direct characterization of interphase, and thermomechanical properties, *Polymer*, Vol. 46, p. 7018, 2005. Stimuli Responsive Polymers.

100. V.P. Cyras, L.B. Manfredi, M.-T. Ton-That, and A. Vázquez, Physical and mechanical properties of thermoplastic starch/montmorillonite nanocomposite films, *Carbohydrate Polymers*, Vol. 73, p. 55, 2008.
101. H.M.C. Azeredo, L.H.C. Mattoso, D. Wood, T.G. Williams, R.J. Avena-Bustillos, and T.H. McHugh, Nanocomposite edible films from mango puree reinforced with cellulose nanofibers, *Journal of Food Science*, Vol. 74, p. N31, 2009.
102. G.A. Burdock, Safety assessment of hydroxypropyl methylcellulose as a food ingredient, *Food and Chemical Toxicology*, Vol. 45, p. 2341, 2007.
103. M.R. De Moura, R.J. Avena-Bustillos, T.H. McHugh, J.M. Krochta, and L.H.C. Mattoso, Properties of novel hydroxypropyl methylcellulose films containing chitosan nanoparticles, *Journal of Food Science*, Vol. 73, p. N31, 2008.
104. Y. He, Y. Zhou, J. Cai, Y. Feng, B. Luo, and M. Liu, Facile fabrication of hydrophobic paper by HDTMS modified chitin nanocrystals coating for food packaging, *Food Hydrocolloids*, p. 107915, 2022.
105. M.S. Andrade, O.H. Ishikawa, R.S. Costa, M.V.S. Seixas, R.C.L.B. Rodrigues, and E.A.B. Moura, Development of sustainable food packaging material based on biodegradable polymer reinforced with cellulose nanocrystals, *Food Packaging and Shelf Life*, Vol. 31, p. 100807, 2022.
106. P. Thivya, S. Akalya, and V.R. Sinija, A comprehensive review on cellulose-based hydrogel and its potential application in the food industry, *Applied Food Research*, p. 100161, 2022.
107. J. Maitra and V.K. Shukla, Cross-linking in hydrogels — A review, *Am. J. Polym. Sci*, Vol. 4, p. 25, 2014.
108. N. Ranganathan, R.J. Bensingh, M.A. Kader, and S.K. Nayak, Synthesis and properties of hydrogels prepared by various polymerization reaction systems, *Online*, pp. 487–511, 2019.
109. S. Thiangtham, J. Runt, and H. Manuspiya, Sulfonation of dialdehyde cellulose extracted from sugarcane bagasse for synergistically enhanced water solubility, *Carbohydrate Polymers*, Vol. 208, p. 314, 2019.
110. A.A. Freitas, C.P. Silva, G.T.M. Silva, A.L. Macanita, and F.H. Quina, Ground- and excited-state acidity of analogs of red wine pyranoanthocyanins, *Photochemistry and Photobiology*, Vol. 94, p. 1086, 2018.
111. V. Gomes, A.S. Pires, N. Mateus, V. de Freitas, and L. Cruz, Pyranoflavylium-cellulose acetate films and the glycerol effect towards the development of pH-freshness smart label for food packaging, *Food Hydrocolloids*, Vol. 127, p. 107501, 2022.
112. F.-J. Wang, L.-Q. Wang, X.-C. Zhang, S.-F. Ma, and Z.-C. Zhao, Study on the barrier properties and antibacterial properties of cel-

lulose-based multilayer coated paperboard used for fast food packaging, *Food Bioscience*, Vol. 46, p. 101398, 2022.

113. F.-J. Wang, L.-Q. Wang, X.-C. Zhang, S.-F. Ma, and Z.-C. Zhao, Enhancement of oil resistance of cellulose packaging paper for food application by coating with materials derived from natural polymers, *Journal of Food Engineering*, p. 111039, 2022.
114. H. Shi, L. Wu, Y. Luo, F. Yu, and H. Li, A facile method to prepare cellulose fiber-based food packaging papers with improved mechanical strength, enhanced barrier, and antibacterial properties, *Food Bioscience*, Vol. 48, p. 101729, 2022.
115. M.Z. Al-Gharrawi, J. Wang, and D.W. Bousfield, Improving water vapor barrier of cellulose based food packaging using double layer coatings and cellulose nanofibers, *Food Packaging and Shelf Life*, Vol. 33, p. 100895, 2022.
116. H. Zhu, Z. Han, J.-H. Cheng, and D.-W. Sun, Modification of cellulose from sugarcane (Saccharum officinarum) bagasse pulp by cold plasma: Dissolution, structure and surface chemistry analysis, *Food Chemistry*, Vol. 374, p. 131675, 2022.
117. Y. Xie, Y. Pan, and P. Cai, Cellulose-based antimicrobial films incroporated with ZnO nanopillars on surface as biodegradable and antimicrobial packaging, *Food Chemistry*, Vol. 368, p. 130784, 2022.
118. B. Qu and Y. Luo, Preparation and characterization of carboxymethyl cellulose capped zinc oxide nanoparticles: A proof-of-concept study, *Food Chemistry*, Vol. 389, p. 133001, 2022.
119. S. LakshmiBalasubramaniam, C. Howell, M. Tajvidi, and D. Skonberg, Characterization of novel cellulose nanofibril and phenolic acid-based active and hydrophobic packaging films, *Food Chemistry*, Vol. 374, p. 131773, 2022.
120. J. Jacob, U. Lawal, S. Thomas, and R.B. Valapa, Biobased polymer composite from poly(lactic acid): Processing, fabrication, and characterization for food packaging in Y. Zhang, (Ed.), *Processing and Development of Polysaccharide-Based Biopolymers for Packaging Applications*, chapter 4, pp. 97–115. Elsevier, 2020.
121. J. Lunt, Large-scale production, properties and commercial applications of polylactic acid polymers, *Polymer Degradation and Stability*, Vol. 59, p. 145, 1998.
122. T.J. Madera-Santana, R. Meléndrez, G. González-García, P. Quintana-Owen, and S.D. Pillai, Effect of gamma irradiation on physicochemical properties of commercial poly (lactic acid) clamshell for food packaging, *Radiation Physics and Chemistry*, Vol. 123, p. 6, 2016.
123. R.G. Sinclair, The case for polylactic acid as a commodity packaging plastic, *Journal of Macromolecular Science, Part A*, Vol. 33, p. 585, 1996.

124. C. Thellen, C. Orroth, D. Froio, D. Ziegler, J. Lucciarini, R. Farrell, N.A. D'Souza, and J.A. Ratto, Influence of montmorillonite layered silicate on plasticized poly(*l*-lactide) blown films, *Polymer*, Vol. 46, p. 11716, 2005.
125. J.-W. Rhim, J.-H. Lee, and S.-I. Hong, Increase in water resistance of paperboard by coating with poly(lactide), *Packaging Technology and Science*, Vol. 20, p. 393, 2007.
126. L. Cabedo, J. Luis Feijoo, M. Pilar Villanueva, J.M. Lagarón, and E. Giménez, Optimization of biodegradable nanocomposites based on PLA/PCL blends for food packaging applications, *Macromolecular Symposia*, Vol. 233, p. 191, 2006.
127. E.P. Winters and D.L. Deardorff, Zein as a film-type coating for medicinal tablets, *Journal of the American Pharmaceutical Association*, Vol. 47, p. 608, 1958.
128. J.W. Lawton, Zein: A history of processing and use, *Cereal Chemistry*, Vol. 79, p. 1, 2002.
129. R. Shukla and M. Cheryan, Zein: The industrial protein from corn, *Industrial Crops and Products*, Vol. 13, p. 171, 2001.
130. X. Liu, Q. Sun, H. Wang, L. Zhang, and J.-Y. Wang, Microspheres of corn protein, zein, for an ivermectin drug delivery system, *Biomaterials*, Vol. 26, p. 109, 2005.
131. N. Jakel, Method and system for producing a zein protein product from a whole stillage byproduct produced in a corn dry-milling process, US Patent 10 875 889, assigned to Fluid Quip Technologies LLC, December 29, 2020.
132. S.S. Fernandes, V.P. Romani, G. da Silva Filipini, and V. G. Martins, Chia seeds to develop new biodegradable polymers for food packaging: Properties and biodegradability, *Polymer Engineering & Science*, Vol. 60, p. 2214, 2020.
133. N. Muñoz-Tebar, A. Molina, M. Carmona, and M.I. Berruga, Use of chia by-products obtained from the extraction of seeds oil for the development of new biodegradable films for the agri-food industry, *Foods*, Vol. 10, p. 620, 2021.
134. A.C. Mendes and G.A. Pedersen, Perspectives on sustainable food packaging:–is bio-based plastics a solution?, *Trends in Food Science & Technology*, Vol. 112, p. 839, 2021.
135. J. Rydz, M. Musiol, B. Zawidlak-Wegrzynska, and W. Sikorska, Present and future of biodegradable polymers for food packaging applications in A.M. Grumezescu and A.M. Holban, (Eds.), *Biopolymers for Food Design*, Handbook of Food Bioengineering, chapter 14, pp. 431–467. Academic Press, 2018.
136. T. Iwata, Biodegradable and bio-based polymers: Future prospects of eco-friendly plastics, *Angewandte Chemie International Edition*, Vol. 54, p. 3210, 2015.

137. R. Singla and R. Mehta, Preparation and characterization of polylactic acid-based biodegradable blends processed under microwave radiation, *Polymer-Plastics Technology and Engineering*, Vol. 51, p. 1014, 2012.
138. Z. Liu, Y. Lei, Z. Hu, W. Kong, C. Zhou, and J. Lei, Preparation, characterization and properties of poly(lactic acid)/poly (1,4-butylene adipate) blends for biodegradable packaging materials, *Macromolecular Research*, Vol. 25, p. 439, 2017.
139. M. Koller, Linking food industry to green plastics–polyhydroxyalkanoate (PHA) biopolyesters from agro-industrial by-products for securing food safety, *Applied Food Biotechnology*, Vol. 6, p. 1, 2018.
140. M.J. Fabra, A. López-Rubio, and J.M. Lagaron, On the use of different hydrocolloids as electrospun adhesive interlayers to enhance the barrier properties of polyhydroxyalkanoates of interest in fully renewable food packaging concepts, *Food Hydrocolloids*, Vol. 39, p. 77, 2014.
141. M.J. Fabra, P. Pardo, M. Martínez-Sanz, A. Lopez-Rubio, and J.M. Lagarón, Combining polyhydroxyalkanoates with nanokeratin to develop novel biopackaging structures, *Journal of Applied Polymer Science*, Vol. 133, 2016.
142. C.S. de Andrade, G.G. Fonseca, L.H. Innocentini Mei, and F.M. Fakhouri, Development and characterization of multilayer films based on polyhydroxyalkanoates and hydrocolloids, *Journal of Applied Polymer Science*, Vol. 134, 2017.
143. M. Seggiani, P. Cinelli, N. Mallegni, E. Balestri, M. Puccini, S. Vitolo, C. Lardicci, and A. Lazzeri, New bio-composites based on polyhydroxyalkanoates and Posidonia oceanica fibres for applications in a marine environment, *Materials*, Vol. 10, p. 326, 2017.
144. L.C. Lins, V. Bugatti, S. Livi, and G. Gorrasi, Ionic liquid as surfactant agent of hydrotalcite: Influence on the final properties of polycaprolactone matrix, *Polymers*, Vol. 10, p. 44, 2018.
145. A.M. Díez-Pascual and A.L. Díez-Vicente, Poly (3-hydroxybutyrate)/ZnO bionanocomposites with improved mechanical, barrier and antibacterial properties, *International Journal of Molecular Sciences*, Vol. 15, p. 10950, 2014.
146. M.P. Arrieta, M.d.M. Castro-Lopez, E. Rayón, L.F. Barral-Losada, J.M. López-Vilarino, J. López, and M.V. González-Rodríguez, Plasticized poly(lactic acid)–poly (hydroxybutyrate)(PLA–PHB) blends incorporated with catechin intended for active food-packaging applications, *Journal of Agricultural and Food Chemistry*, Vol. 62, p. 10170, 2014.
147. C. Swapna Joseph, K.V. Harish Prashanth, N.K. Rastogi, A.R. Indiramma, S. Yella Reddy, and K.S.M.S. Raghavarao, Optimum blend of chitosan and poly-(ε-caprolactone) for fabrication of films

for food packaging applications, *Food and Bioprocess Technology*, Vol. 4, p. 1179, 2011.

148. T. Børresen, Microplastics–a new threat to aquatic food safety?, *Journal of Aquatic Food Product Technology*, Vol. 24, p. 415, 2015.
149. K. Zhang, A.H. Hamidian, A. Tubić, Y. Zhang, J.K. Fang, C. Wu, and P.K. Lam, Understanding plastic degradation and microplastic formation in the environment: A review, *Environmental Pollution*, Vol. 274, p. 116554, 2021.
150. J.R. Jambeck, R. Geyer, C. Wilcox, T.R. Siegler, M. Perryman, A. Andrady, R. Narayan, and K.L. Law, Plastic waste inputs from land into the ocean, *Science*, Vol. 347, p. 768, 2015.
151. R. Geyer, J.R. Jambeck, and K.L. Law, Production, use, and fate of all plastics ever made, *Science Advances*, Vol. 3, p. e1700782, 2017.
152. C. Hogstrand, Presence of microplastics and nanoplastics in food, with particular focus on seafood, *EFSA Journal*, Vol. 14, p. e04501, 2016.
153. L. Van Cauwenberghe and C.R. Janssen, Microplastics in bivalves cultured for human consumption, *Environmental Pollution*, Vol. 193, p. 65, 2014.
154. A.A. Koelmans, E. Besseling, and W.J. Shim, Nanoplastics in the aquatic environment. Critical review in M. Bergmann, M. Klages, and L. Gutow, (Eds.), *Marine Anthropogenic Litter*, chapter 12, pp. 325–340. Springer International Publishing, 2015.
155. D. Santillo, K. Miller, and P. Johnston, Microplastics as contaminants in commercially important seafood species, *Integrated Environmental Assessment and Management*, Vol. 13, p. 516, 2017.
156. J. Li, C. Green, A. Reynolds, H. Shi, and J.M. Rotchell, Microplastics in mussels sampled from coastal waters and supermarkets in the United Kingdom, *Environmental Pollution*, Vol. 241, p. 35, 2018.
157. S. Rainieri and A. Barranco, Microplastics, a food safety issue?, *Trends in Food Science & Technology*, Vol. 84, p. 55, 2019.
158. D.A. Horn, E.F. Granek, and C.L. Steele, Effects of environmentally relevant concentrations of microplastic fibers on Pacific mole crab (Emerita analoga) mortality and reproduction, *Limnology and Oceanography Letters*, Vol. 5, p. 74, 2020.
159. H. Tong, Q. Jiang, X. Hu, and X. Zhong, Occurrence and identification of microplastics in tap water from China, *Chemosphere*, Vol. 252, p. 126493, 2020.
160. M. Al-Sid-Cheikh, S.J. Rowland, K. Stevenson, C. Rouleau, T.B. Henry, and R.C. Thompson, Uptake, whole-body distribution, and depuration of nanoplastics by the scallop Pecten maximus at environmentally realistic concentrations, *Environmental Science & Technology*, Vol. 52, p. 14480, 2018.

161. T. Özgür Sökmen, E. Sulukan, M. Türkoglu, A. Baran, M. Özkaraca, and S.B. Ceyhun, Polystyrene nanoplastics (20 nm) are able to bioaccumulate and cause oxidative DNA damages in the brain tissue of zebrafish embryo (Danio rerio), *NeuroToxicology*, Vol. 77, p. 51, 2020.
162. I. Ferreira, C. Venâncio, I. Lopes, and M. Oliveira, Nanoplastics and marine organisms: What has been studied?, *Environmental Toxicology and Pharmacology*, Vol. 67, p. 1, 2019.
163. E.S. Okeke, C.O. Okoye, E.O. Atakpa, R.E. Ita, R. Nyaruaba, C.L. Mgbechidinma, and O.D. Akan, Microplastics in agroecosystems-impacts on ecosystem functions and food chain, *Resources, Conservation and Recycling*, Vol. 177, p. 105961, 2022.
164. L. Nizzetto, G. Bussi, M.N. Futter, D. Butterfield, and P.G. Whitehead, A theoretical assessment of microplastic transport in river catchments and their retention by soils and river sediments, *Environmental Science: Processes & Impacts*, Vol. 18, p. 1050, 2016.
165. A.A. Horton, A. Walton, D.J. Spurgeon, E. Lahive, and C. Svendsen, Microplastics in freshwater and terrestrial environments: Evaluating the current understanding to identify the knowledge gaps and future research priorities, *Science of the Total Environment*, Vol. 586, p. 127, 2017.
166. M. Vithanage, S. Ramanayaka, S. Hasinthara, and A. Navaratne, Compost as a carrier for microplastics and plastic-bound toxic metals into agroecosystems, *Current Opinion in Environmental Science & Health*, Vol. 24, p. 100297, 2021.
167. A. Dey, C.V. Dhumal, P. Sengupta, A. Kumar, N.K. Pramanik, and T. Alam, Challenges and possible solutions to mitigate the problems of single-use plastics used for packaging food items: A review, *Journal of Food Science and Technology*, Vol. 58, p. 3251, 2021.
168. E.A. Baldwin, R. Hagenmaier, J. Bai, and J.M. Krochta, (Eds.), *Edible Coatings and Films to Improve Food Quality*, CRC Press, Boca Raton, 1994.
169. R. Priyadarshi, S.-M. Kim, and J.-W. Rhim, Pectin/pullulan blend films for food packaging: Effect of blending ratio, *Food Chemistry*, Vol. 347, p. 129022, 2021.
170. X. Hu, Y. Liu, D. Zhu, Y. Jin, H. Jin, and L. Sheng, Preparation and characterization of edible carboxymethyl cellulose films containing natural antibacterial agents: Lysozyme, *Food Chemistry*, Vol. 385, p. 132708, 2022.
171. S. Khajeh-Ali, F. Shahidi, and N. Sedaghat, Evaluation of the effect of carboxy methyl cellulose edible coating containing Astragalus honey (Astragalus gossypinus) on the shelf-life of pistachio kernel, *Food Control*, Vol. 139, p. 109094, 2022.

6

Additives

There is a monograph dealing with food additives (1). This book presents research on food phosphates, commercial starches, antibrowning agents, essential fatty acids, and fat substitutes, as well as studies on consumer perceptions of food additives.

Furthermore, *The Chemistry of Food Additives and Preservatives* is an up-to-date reference guide on the range of different types of additives, both natural additives and synthetic additives, that are used in the food industry (2). This book contains full details about the chemistry of each major class of food additives.

6.1 Chemical Additives

With the increased production of plastic materials, an increase in the amount of plastic material wastage has occurred, which is creating a number of challenges as well as opportunities in the waste management industry (3). There are various chemical additives in plastic products for enhancing the properties of the polymer and prolonging its life.

Commonly used additives are shown in Table 6.1 and some materials are shown in Figure 6.1.

Despite the usefulness of these additives, their potential to contaminate soil, air, water and food has been widely documented. These additives can potentially migrate and lead to undesirable human exposure via food contact materials such as packaging (3).

Table 6.1 Most commonly used additives in plastic materials (4).

Category and Substances	
Plasticizers	Abbreviation
Diisoheptyl phthalate	DIHP
Benzyl butyl phthalate	BBP
Bis(2-ethylhexyl)phthalate	DEHP
Bis(2-methoxyethyl) phthalate	DMEP
Dibutyl phthalate	DBP
Dipentyl phthalate	DPP
Di(2-ethylhexyl) adipate	DEHA
Di-octyl adipate	DOA
Diisobutyl phthalate	DiBP
Tris(2 chloroethyl)phosphate	TCEP
Dicyclohexyl phthalate	DCHP
Butyl benzyl phthalate	BBP
Diheptyl adipate	DHA
Heptyl adipate	HAD
Heptyl octyl adipate	HOA
Flame retardants	Abbreviation
Boric acid	
Decabromodiphenylethane	
Tetrabromobisphenol A	TBBPA
Tris(2-chloroethyl)phosphate	
Tris(2-chlorisopropyl)phosphate	TCPP
Hexabromocyclohexane	HBCCD
Stabilizers	Abbreviation
Bisphenol A	
Octylphenol	
1,3,5-Tris(oxiran-2-ylmethyl)-1,3,5-triazinane-2,4,6-trione	
Butylated hydroxytoluene	BHT
2-*tert*-Butyl-4-hydroxyanisole	
3-*tert*-Butyl-4-hydroxyanisole	
Tris-nonyl-phenyl phosphate	
Tris(2, 4-di-*tert*-butylphenyl) phosphite	
Tetrakismethylene-(3,5-di-*tert*-butyl-4-hydroxyhydrocinnamate) methane	

Table 6.1 (cont) Most commonly used additives in plastic materials (4).

Category and Substances	
Curing agents	
4,4′-Diaminodiphenylmethane 2,2′-Dichloro-4,4′-methylene dianiline 1,3,5-Tris(oxiran-2-ylmethyl)-1,3,5-triazinane-2,4,6-trione 1,3,5-Tris[(2S and 2R)-2,3-epoxypropyl]-1,3,5-triazine-2,4,6-(1H,3H,5H)-trione	 MOCA
Heat stabilizers	Heat stabilizers
Nonylphenol salts	
Slip agents	Slip agents
Erucamide Zinc stearate	Oleamide Waxes
Blowing agents	Blowing agents
Azodicarbonamide Benzene disulfonyl hydrazide	ADC Pentane
Biocides	Biocides
Triclosan 10,10′-Oxybisphenoarsine	Tributyltin
Pigments	
Cobalt(II) diacetate Zinc oxide	Zinc sulfide Iron oxide
Fillers	
Calcium carbonate Clay Glimmer Wood powder Barium sulfate	Talc Zinc oxide Metal powder Asbestos Glass microspheres

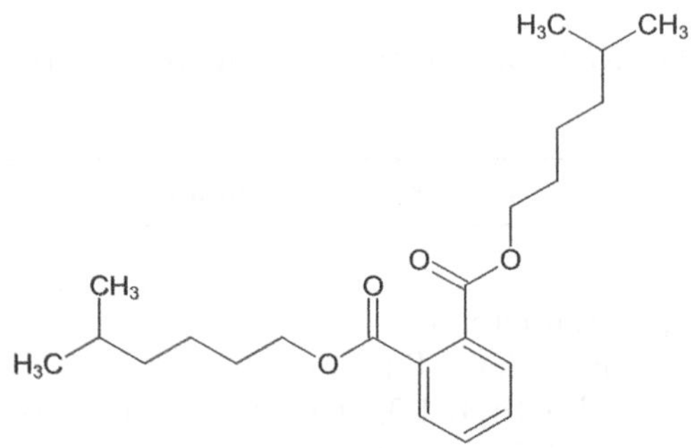

Diisoheptyl phthalate

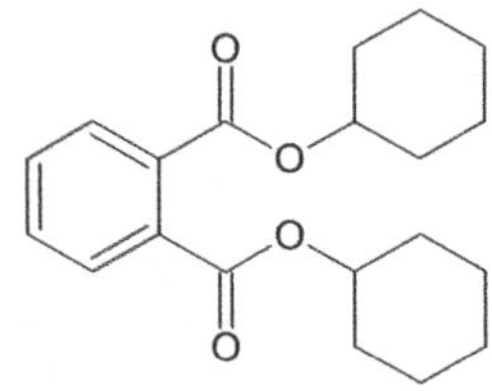

Dicyclohexyl phthalate

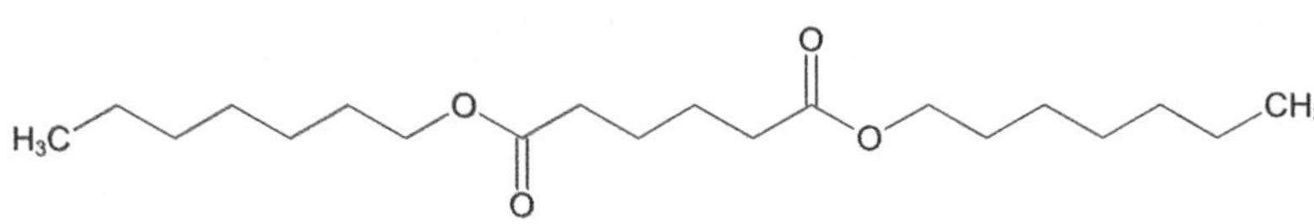

Diheptyl adipate

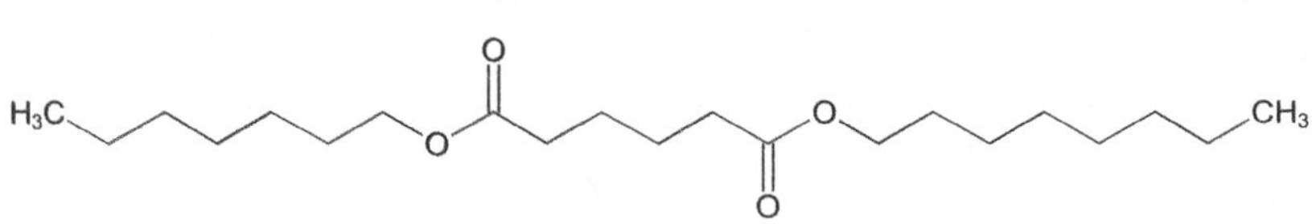

Heptyl octyl adipate

Figure 6.1 Plasticizers.

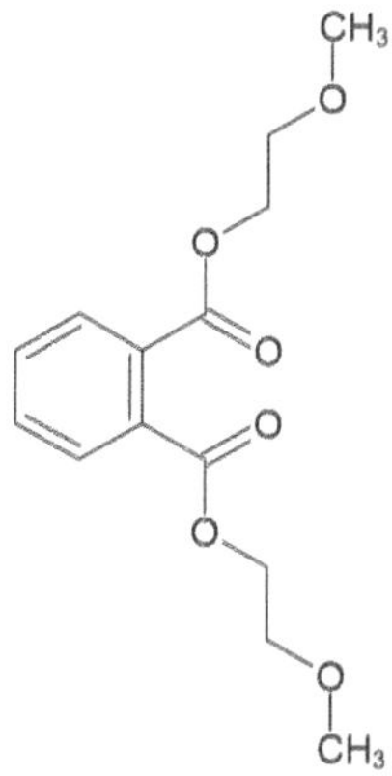

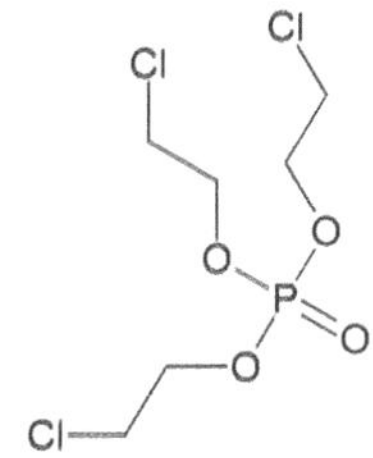

Bis(2-methoxyethyl) phthalate Tris(2-chloroethyl)phosphate

Figure 6.1 (cont) Plasticizers.

6.1.1 *Hazardous Metal Additives*

Many additives and catalysts that have been used in plastics were based on compounds of toxic metals and metalloids, like arsenic, cadmium, chromium(VI), and lead (5).

Despite current restrictions, hazardous additives remain in plastics because of the pervasiveness of many products and the more general contamination of recycled goods. Little is understood about their presence and impact on the environment.

A review of the uses of hazardous, metal-based additives in plastics has been given (5). Also, the relevant European regulations have been explained that have been introduced to restrict or prohibit usage in various sectors, and the likely environmental impacts of hazardous additives once plastics are lost in nature.

A widespread occurrence of hazardous metals in environmental plastics has been shown, with impacts ranging from contamination of the waste stream to an increase in the density and settling rates of material in aquatic systems (5).

Because metal additives are generally not chemically bound to polymers, they have the propensity to mobilize from the matrix via diffusion and mass transfer into the surrounding aqueous phase (6, 7).

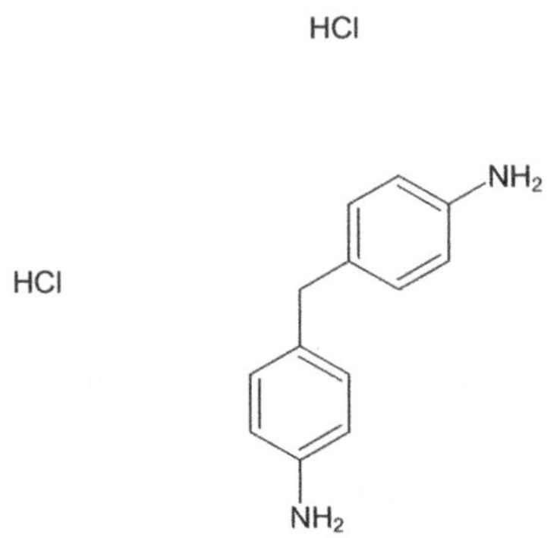

4,4′-Diaminodiphenylmethane

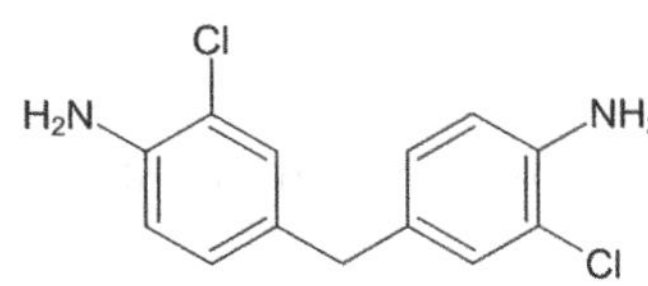

2,2′-Dichloro-4,4′-methylene dianiline

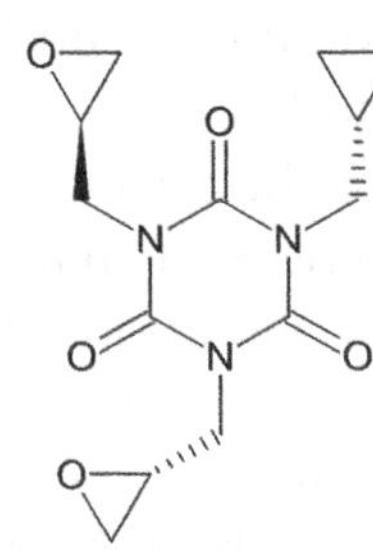

1,3,5-Tris[(2S and 2R)-2,3-epoxypropyl]-1,3,5-triazine-2,4,6-(1H,3H,5H)-trione

Figure 6.1 (cont) Curing agents.

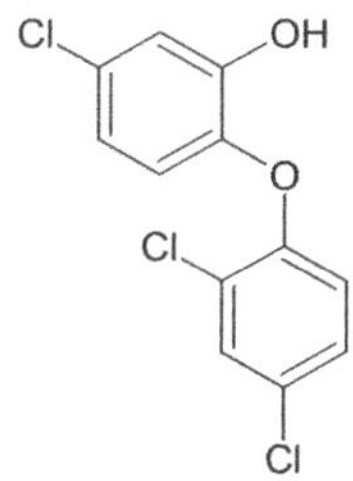

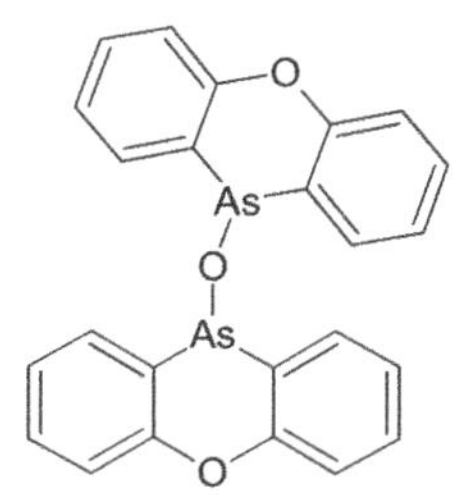

Triclosan 10,10′-Oxybisphenoarsine

Figure 6.1 (cont) Biocides.

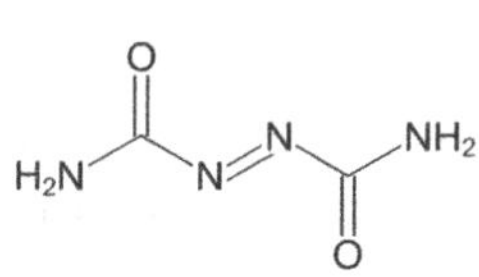

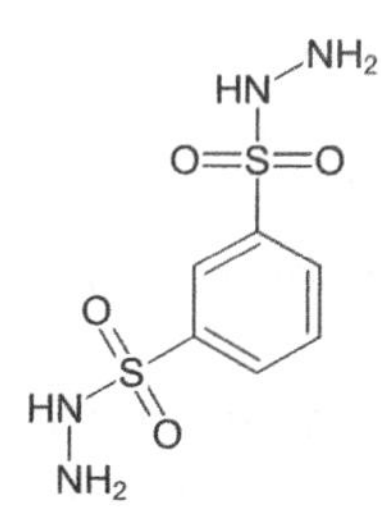

Azodicarbonamide Benzene disulfonyl hydrazide

Figure 6.1 (cont) Blowing agents.

6.2 Mathematical Models

The mathematical modeling of migration of additives from plastics to food simulants was studied using experimental data published in the literature, following two routes (8):

1. The conventional approach using the solution of Fick's second law, and
2. A kinetic model based on the Weibull distribution function.

The objective of this comparison was to permit using a mathematically simpler model equally able to describe migration data, and that could have a generally wider applicability by describing situations more complex than those that simple diffusional phenomena can describe.

The relationship between the parameters of the two models was analyzed by regression of data generated by Fick's law using the Weibull model.

The results showed that the time constant parameter is related to the diffusion coefficient and the material thickness. It depends on temperature and on the molecular weight of the migrant in a similar manner to the diffusion coefficient. The activation energy presented values from 72 to 125 $kJ\,mol^{-1}$. The shape parameter is a function of the contribution of the mass transfer resistance at the interface. It was independent of temperature and of the migrant, assuming a global constant value of 0.67, for the systems and conditions studied.

The results indicate that the Weibull model can be used to describe and analyze the migration of additives from plastics to foods with a meaning of the parameters in terms of the underlying physical phenomena (8).

The model can be applied in packaging systems with mass transfer processes more complex than those that simple diffusional phenomena can describe. In the migration of components from paper-based materials (instead of plastics) into solid foods, mass transfer occurs in a heterogeneous medium of paper fibers and air. Migration curves commonly present an initial lag phase for heavier migrants. The sigmoidal shape of the migration curves is not adequately described by the established models based on Fick's second law solutions, and the Weibull model is a good alternative. Additionally, this simple model can potentially be used in probabilistic approaches for exposure assessments of consumers to migrants originating from packaging materials (8).

The potential for the use of migration modeling for studying polyolefin packaging materials, such as low-density and high-density poly(ethylene) and poly(propylene), was summarized and demonstrated with practical examples (9).

For these polymers, an upper limit of migration into foodstuffs can be predicted with a high degree of statistical confidence. The only analytical information needed for modeling in such cases is the initial concentration of the migrant in the polymer matrix.

For polyolefins of unknown origin or newly developed materials with new properties, a quick experimental method has been described for obtaining the characteristic matrix parameter needed for the process of migration modeling.

For an easy handling of both the experimental results and the diffusion model, a user-friendly software has been developed. An additional aim of the described method is the determination of the migrant partition between polymer and food or food simulant and the specific contribution of the migrant molecular structure on the diffusion coefficient. For migration modeling of packaging materials with multilayer structures, a numerical solution of the diffusion equation is described. This procedure has also been applied for modeling the migration into solid or high viscous foodstuffs (9).

The permeation through and diffusion/migration from high-density poly(ethylene), poly(butylene), poly(propylene) and crosslinked poly(ethylene) films was investigated experimentally with three different methods to determine diffusion coefficients in these polyolefins for a series of additives, their degradation products and other organic substances in the 20°C to 60°C temperature range (10).

The experimental methods used were dynamic permeation through additive-free polymer films, kinetic desorption from additivated films into water and kinetic migration from additivated into additive-free polymer films. It was found that in general the temperature dependence of the obtained diffusion coefficients was well represented by the Arrhenius law.

Some of these results also suggested that the contact of the polyolefins with water had an influence on the magnitude of the diffusion coefficients and on their apparent activation energy of diffusion (10).

The migration of phthalate from poly(ethylene terephthalate) bottles containing non-alcoholic beer was done by performing an adaptive neuro-fuzzy inference system analysis (11).

The data showed that the storage temperature, contact surface and storage period correlates with the rate of migration. The migration of phthalate increases with storage duration gradually and reduces under different temperatures and contact surface. Moreover, increased temperature and storage duration resulted in an increase in migration level ranging from 0.6 $\mu g\, l^{-1}$ to 2.9 $\mu g\, l^{-1}$ (11).

The study used an adaptive neuro-fuzzy inference system analysis architecture, which consists of three inputs (temperature, surface and storage period), Gaussian bell membership functions for each input variable and one output layer, which represent the migration level. The validation and training models showed an excellent match between the experimental and predicted values of adaptive neuro-fuzzy inference system analysis.

The analysis of the model showed that adaptive neuro-fuzzy inference system analysis is a powerful tool for predicting phthalate migration from bottles containing non-alcoholic beer (11).

6.3 Evaluation of Models

The European Commission charged an international group of experts with demonstrating that migration modeling can be regarded as a valid and reliable tool to calculate reasonable worst-case migration rates from the most important food contact plastics into the European Union official food simulants (12).

The main steps to follow to build up and validate a migration estimation model that can be used for a series of plastic food contact materials and migrants for regulatory purposes are shown (12).

The additives used for migration measurements are shown in Table 6.2. Some of the compounds are shown in Figure 6.2

6.4 Plastic Additives by Swelling

The plasticizing effect of swelling and its effect of promoting the migration of polymer additives is widely known (13). It is usual to explain the relatively high diffusion coefficients obtained using

Table 6.2 Additives used for migration measurements (12).

Material
2-Hydroxy-4-*n*-octylbenzophenone
3,5-Di-*tert*-butyl-4-hydroxybenzoic acid, hexadecyl ester
2,2′,2″-Nitrilo(triethyl-tris(3,3′,5,5′-tetra-*tert*-butyl-1,1′-biphenyl- 2,2′-diyl)phosphite)
Phosphorous acid, tris(2,4-di-*tert*-butylphenyl)ester
Phosphorous acid, bis(2,4-di-*tert*-butyl-6-methylphenyl)ethyl ester
Pentaerythritol tetrakis(3-(3,5-di-*tert*-butyl-4-hydroxyphenyl)propionate)
Octadecyl 3-(3,5-di-*tert*-butyl-4-hydroxyphenyl)propionate
1,3,5-Trimethyl-2,4,6-tris(3,5-di-*tert*-butyl-4-hydroxybenzyl)benzene
Triethyleneglycol-bis(3-(3-*tert*-butyl-4-hydroxy-5-methylphenyl)propionate)
Acrylic acid, 2-*tert*-butyl-6-(3-*tert*-butyl-2-hydroxy-5-methylbenzyl)-4-methylphenyl ester
d,l-α-Tocopherol
N,*N*′-bis(3-(3,5-di-*tert*-butyl-4-hydroxyphenyl)propionyl)-hydrazide
1,1-Bis(2-hydroxy-3,5-di-*tert*-butylphenyl)ethane
Bis(2,6-di-*tert*-butyl-4-methylphenyl)pentaerithritol diphosphite
2-(2-Hydroxy-3,5-bis(1,1-dimethylbenzyl)phenyl)benzotriazole
2-(2′-Hydroxy-3′-*tert*-butyl-5′-methylphenyl)-5-chlorobenzotriazole
Bis(2,2,6,6-Tetramethyl-4-piperidyl) sebacate
Triisopropanolamine
1,1,3-Tris(2-methyl-4-hydroxy-5-*tert*-butylphenyl)butane
Bis(2,4-di-*tert*-butylphenyl)pentaerythritol diphosphite
2,4,6-Tris(*tert*-butyl)phenyl 2-butyl-2-ethyl-1,3-propanediol phosphite
2,5-Bis(5-*tert*-butyl-2-benzoazolyl)thiophene
Acid adipic, bis(2-ethylhexyl)ester
Amines, bis(hydrogenated tallow alkyl) oxidized (mixture)
2,2′-Ethylidene bis(4,6-di-*tert*butyl-phenyl)fluorophosphonite
Bis(4-ethylbenzylidene)sorbitol
4-Ethoxy ethylbenzoate
N,*N*-Bis(2-hydroxyethyl)lauramide
Diethyl phthalate
Benzyl butyl phthalate

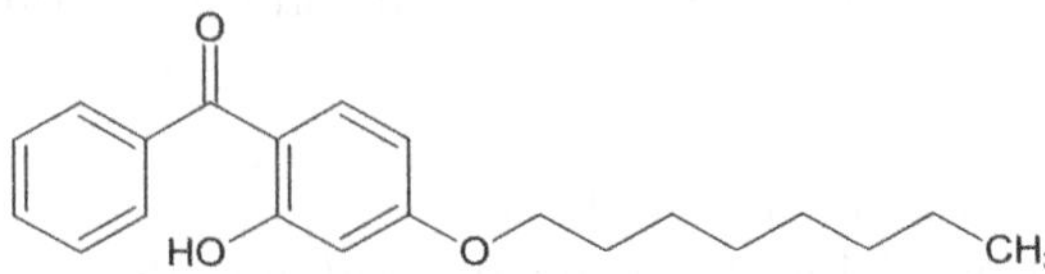

2-Hydroxy-4-*n*-octylbenzophenone

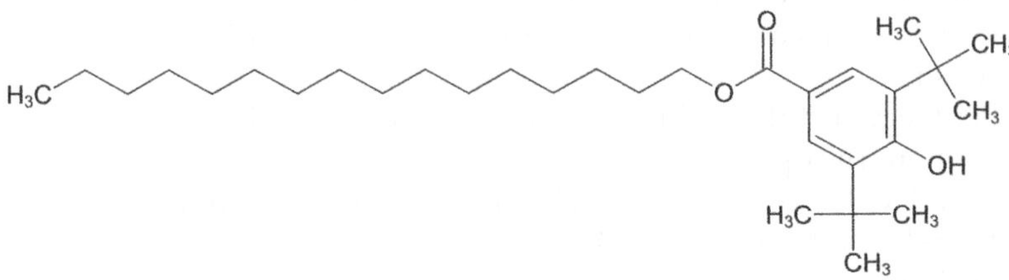

3,5-Di-*tert*-butyl-4-hydroxybenzoic acid, hexadecyl ester

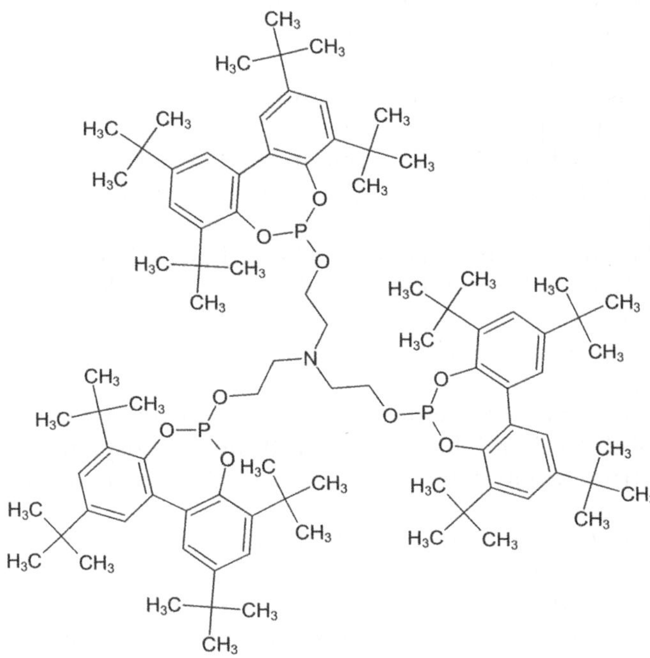

2,2′,2″-Nitrilo(triethyl-tris(3,3′,5,5′-tetra-*tert*-butyl-1,1′-biphenyl- 2,2′-diyl)phosphite)

Figure 6.2 Additives for migration measurements.

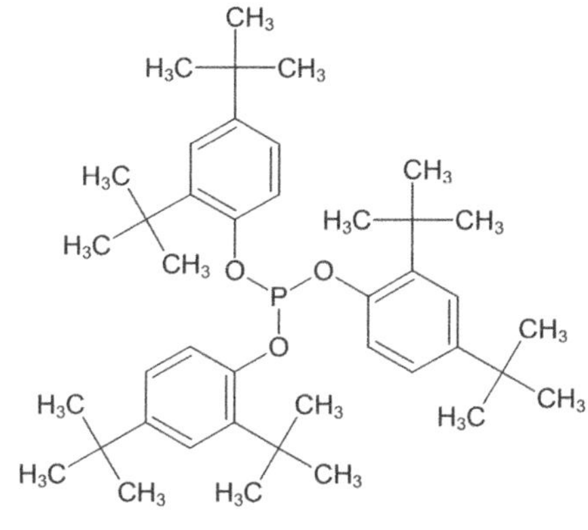

Phosphorous acid, tris(2,4-di-*tert*-butylphenyl)ester

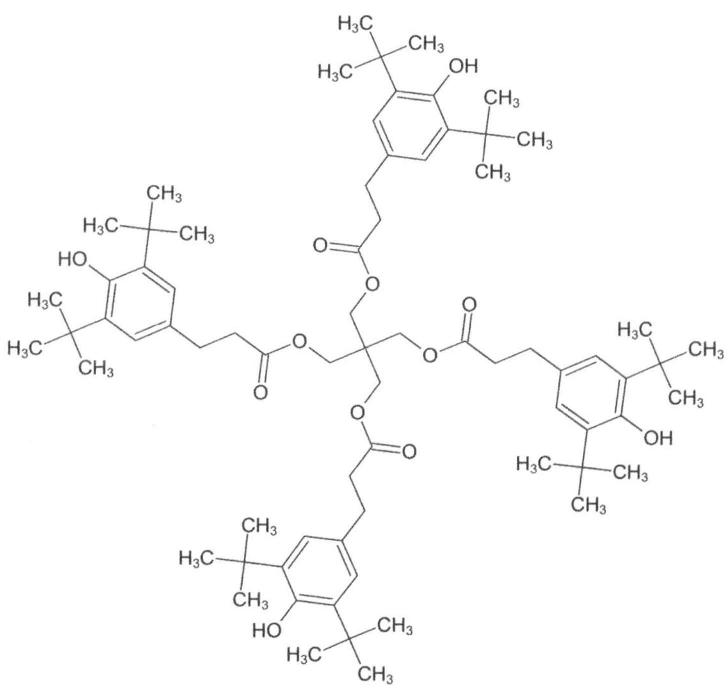

Pentaerythritol tetrakis(3-(3,5-di-*tert*-butyl-4-hydroxyphenyl)propionate)

Figure 6.2 (cont) Additives for migration measurements.

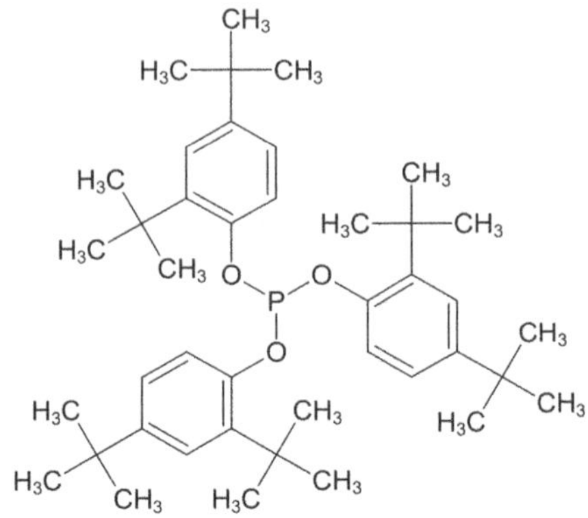

Phosphorous acid, tris(2,4-di-*tert*-butylphenyl)ester

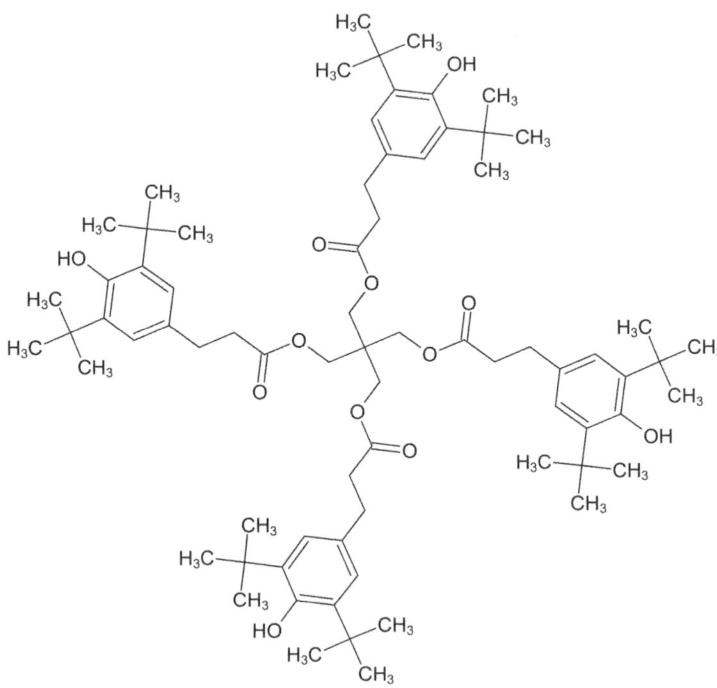

Pentaerythritol tetrakis(3-(3,5-di-
tert-butyl-4-hydroxyphenyl)propionate)

Figure 6.2 (cont) Additives for migration measurements.

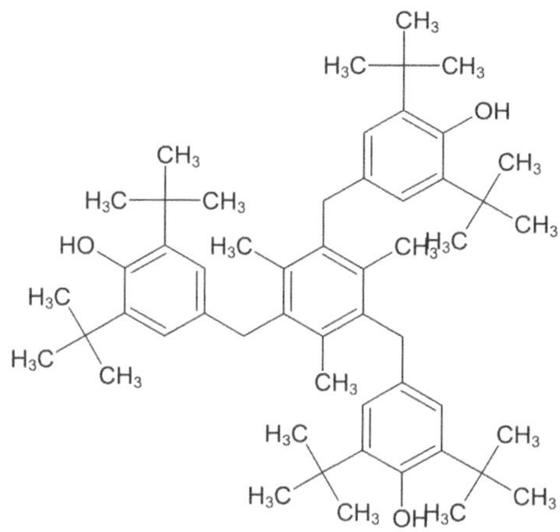

1,3,5-Trimethyl-2,4,6-tris(3,5-di-*tert*-butyl-4-hydroxybenzyl)benzene

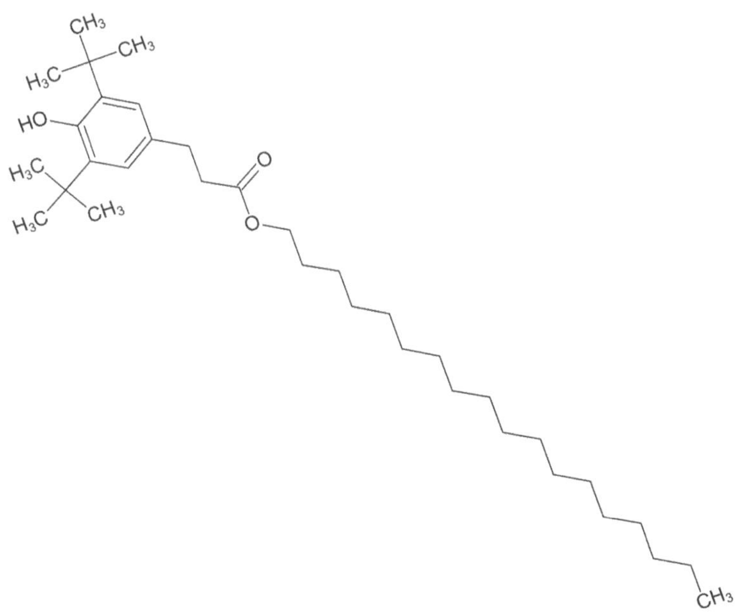

Octadecyl 3-(3,5-di-*tert*-butyl-4-hydroxyphenyl)propionate

Figure 6.2 (cont) Additives for migration measurements.

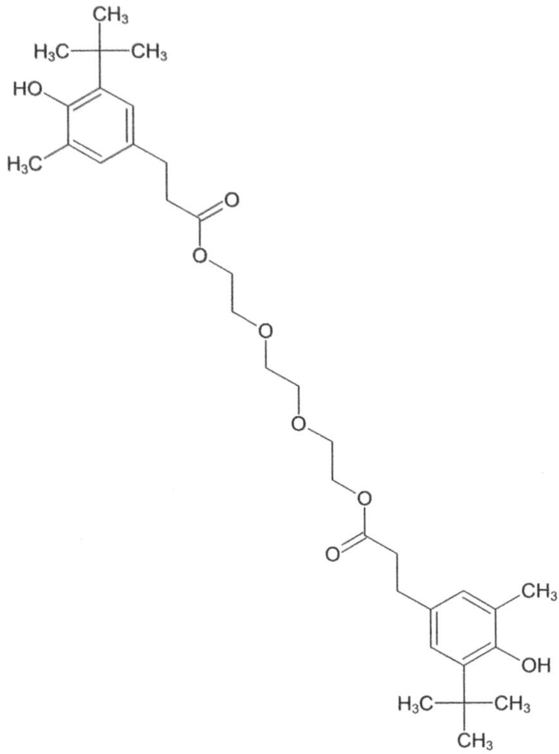

Triethyleneglycol-bis(3-(3-*tert*-butyl-4-hydroxy-5-methylphenyl)propionate)

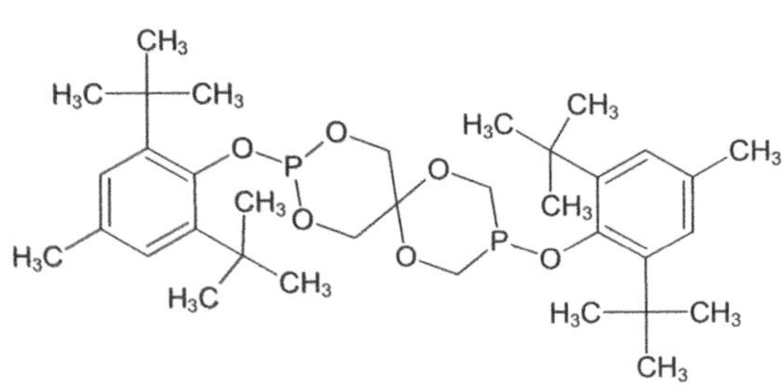

Bis(2,6-di-*tert*-butyl-4-methylphenyl)pentaerithritol diphosphite

Figure 6.2 (cont) Additives for migration measurements.

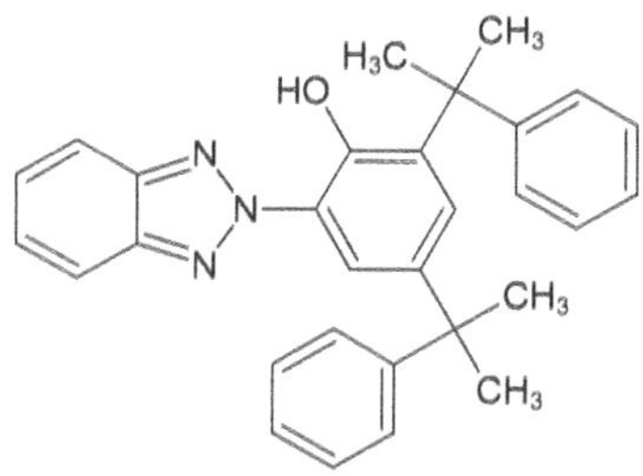

2-(2-Hydroxy-3,5-bis(1,1-dimethylbenzyl)phenyl)benzotriazole

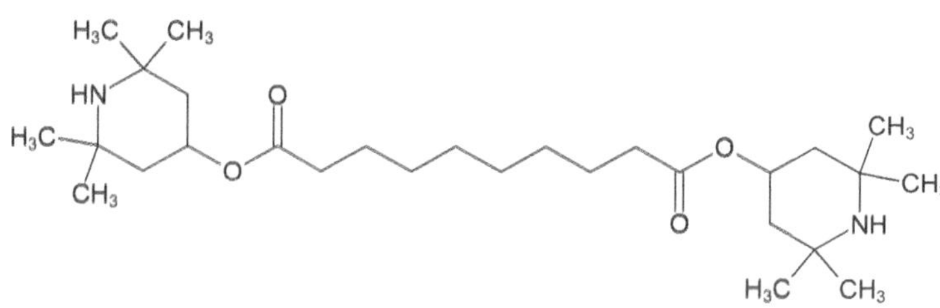

Bis(2,2,6,6-Tetramethyl-4-piperidyl) sebacate

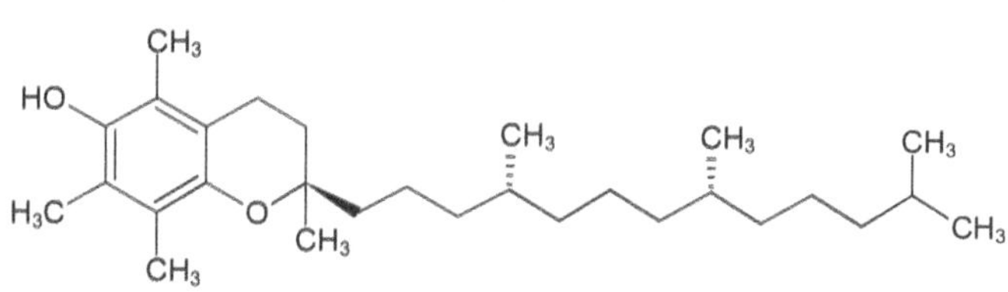

Tocopherol

Figure 6.2 (cont) Additives for migration measurements.

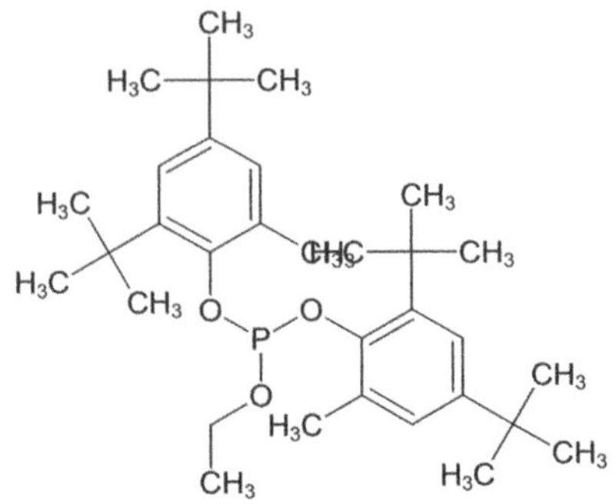

Phosphorous acid, tris(2,4-di-*tert*-butylphenyl)ester

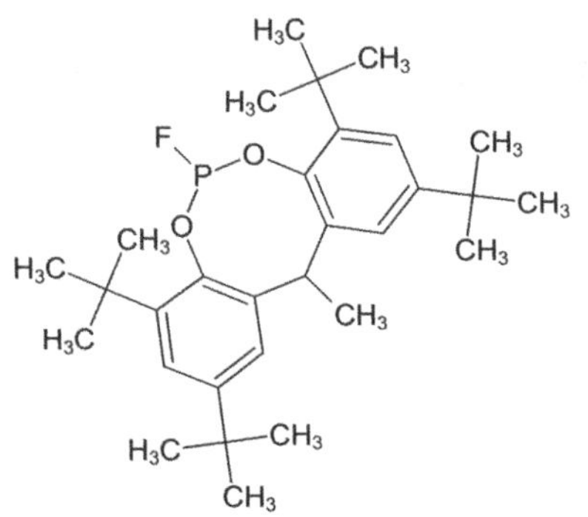

2,2′-Ethylidene bis(4,6-di-*tert*butyl-phenyl)fluorophosphonite

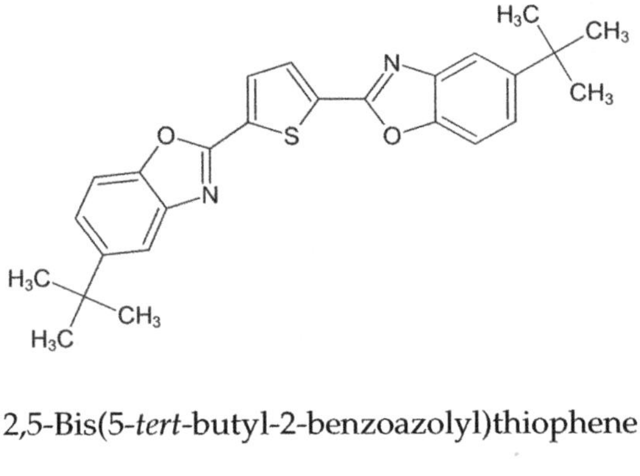

2,5-Bis(5-*tert*-butyl-2-benzoazolyl)thiophene

Figure 6.2 (cont) Additives for migration measurements.

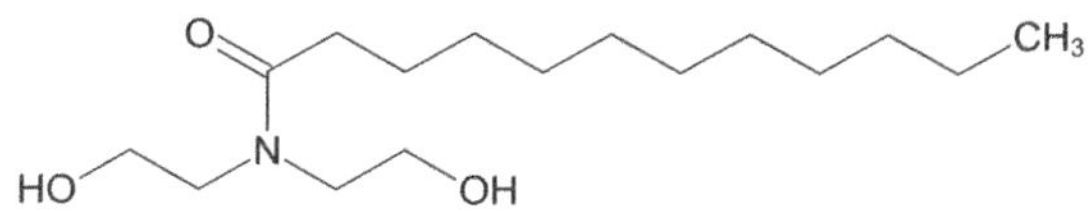

N,N-Bis(2-hydroxyethyl)lauramide

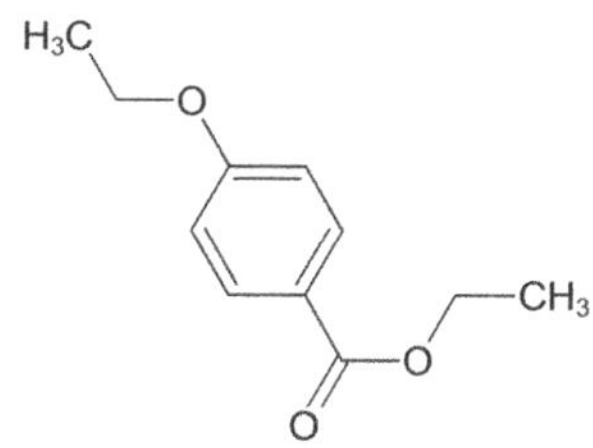

4-Ethoxy ethylbenzoate

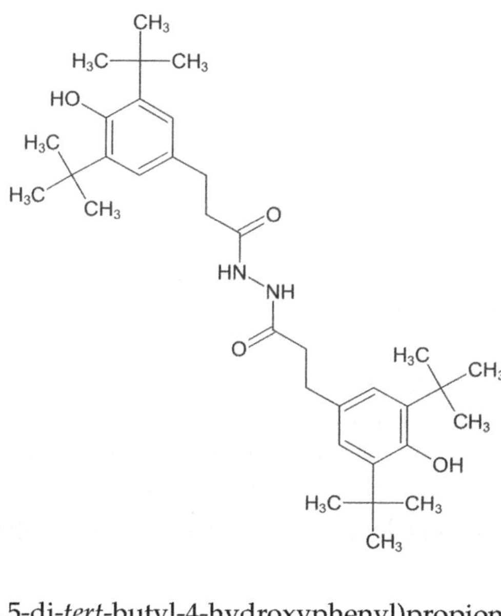

N,N'-Bis(3-(3,5-di-*tert*-butyl-4-hydroxyphenyl)propionyl)-hydrazide

Figure 6.2 (cont) Additives for migration measurements.

food simulants mimicking fatty food by the swelling of the polymer (14, 15).

The migration of various plastic additives, such as antioxidants, UV absorbers, and plasticizers poly(propylene) (PP) and poly(lactic acid) was investigated in a series of experiments conducted over a period of 13 *d*.

In the migration tests, isooctane and ethanol 95 v/v% were applied as food simulants of fatty food.

The tested materials are collected in Table 6.3, and some compounds are shown in Figure 6.3.

Table 6.3 Migration materials.

Antioxidants	
BHT	2,6-Di-*tert*-butyl-4-methylphenol
Ionox 220	4,4′-Methylene-bis(2,6-di-*tert*-butylphenol
Irgafos 168	(Tris(2,4-di-*tert*-butylphenyl) phosphite
UV Absorber	
Uvinul 3039	2-Ethylhexyl (Z)-3-cyano-2,3-diphenylprop-2-enoate
Plasticizers	
TBAC	Tributyl acetyl citrate
TOTM	Tris(2-ethylhexyl) trimellitate

As fatty food simulants, both ethanol 95 v/v% and isooctane were used. A non-Fickian behavior was observed on multiple occasions. The kinetic curves of both migration concentrations and swelling were evaluated using variography, to objectively determine the starting points of long-lasting plateaus as well as short halts in the increase.

A strong correlation between migration and swelling was observed; the kinetic curves showed that migration always followed swelling. Also, more intensive swelling results in the increased migration of the additives. Consequently, migration testing can be improved by considering the swelling of the plastic (13).

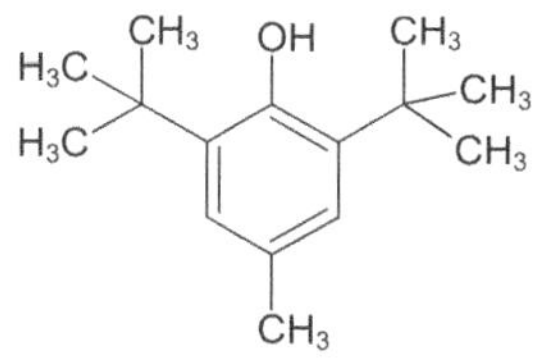

2,6-Di-*tert*-butyl-4-methylphenol

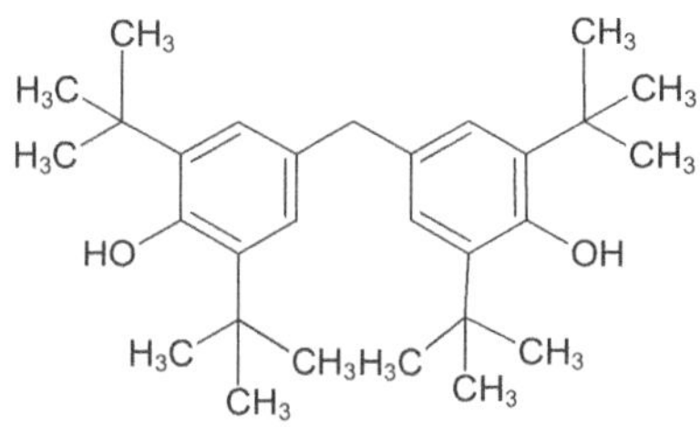

4,4′-Methylene-bis(2,6-di-*tert*-butylphenol

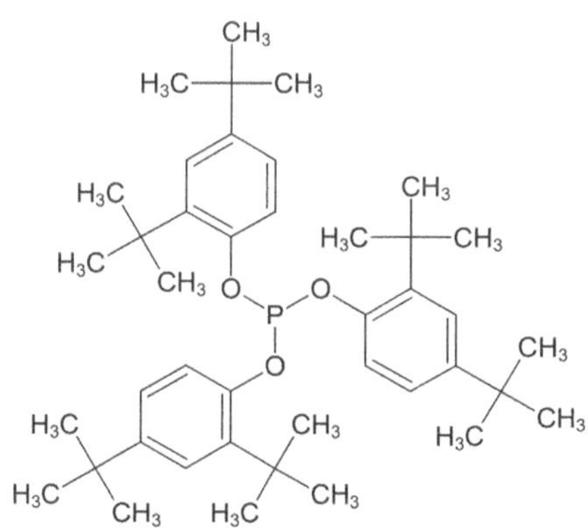

Tris(2,4-di-*tert*-butylphenyl) phosphite

Figure 6.3 Compounds for migration tests.

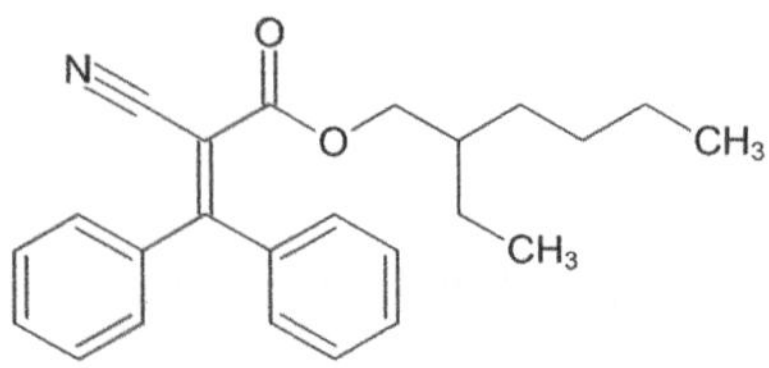

2-Ethylhexyl (Z)-3-cyano-2,3-diphenylprop-2-enoate

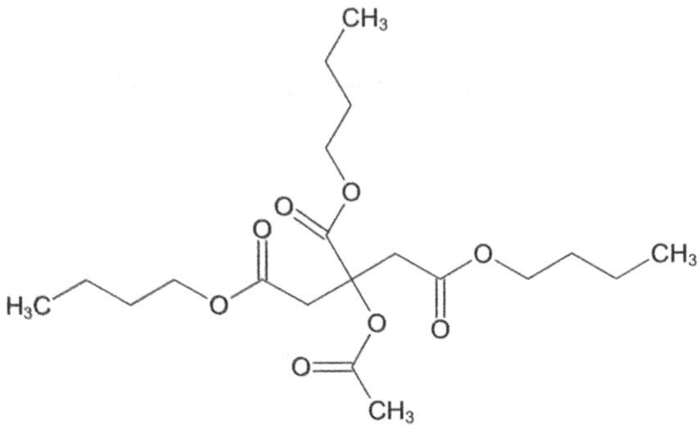

Tributyl acetyl citrate

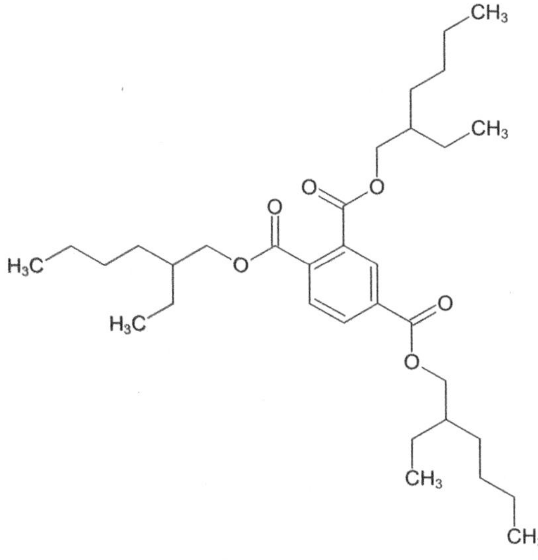

Tris(2-ethylhexyl) trimellitate

Figure 6.3 (cont) Compounds for migration tests.

6.5 Fatty Acid Amide

Specific migration studies using food simulants have been conducted on a range of polymers which are used in food packaging and contain commonly used fatty acid amide slip additives (16). Migration levels using the test conditions (10 *d* at 40°C) were found to depend primarily upon the food simulant and the polymer type, with the fat simulant and low-density poly(ethylene) (LDPE) giving the highest values in the range of 1.8 *mg* to 3.1 *mg* of fatty acid amide per kilogram of simulant.

Extremely low levels of migration, of less than 0.05 *mg* of fatty acid amide per kilogram of simulant, were found with aqueous food simulants for all plastics tested, and fat simulant with poly(vinyl chloride) (PVC) and poly(styrene) (PS) plastics (16).

6.5.1 *Plasticizers*

Plasticizers, with a comparatively low molecular weight of 300 $g\,mol^{-1}$ to 600 $g\,mol^{-1}$, could potentially migrate from packaging materials into food, thereby becoming indirect food additives.

The most commonly used plasticizers in PVC, poly(vinyl alcohol), and poly(ethylene) are phthalate- and adipate-based compounds. Their migration to food under various conditions has been documented in the literature (17–20).

In a study using olive oil, phthalates exhibited an extremely high transfer (migration) rate of 350% when used in gasket material for closures (17). This indicates that the transfer was not only stemmed from the gasket, but also from underneath the seal or rim.

Also, the migration of phthalates, alkylphenols, bisphenol A and di(2-ethylhexyl) adipate from food packaging was investigated (21).

The migration tests were carried out as suggested by the Directive 85/572/CEE (22) and 82/711/EEC (23).

6.5.2 *Antioxidants*

The quantification of migrated antioxidants and their degraded products from different polymers under various conditions was assessed in several studies (24–29).

The migration of antioxidants has been studied mostly in the cases of PP (27) and LDPE (29).

The migration of antioxidants from PP films of several thicknesses into *n*-heptane and 95% ethanol at 20°C, 37°C and 60°C was evaluated (25). It was found that the thicker the polymer in *n*-heptane, the higher the migration rates, compared to ethanol.

It was observed that continuous microwave (MW) heating for up to 1 *h* favors degradation of antioxidant in FSs when compared to conventional heating using an oil bath (26, 27). The experimental temperature was kept at 80°C for both MW and conventional heating. However, this 1 *h* of MW heating in contact with food is not relevant for industrial or domestic applications. It was observed that high temperature caused more swelling of PP in isooctane during MW heating and increased the diffusion coefficient by factors of 100–1000 in aqueous food (26).

6.5.3 *Monomers and Oligomers*

Several studies have reported the migration of styrene into food (30–32).

Here, the daily styrene exposure ws estimated at 18.2–55.2 μg for individuals, with an annual exposure of 6.7–20.2 mg. This level of exposure causes an irritation to the human organs and skin, as well as neurological disorders (33).

6.5.4 *Light Stabilizers*

Light stabilizers are used for the protection of plastics from sun and weather exposure. Polyolefins are susceptible to UV light, O_2, moisture, and heat, resulting in polymer brittleness, surface crazing, color change and product failure. Polyolefins usually contain hindered amine light stabilizers such as Tinuvin 622, Tinuvin 765 and Chimassorb 944 (34, 35).

6.5.5 *Slip Additives*

The most commonly used slip additives in plastics are fatty acid amides such as oleamides, stearyl erucamide, stearamide, erucamide and oleyl palmitamide. They usually act as lubricants, thus preventing films from sticking together (34).

The migration of 5 fatty acid amides was studied. Oleamide, erucamide, stearamide, stearyl amide, and oleyl palmitamide from 4 polymer materials (LDPE, PP, PS, and PVC) was determined at 40°C after 10 *d* of storage (16). The fatty acid amides are shown in Figure 6.4.

The highest migration occurred from LDPE to olive oil. Additive migrations from LDPE were: 88% for oleamide, 98% for erucamide, and 95% for stearamide, respectively. The highest slip additive migration was measured in 65 μm LDPE film, compared to PVC or PS films, i.e., less than 1% of the total compounds.

6.6 Perfluorinated Compounds

Perfluorinated compounds are used in food packaging materials as coatings/additives for oil and moisture resistance (36). In a study, foodstuff packaging materials collected from the Greek market, made of paper, paperboard or aluminum foil, were analyzed for the determination of perfluorinated compounds.

For the analysis of the samples, pressurized liquid extraction, liquid chromatography-tandem mass spectrometry and an isotope dilution method were applied to develop a specific and sensitive method of analysis for the quantification of 12 perfluorinated compounds.

The compounds are listed in Table 6.4.

No perfluorinated compounds were quantified in aluminum foil wrappers, baking paper materials or beverage cups.

Perfluorotridecanoic acid, perfluorotetradecanoic acid and perflyohexadecanoic acid were detected in fast-food boxes. In the ice cream cup sample only perfluorohexanoic acid was found. On the other hand, several perfluorinated compounds were quantified and detected in fast-food wrappers, while the highest levels of perfluorinated compounds were found in the microwave popcorn bag. Perfluorooctanoic acid and perfluorooctane sulfonate were not detected in any of the samples. Some of these compounds are shown in Figure 6.5.

Compared to other studies from different countries, very low concentrations of perfluorinated compounds were detected in the packaging materials analyzed (36). The results of the study suggest

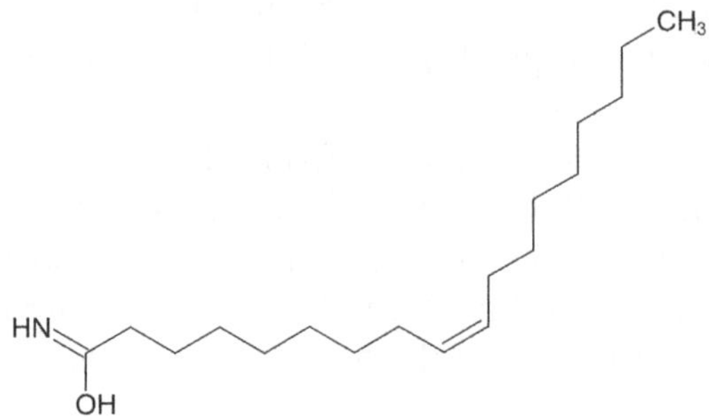

Oleamide

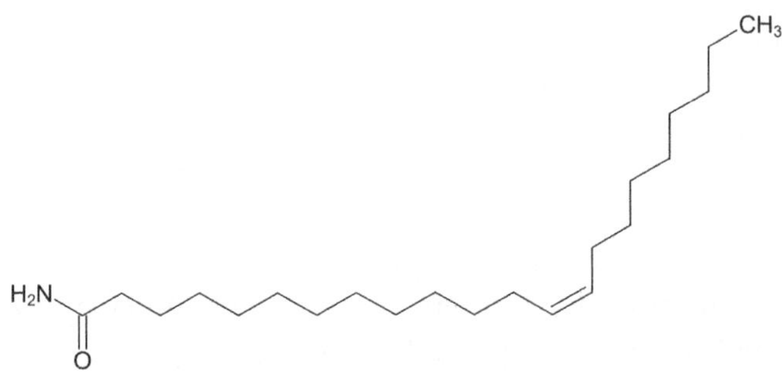

Erucamide

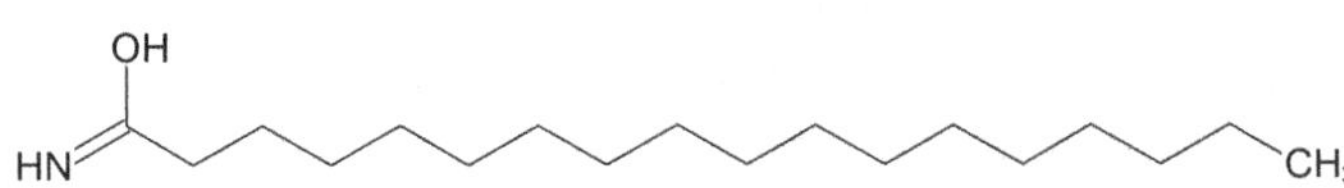

Stearamide

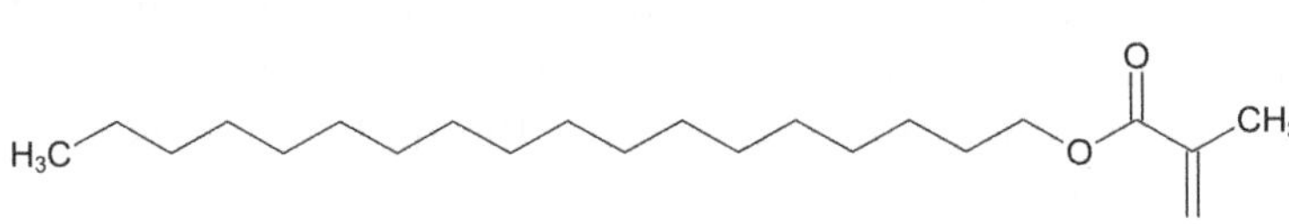

Stearyl amide

Figure 6.4 Fatty acid amides.

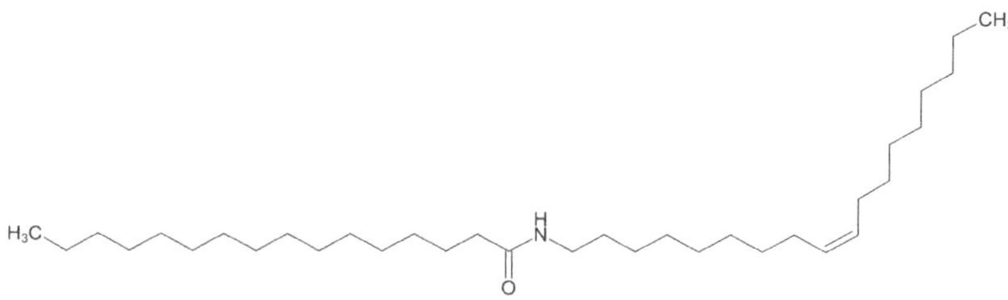

Oleyl palmitamide

Figure 6.4 (cont) Fatty acid amides.

Table 6.4 Perfluorinated compounds.

Compound	Abbreviation
Perfluorobutanoic acid	PFBA
Perfluoropentanoic acid	PFPeA
Perfluorohexanoic acid	PFHxA
Perfluoroheptanoic acid	PFHpA
Perfluorooctanoic acid	PFOA
Perfluorononanoic acid	PFNA
Perfluorodecanoic acid	PFDA
Perfluoroundecanoic acid	PFUnDA
Perfluorododecanoic acid	PFDoA
Perfluorobutane sulfonate	PFBS
Perfluorohexane sulfonate	PFHxS
Perfluorooctane sulfonate	PFOS
Perfluorotridecanoic acid	PFTrDA
Perfluorotetradecanoic acid	PFTeDA
Perflyohexadecanoic acid	PFHxDA
Perfluorooctadecanoic acid	PFODA
Perfluorodecane sulfonate	PFDS

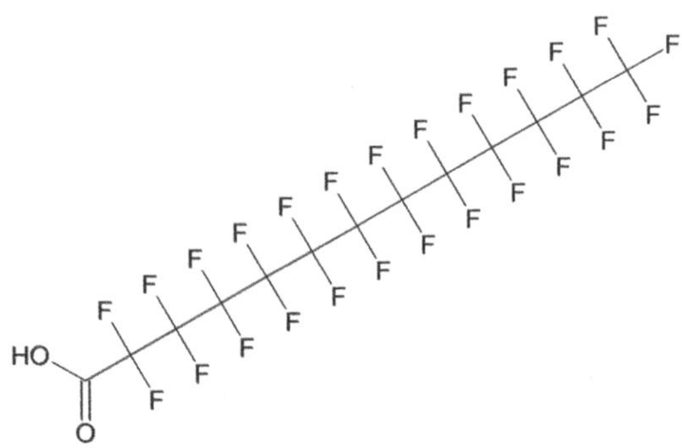

Perfluorotridecanoic acid

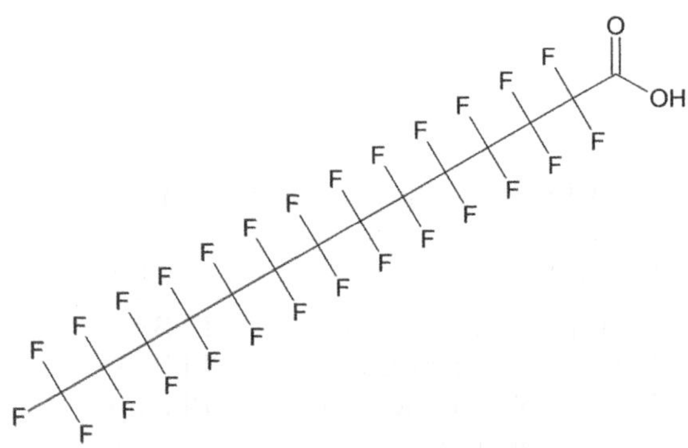

Perfluorotetradecanoic acid

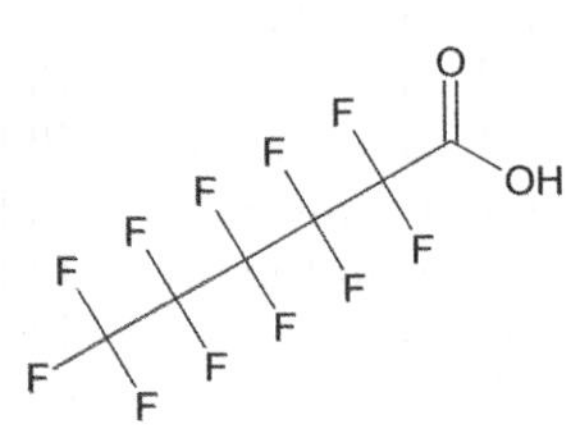

Perfluorohexanoic acid

Figure 6.5 Perfluorinated compounds.

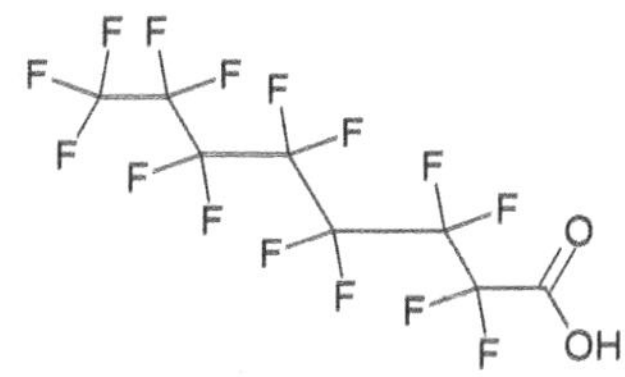

Perfluorooctanoic acid

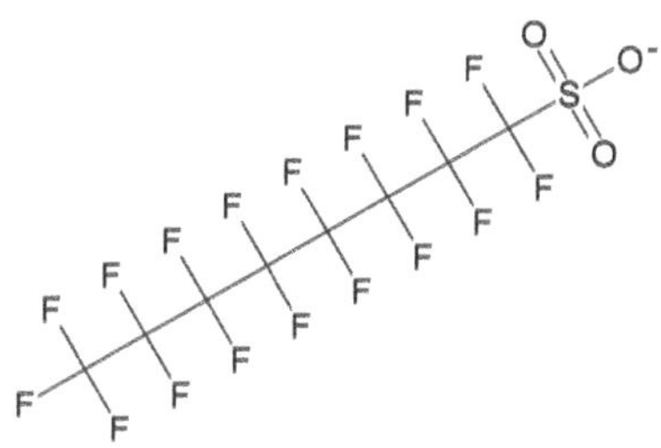

Perfluorooctane sulfonate

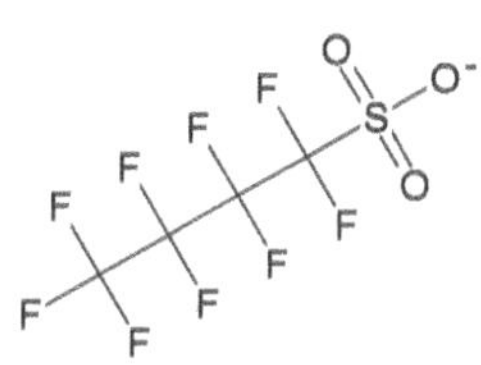

Perfluorobutane sulfonate

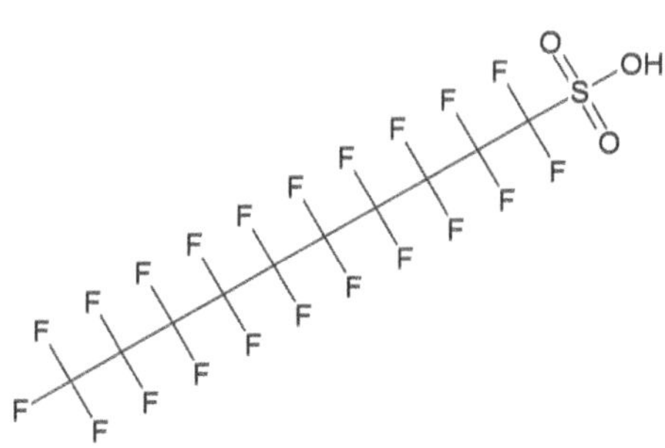

Perfluorodecane sulfonate

Figure 6.5 (cont) Perfluorinated compounds.

that probably no serious danger to consumer health can be associated with perfluorinated compounds contamination of packaging materials that are used in Greece.

6.7 Toxic Materials in Plastics

6.8 Aflatoxins

Various acidified clays and minerals can be used as food additive to kill or inhibit the growth of harmful microorganisms and to inactivate mycotoxins, such as aflatoxins, present as contaminants in human foods and animal feeds (37).

Aflatoxins are produced by *Aspergillus flarus* or *A. parasiticus*. There are at least four naturally occurring aflatoxins, namely aflatoxin B1, aflatoxin B2, aflatoxin G1 and aflatoxin G2, as shown in Figure 6.6.

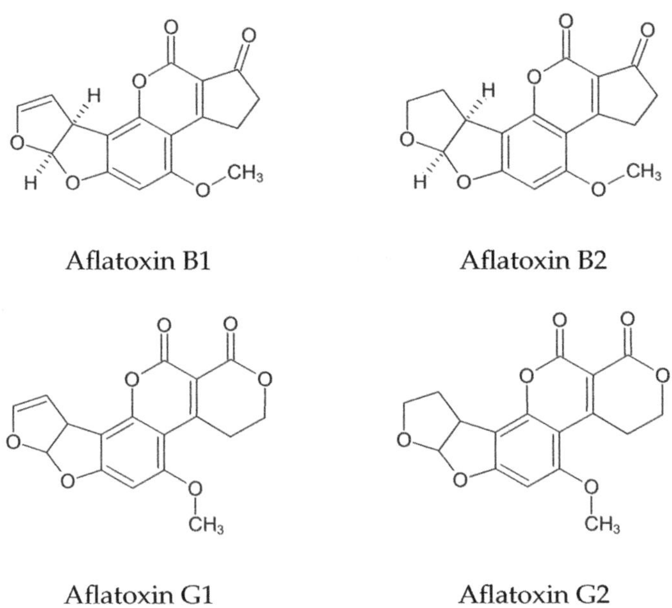

Figure 6.6 Naturally occurring aflatoxins.

Many aflatoxins occur as natural contaminants in a variety of foods and feeds such as corn, wheat, barley, beans, sorghum, moldy peanuts, mixed feed, and some coffee beans. They are also found as residues in the liver and kidneys of pigs. Many of these aflatoxins have been known to be strong carcinogens, thus causing cancers in humans, and they are also capable of eliciting other toxic effects, such as teratogenesis.

Microorganism and aflatoxin contamination of human foods and animal feeds constitutes a serious threat to the health of humans and animals. Generally, microorganisms are considered to be any microscopic or ultramicroscopic animal, plant, bacterium, virus, etc., and aflatoxins are harmful byproducts of mold growth, which both are potentially fatal to humans and animals. Furthermore, such contamination in human food and animal feeds also can lead to severe economic losses.

Many strategies have been developed to inactivate aflatoxins in foods and feeds. These strategies, which have shown varying degrees of success, include processing the food and feed, biocontrol and microbial inactivation, chemical treatment after structural degradation, dietary modification of toxicity and absorption method to reduce bioavailable aflatoxin.

One popular method is to add an non-nutritive adsorbent in contaminated feeds to inactivate the aflatoxin. Various adsorbents have been used such as aluminas, zeolites, silicas, phyllosilicates, bentonite, activated charcoal, and montmorillonite (38, 39). In particular, NovaSil Plus™ (hydrated sodium calcium aluminosilicate produced by the Engelhard Corporation) has been used to absorb and inactivate aflatoxin (40).

6.8.1 *Sorbents for Aflatoxin*

Sorption with plant-derived materials has emerged as a potential alternative over the aflatoxin B1 binding agents. This technology is based on a set of chemical or physical mechanisms including (41):

1. Hydrogen bonding,
2. Electrostatic attraction,
3. Ion exchange,
4. Chelation,
5. Precipitation, and

6. Complexation.

A widely used physical decontamination strategy is the adsorption of aflatoxins onto different materials. Inorganic binders such as phyllosilicates (bentonites, montmorillonites, smectites, kaolinites, and illites) and tectosilicates (zeolites); organic binders (yeast cell wall, lactic acid bacteria, micronized fibers, and activated carbon), and polymers (cholestyramine, divinylbenzene-styrene, poly(vinylpyrrolidone), chitosan, hydroxypropyl methyl cellulose, sodium carboxymethyl cellulose, and microcrystalline cellulose) have been reported to remove aflatoxin B1 (42, 43).

6.8.1.1 *Biosorbents for Aflatoxin*

In-vitro experiments were conducted to evaluate the effectiveness of two biosorbents, i.e., lettuce and field horsetail, for removing aflatoxin B1 (41). Formosa firethorn was used as reference material.

The adsorption of aflatoxin B1 (190 $ng\,ml^{-1}$) was investigated using two sorbent contents of 0.5% and 0.1% w/v, and three pH values of 2, 5, and 7.

Batch experiments were performed at 40°C for 2 h. Several methodologies were used to characterize the nature of the biosorbent-aflatoxin B1 interaction.

In general, when using biosorbents at 0.5% w/v, aflatoxin B1 was well adsorbed by the three tested biomaterials (70 to 100%). Furthermore, with the lowest biosorbent content of 0.1% w/v, significant aflatoxin B1 adsorption efficiencies of 33 to 50% could be found at pH 5. At a pH of 7, lettuce showed the highest ability for the removal of aflatoxin B1 with 95%. Furthermore, the aflatoxin B1-loaded biosorbents showed that chemical and physical mechanisms were involved in the adsorption process (41).

6.9 Migration of Additives from Plastics into Food

The migration of additives from plastics into food has been presented in a monograph (44).

It examines the intrusion of foreign chemicals into food via additives present in plastics packaging and the toxic hazards they pose to consumers.

This book shows how direct contact between the packed commodity and the plastic is likely to result in the transfer of polymer additives, adventitious impurities such as monomers, catalyst remnants, and residual polymerization solvents, and low-molecular-weight polymer fractions from the plastic into the packaged material.

The determination of various types of polymer additives and residual monomers in extractants of liquid foodstuffs and beverages, solid foods, edible oils, and fatty foodstuffs is considered (44).

6.10 Marine Environments

Marine litter plastics and microplastics and their toxic chemicals components have been described in a chapter of a monograph (45).

Plastic pollution is a fast-rising environmental catastrophe. Microplastics and nanoplastics are ubiquitous components of most aquatic environments, and their burgeoning prevalence is endangering aquatic organisms (46). Recent studies have documented the entanglement of marine and freshwater biota with plastic litters, particularly ghost fishing gear, resulting in suffocation, drowning, or starving to death.

Numerous reports have shown that aquatic organisms readily ingest and accumulate these emerging contaminants in their digestive systems. Given experimental evidence that contaminants-laden microplastics and nanoplastics can persist in the gastrointestinal tract for considerable durations, investigations have documented a high probability of lethal and sublethal toxicological effects associated with direct and indirect microplastics and nanoplastics ingestions. These include chronic protein modulation, DNA damage, embryotoxicity, gastrointestinal toxicity, genotoxicity, growth inhibition toxicity, histopathotoxicity, liver toxicity, neurotoxicity, oxidative stress, reproductive toxicity, and tissue damage. Today, reports have proven the transfer of microplastics and nanoplastics across the aquatic food web to humans. However, the mechanisms of multiple contaminants-laden microplastics and nanoplastics-induced toxicities, size-dependent toxicity, and the comprehensive mode-of-action and alterations of digestive, reproductive, and neurological systems' functionality in marine organisms are still unclear.

A review has been presented that mainly addresses the prevalence, food web interactions, and toxicity assessment of micro(nano)plastics in marine and freshwater organisms. It summarizes documented studies based on the following broad objectives (46):

1. The occurrence and prevalence of micro(nano)plastic particles in marine and freshwater environments,
2. The ingestion of microplastics and nanoplastics by aquatic biota and the food web exposure routes and bioaccumulation of contaminated microplastics and nanoplastics by higher trophic entities,
3. The adsorption and desorption of persistent organic pollutants, metals, and chemical additives on/from micro(nano)plastics, and
4. The probable ecotoxicological effects of micro(nano)plastics ingestion on aquatic biota.

6.11 Surfactants

The migration from a multilayer plastic material intended for food contact showed that 2,4,7,9-tetramethyl-5-decyne-4,7-diol, c.f. Figure 6.7, used as a surfactant in the adhesive employed to build the multilayer, was transferred to water and other food simulants in contact with the plastic (47).

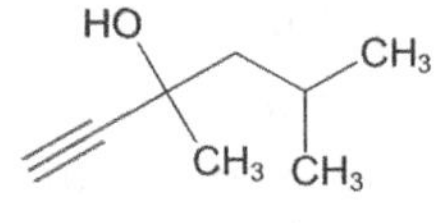

Figure 6.7 2,4,7,9-Tetramethyl-5-decyne-4,7-diol.

When these multilayer plastics were used for containing seminal doses for artificial insemination, it was found that fertility was seriously damaged in terms of motility, acrosome integrity, mitochondrial activity and penetration capacity in the cells, thus affecting male fertility.

Quantitative proteomic analysis of exposed germinal cells demonstrated the inhibition of key proteins involved in the fertilization capacity by affecting the cytoskeleton, sperm motility, the energy machinery and sperm defense mechanisms against oxidation, therefore confirming the surfactant-induced male infertility (47).

6.12 Salting-Out Agents

6.12.1 Determination of Food Stimulants

A highly efficient salting-out agent, potassium tri(poly phosphate), used to assist the liquid–liquid extraction sample preparation method was developed for the determination of the polar ethylene glycol and diethylene glycol in liquid samples (food stimulants and liquid foods) (48).

To obtain satisfactory extraction efficiencies for the studied analytes, several parameters affecting this procedure were optimized.

The effects of the salting-out reagent, type of extraction solvent, volumes of ethanol, amount of potassium triphosphate, vortexing and standing times on the extraction efficiency were investigated.

Analytical performances of the method were satisfactory, they showed a good linearity in the selected range ($R^2 > 0.9998$). The limits of detection were 0.2 $mg\,l^{-1}$ in all cases. Average recoveries in the ranges of 81.2% to 109.3% for ethylene glycol and 88.1% to 109.1% for diethylene glycol were obtained (48).

In relation to the green and low solvent consumption, the proposed methodology could be considered rapid and environmentally friendly. The here developed method can be used not only for the determination of ethylene glycol and diethylene glycol in food simulation solution, but also for liquid foods to obtain acceptable results (48).

6.12.2 Determination of Biogenic Amines

Biogenic amines are compounds present in foodstuff of animal origin that not only can be harmful to the consumer but can also serve as indicators of the product's edibility status (49). However, the analysis of biogenic amines is rather complicated.

Derivatization methods have been employed to make their analysis less challenging. Salting-out assisted liquid-liquid extraction was herein applied to decrease the time that the derivatizing agent dansyl chloride requires to be effective (49). Dansyl chloride is shown in Figure 6.8.

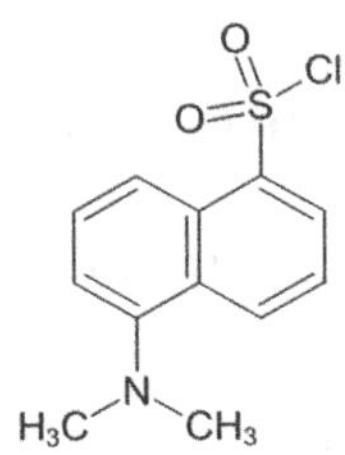

Figure 6.8 Dansyl chloride.

The dansyl chloride derivates were separated and analyzed by high-performance liquid chromatography (HPLC) with UV and fluorescence detection.

So, it was possible to analyze methylamine, dimethylamine, ethylamine, β-phenylethylamine, isopentylamine, putrescine, cadaverine, histamine, tyramine, spermidine and spermine in a straightforward manner in several fish-based and meat-based samples. Biogenic amines are shown in Figure 6.9 and in Figure 1.3.

The results were obtained in the order of magnitude of about a few $mg\,kg^{-1}$. The limits of detection ranged from 0.0075 $mg\,l^{-1}$ to 1.6 $mg\,l^{-1}$ (49).

6.12.3 *Liquid-Liquid Microextraction*

Salting-out assisted liquid-liquid microextraction was integrated with the derivatization procedure to establish a one-step sample pretreatment approach for the rapid analysis of 14 biogenic amines in fruit juices (50).

This methodology consists of salting-out of analytes, derivatization with ethyl chloroformate, extraction with ethyl acetate, and the

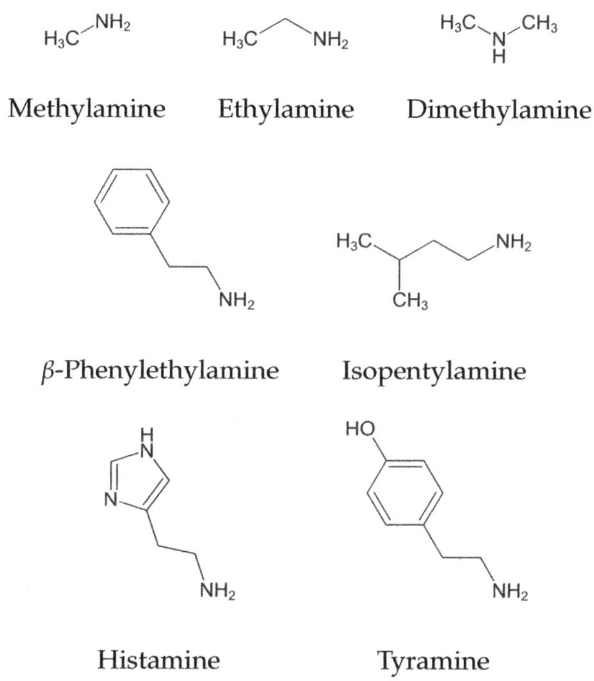

Figure 6.9 Biogenic amines.

analysis of the derivatized biogenic amines using gas chromatography-mass spectroscopy spectrometry. Ethyl chloroformate is shown in Figure 2.2.

Optimization of the salting-out assisted liquid-liquid microextraction parameters, including the amount of sample, NaOH, and ethyl chloroformate was carried out through a Box-Behnken response surface design. The developed method exhibits satisfactory limits of detection from 1.5 $\mu g\, l^{-1}$ to 8.1 $\mu g\, l^{-1}$ and quantification from 5.0 $\mu g\, l^{-1}$ to 26.7 $\mu g\, l^{-1}$. Average recoveries were between 84% and 108%.

The here developed procedure was used for the determination of biogenic amines in juices of different berries. The highest concentrations were found for cadaverine, putrescine, tryptamine, and tyramine (50).

In a similar study, monitoring of the biogenic amine content in food products, including wine, was presented (51).

6.12.4 *Determination of Aromatic Amines*

Carcinogenic primary aromatic amines can be released from improperly manufactured food packaging materials (52). The limit for the sum of primary aromatic amines is set to 10 $\mu g\, kg^{-1}$ in accordance with Commission Regulation No. 10/2011 (FCM Regulation).

However, a lower individual limit, 2 $\mu g\, kg^{-1}$, has been recently introduced for the carcinogenic primary aromatic amines in Commission Regulation No. 2020/1245. As the majority of the previously published methods are no longer compliant with the current regulation, a UHPLC-MS/MS method was developed to enable food packaging compliance testing for primary aromatic amines not only from 3% (w/v) acetic acid, but also from 10% (v/v) ethanol food simulant.

Since the latest amendment of the FCM Regulation refers to the list of the 22 restricted primary aromatic amines of EU Regulation No. 1907/2006, these primary aromatic amines were selected as target compounds along with aniline and *p*-toluidine, the most common impurities of azo colorants and isocyanates (52).

A list of primary aromatic amines and their carcinogenic categorization is given in Table 6.5. Some of these compounds are shown in Figure 6.10.

Table 6.5 List of primary aromatic amines.

Abbreviation	Compound	IARC Carcenogenic category
DAANI	2,4-Diaminoanisole	2B
DATOL	2,4-Diaminotoluene	2B
ANL	Aniline	3
PTOL	*p*-Toluidine	–
ODIA	4,4′-Oxydianiline	2B
BNZ	Benzidine	1
OANI	*o*-Anisidine	2B
DDPM	4,4′-Diaminodiphenylmethane	2B
OTOL	*o*-Toluidine	2A
PCRES	*p*-Cresidine	2B
DMXB	3,3′-Dimethoxybenzidine	2B
DMTB	3,3′-Dimethylbenzidine	2B
DMAPM	3,3′-Dimethyl-4,4′-diaminodiphenylmethane	2B
4CLA	4-Chloroaniline	2B
TMA	2,4,5-Trimethylaniline	3
TDIA	4,4′-Thiodianiline	2B
NAP	2-Naphthylamine	1
2M5NA	2-Methyl-5-nitroaniline	3
4ABP	4-Aminobiphenyl	1
4CLOT	4-Chloro-*o*-toluidine	2A
DCLB	3,3′-Dichlorobenzidine	2B
MOCA	4,4′-Diamino-3,3′-dichlorodiphenylmethane	1
4AZB	4-Aminoazobenzene	2B
OAZT	*o*-Aminoazotoluene	2B

IARC Categorization of the International Agency for Research on Cancer (IARC) (53)
1 Carcinogenic to humans
2A Probably carcinogenic to humans
2B Possibly carcinogenic to humans
3: Not classifiable as to its carcinogenicity to humans

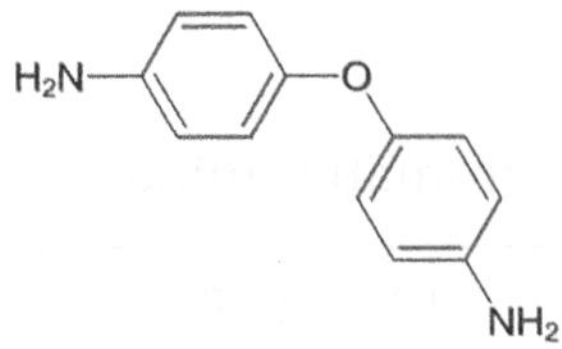

4,4′-Oxydianiline

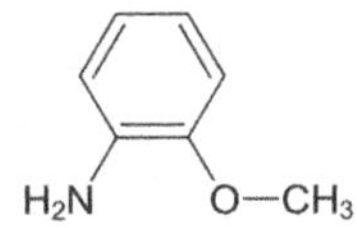

o-Anisidine

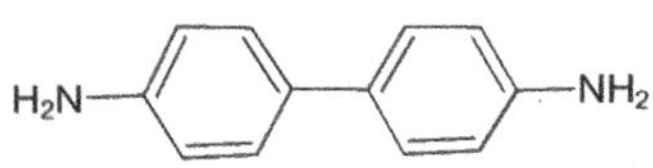

Benzidine

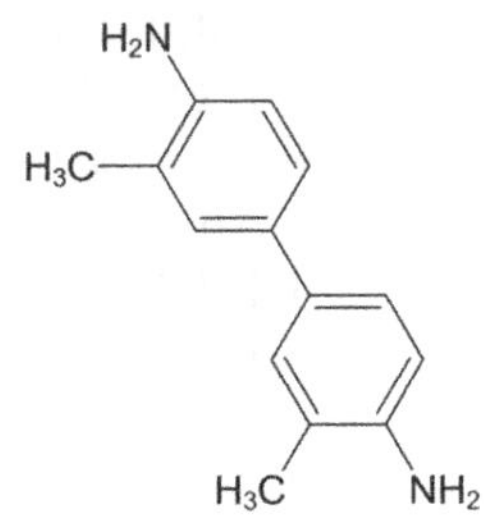

3,3′-Dimethylbenzidine

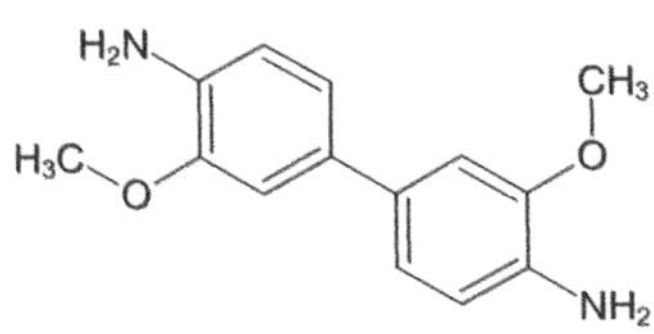

3,3′-Dimethoxybenzidine

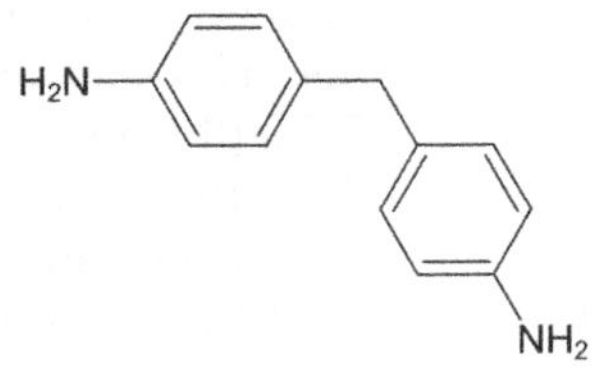

4,4′-Diaminodiphenylmethane

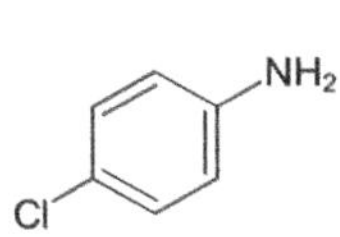

4-Chloroaniline

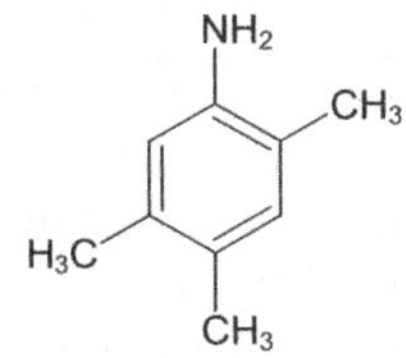

2,4,5-Trimethylaniline

Figure 6.10 Aromatic amines.

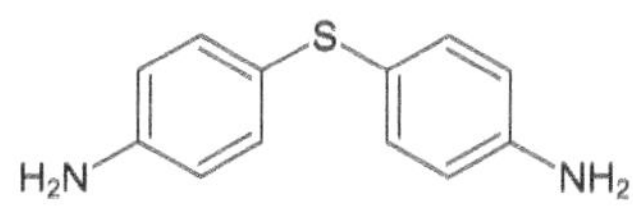

4,4′-Thiodianiline

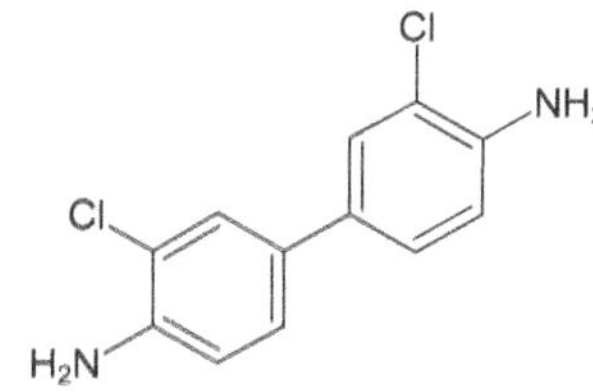

3,3′-Dichlorobenzidine

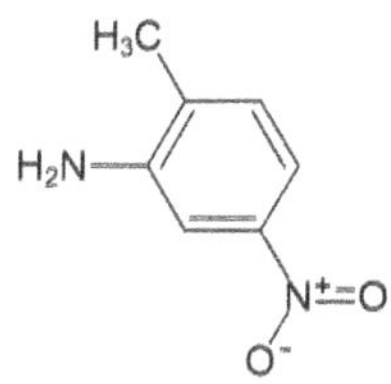

2-Methyl-5-nitroaniline

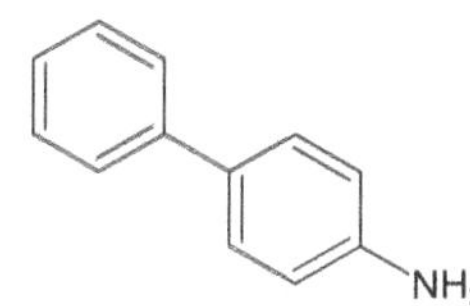

4-Aminobiphenyl

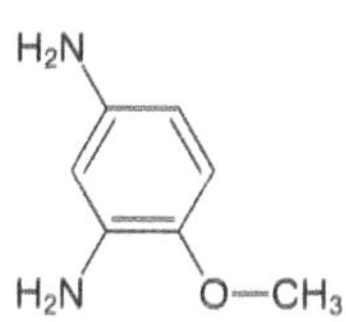

4-Methoxy-1,3-benzenediamine

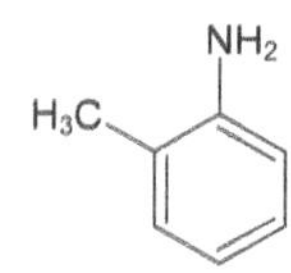

o-Aminoazotoluene

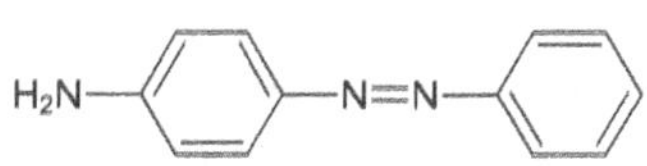

4-Aminoazobenzene

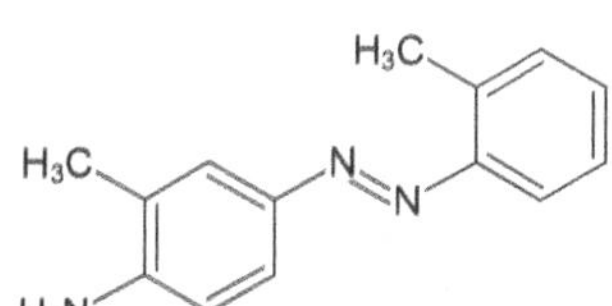

o-Aminoazotoluene

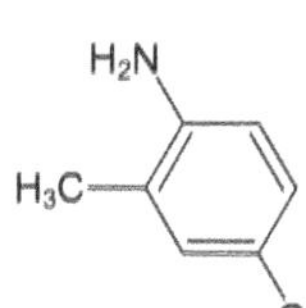

4-Chloro-*o*-toluidine

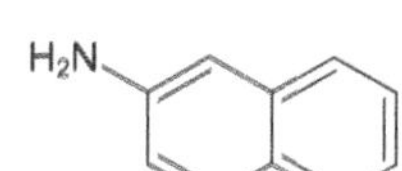

2-Naphthylamine

Figure 6.10 (cont) Aromatic amines.

An enrichment factor of 20 could be achieved combining solid phase extraction with salting-out assisted liquid-liquid extraction. The method was successfully validated and applied on real samples. Limit of quantitation and limit of detection values were 0.15 $\mu g\, l^{-1}$ and 0.05 $\mu g\, l^{-1}$ for both food simulants, respectively, except for 2,4-diaminotoluene, aniline and 4,4′-oxydianiline.

However, even these compounds had a lower limit of detection values than the new individual limit of 2 $\mu g\, kg^{-1}$. The cumulative limit of detection values for both food simulants (1.6 $\mu g\, l^{-1}$ and 1.5 $\mu g\, l^{-1}$ for 3% (w/v) acetic acid and 10% (v/v) ethanol, respectively) was lower than the 10 $\mu g\, kg^{-1}$ specified in the FCM Regulation. Accuracy values were between 70 and 118% for both food simulants for the majority of primary aromatic amines. Both within-day and between-day precision values were below 20%.

This method proved to be suitable for daily routine analysis enabling compliance testing of food packaging materials according to the latest regulations. Also, the method was successfully applied for the analysis of plastic kitchenware samples (52).

An extraction experiment of primary aromatic amines was carried out by using different solvents to seek out the best solvent for extraction among methylene chloride, *n*-hexane, ethyl acetate and methanol (54). Migration tests of primary aromatic amines from different polymer flexible packaging composite materials under both conditions of 40°C in 10 days and 70°C in 2 hours in neutral and oil simulants were conducted.

An effective HPLC method in ESI+ mode was proposed to separate and screen 25 kinds of primary aromatic amines listed in the Chinese Standard SN/T 2893-2011 (55). The extraction test clearly can confirm the existence of primary aromatic amines in both a poly(urethane) ink and UV ink (54).

References

1. A.L. Branen, P.M. Davidson, S. Salminen, and J. Thorngate, (Eds.), *Food Additives*, Food Science and Technology, CRC Press, Boca Raton, Florida, 2nd edition, 2001.
2. T.A.M. Msagati, *The Chemistry of Food Additives and Preservatives*, Wiley, 2012.

3. J.N. Hahladakis, C.A. Velis, R. Weber, E. Iacovidou, and P. Purnell, An overview of chemical additives present in plastics: Migration, release, fate and environmental impact during their use, disposal and recycling, *Journal of Hazardous Materials*, Vol. 344, p. 179, 2018.
4. E. Hansen, N.H. Nilsson, D. Lithner, and C. Lassen, Hazardous substances in plastic materials, *COWI in cooperation with Danish Technological Institute*, pp. 7–8, 2013.
5. A. Turner and M. Filella, Hazardous metal additives in plastics and their environmental impacts, *Environment International*, Vol. 156, p. 106622, 2021.
6. C. Chen, L. Chen, Y. Yao, F. Artigas, Q. Huang, and W. Zhang, Organotin release from polyvinyl chloride microplastics and concurrent photodegradation in water: Impacts from salinity, dissolved organic matter, and light exposure, *Environmental Science & Technology*, Vol. 53, p. 10741, 2019.
7. S. Mao, W. Gu, J. Bai, B. Dong, Q. Huang, J. Zhao, X. Zhuang, C. Zhang, W. Yuan, and J. Wang, Migration of heavy metal in electronic waste plastics during simulated recycling on a laboratory scale, *Chemosphere*, Vol. 245, p. 125645, 2020.
8. M.F. Pocas, J.C. Oliveira, R. Brandsch, and T. Hogg, Analysis of mathematical models to describe the migration of additives from packaging plastics to foods, *Journal of Food Process Engineering*, Vol. 35, p. 657, 2012.
9. J. Brandsch, P. Mercea, M. Rüter, V. Tosa, and O. Piringer, Migration modelling as a tool for quality assurance of food packaging, *Food Additives & Contaminants*, Vol. 19, p. 29, 2002.
10. P.V. Mercea, A. Kalisch, M. Ulrich, H. Benz, O.G. Piringer, V. Toşa, R. Schuster, S. Aranyi, and P. Sejersen, Modelling migration of substances from polymers into drinking water. Part 1 - Diffusion coefficient estimations, *Polymer Testing*, Vol. 65, p. 176, 2018.
11. R. Rafiei Nazari, S. Noorian, and M. Arabameri, Migration modelling of phthalate from non-alcoholic beer bottles by adaptive neuro-fuzzy inference system, *Journal of the Science of Food and Agriculture*, Vol. 98, p. 2113, 2018.
12. T. Begley, L. Castle, A. Feigenbaum, R. Franz, K. Hinrichs, T. Lickly, P. Mercea, M. Milana, A. O'Brien, S. Rebre, et al., Evaluation of migration models that might be used in support of regulations for food-contact plastics, *Food Additives and Contaminants*, Vol. 22, p. 73, 2005.
13. C. Kirchkeszner, N. Petrovics, T. Tábi, N. Magyar, J. Kovács, B.S. Szabó, Z. Nyiri, and Z. Eke, Swelling as a promoter of migration of plastic additives in the interaction of fatty food simulants with polylactic acid- and polypropylene-based plastics, *Food Control*, Vol. 132, p. 108354, 2022.

14. A.E. Feigenbaum, A.M. Riquet, and D. Scholler, Fatty food simulants: Solvents to mimic the behavior of fats in contact with packaging plastics in S.J. Risch, (Ed.), *Food Packaging*, Vol. 753, chapter 7, pp. 71–81. ACS Publications, 2000.
15. A. Nasiri, S. Peyron, E. Gastaldi, and N. Gontard, Effect of nanoclay on the transfer properties of immanent additives in food packages, *Journal of Materials Science*, Vol. 51, p. 9732, 2016.
16. I. Cooper and P.A. Tice, Migration studies on fatty acid amide slip additives from plastics into food simulants, *Food Additives & Contaminants*, Vol. 12, p. 235, 1995.
17. A. Fankhauser-Noti and K. Grob, Migration of plasticizers from PVC gaskets of lids for glass jars into oily foods: Amount of gasket material in food contact, proportion of plasticizer migrating into food and compliance testing by simulation, *Trends in Food Science & Technology*, Vol. 17, p. 105, 2006.
18. A.E. Goulas, P. Zygoura, and A. Karatapanis, Migration of di(2-ethylhexyl) adipate and acetyltributyl citrate plasticizers from food-grade PVC film into sweetened sesame paste (halawa tehineh): Kinetic and penetration study, *Food Chem. Toxicol.*, Vol. 45, 2007.
19. M. Biedermann, K. Fiselier, and K. Grob, Testing migration from the PVC gaskets in metal closures into oily foods, *Trends in Food Science & Technology*, Vol. 19, p. 145, 2008.
20. D.Y. Wei, M.L. Wang, Z.Y. Guo, S. Wang, H.L. Li, H.N. Zhang, and P.P. Gai, GC/MS method for the determination of adipate plasticizers in ham sausage and its application to kinetic and penetration studies, *Journal of Food Science*, Vol. 74, p. C392, 2009.
21. E. Fasano, F. Bono-Blay, T. Cirillo, P. Montuori, and S. Lacorte, Migration of phthalates, alkylphenols, bisphenol A and di (2-ethylhexyl) adipate from food packaging, *Food Control*, Vol. 27, p. 132, 2012.
22. E. Union, List of simulants to be used for testing migration of constituents of plastic materials and articles intended to come into contact with foodstuffs, Legislation 85/572/EEC, Commission Directive 2007/19/EC, 1985.
23. E. Union, Basic rules necessary for testing migration of the constituents of plastic materials and articles intended to come into contact with foodstuffs, Legislation 82/711/EEC, Commission Directive, 1982.
24. Y. Gao, Y. Gu, and Y. Wei, Determination of polymer additives-antioxidants and ultraviolet (UV) absorbers by high-performance liquid chromatography coupled with UV photodiode array detection in food simulants, *Journal of Agricultural and Food Chemistry*, Vol. 59, p. 12982, 2011.
25. M.S. Dopico-Garcıa, J.M. Lopez-Vilarino, and M.V. González-Rodrıguez, Determination of antioxidant migra-

tion levels from low-density polyethylene films into food simulants, *Journal of Chromatography A*, Vol. 1018, p. 53, 2003.

26. J. Alin and M. Hakkarainen, Type of polypropylene material significantly influences the migration of antioxidants from polymer packaging to food simulants during microwave heating, *Journal of Applied Polymer Science*, Vol. 118, p. 1084, 2010.
27. J. Alin and M. Hakkarainen, Microwave heating causes rapid degradation of antioxidants in polypropylene packaging, leading to greatly increased specific migration to food simulants as shown by ESI-MS and GC-MS, *Journal of Agricultural and Food Chemistry*, Vol. 59, p. 5418, 2011.
28. G. Beldi, S. Pastorelli, F. Franchini, and C. Simoneau, Time- and temperature-dependent migration studies of Irganox 1076 from plastics into foods and food simulants, *Food Additives & Contaminants: Part A*, Vol. 29, p. 836, 2012.
29. D.H. Jeon, G.Y. Park, I.S. Kwak, K.H. Lee, and H.J. Park, Antioxidants and their migration into food simulants on irradiated LLDPE film, *LWT-Food Science and Technology*, Vol. 40, p. 151, 2007.
30. M.S. Tawfik and A. Huyghebaert, Polystyrene cups and containers: Styrene migration, *Food Additives & Contaminants*, Vol. 15, p. 592, 1998.
31. J.O.K. Choi, F. Jitsunari, F. Asakawa, and D.S. Lee, Migration of styrene monomer, dimers and trimers from polystyrene to food simulants, *Food Additives and Contaminants*, Vol. 22, p. 693, 2005.
32. M.-R. Khaksar and M. Ghazi-Khansari, Determination of migration monomer styrene from GPPS (general purpose polystyrene) and HIPS (high impact polystyrene) cups to hot drinks, *Toxicology Mechanisms and Methods*, Vol. 19, p. 257, 2009.
33. I.S. Arvanitoyannis and L. Bosnea, Migration of substances from food packaging materials to foods, *Critical Reviews in Food Science and Nutrition*, Vol. 44, p. 63, 2004.
34. O.-W. Lau and S.-K. Wong, Contamination in food from packaging material, *Journal of Chromatography A*, Vol. 882, p. 255, 2000.
35. G. Przybytniak, K. Mirkowski, A. Rafalski, A. Nowicki, I. Legocka, and Z. Zimek, Effect of hindered amine light stabilizers on the resistance of polypropylene towards ionizing radiation, *Nukleonika*, Vol. 50, p. 153, 2005.
36. E. Zafeiraki, D. Costopoulou, I. Vassiliadou, E. Bakeas, and L. Leondiadis, Determination of perfluorinated compounds (PFCs) in various foodstuff packaging materials used in the Greek market, *Chemosphere*, Vol. 94, p. 169, 2014.
37. R.H. Carpenter, M.C. Kemp, and K.S. McKenzie, Preservative and additive for food and feed, US Patent 9 005 673, assigned to Texas A&M University System, Mionix Corp., April 14, 2015.

38. T.D. Phillips, B.A. Clement, N.D. Heidelbaugh, W.C. Gordon, and L.J. Gordon, Selective immobilization and detection of mycotoxins in solution, US Patent 5 178 832, assigned to Texas A&M University System, January 12, 1993.
39. B. Tengjaroenkul, U. Tengjaroenku, and S. Wongtangtintharn, Study on efficacy of natural Thai clays for adsorption of aflatoxin B1, *Livestock Research for Rural Development*, Vol. 23, p. 232, 2011.
40. D.R. Taylor, K.A. Delaney, and T.D. Phillips, Animal feed additive and method for inactivating mycotoxins present in animal feeds, US Patent 5 165 946, assigned to Texas A&M University System, Engelhard Corp., November 24, 1992.
41. M. de Jesús Nava-Ramírez, A.M. Salazar, M. Sordo, C. López-Coello, G. Téllez-Isaías, A. Méndez-Albores, and A. Vézquez-Durán, Ability of low contents of biosorbents to bind the food carcinogen aflatoxin B1 in vitro, *Food Chemistry*, Vol. 345, p. 128863, 2021.
42. B. Solís-Cruz, D. Hernández-Patlán, E. Beyssac, J.D. Latorre, X. Hernandez-Velasco, R. Merino-Guzman, G. Tellez, and R. López-Arellano, Evaluation of chitosan and cellulosic polymers as binding adsorbent materials to prevent aflatoxin B1, fumonisin B1, ochratoxin, trichothecene, deoxynivalenol, and zearalenone mycotoxicoses through an in vitro gastrointestinal model for poultry, *Polymers*, Vol. 9, p. 529, 2017.
43. P. Vila-Donat, S. Marín, V. Sanchis, and A.J. Ramos, A review of the mycotoxin adsorbing agents, with an emphasis on their multi-binding capacity, for animal feed decontamination, *Food and Chemical Toxicology*, Vol. 114, p. 246, 2018.
44. T.R. Crompton, *Additive Migration from Plastics into Food*, Elsevier, 2013.
45. F. Gallo, C. Fossi, R. Weber, D. Santillo, J. Sousa, I. Ingram, A. Nadal, and D. Romano, Marine litter plastics and microplastics and their toxic chemicals components in L.M. Nollet and K.S. Siddiqi, (Eds.), *Analysis of Nanoplastics and Microplastics in Food*, chapter 10, p. 159. CRC Press, 2020.
46. N.U. Benson, O.D. Agboola, O.H. Fred-Ahmadu, G.E. De-la-Torre, A. Oluwalana, and A. Williams, Micro (nano) plastics prevalence, food web interactions and toxicity assessment in aquatic organisms: A review, *Frontiers in Marine Science*, p. 291, 2022.
47. C. Nerin, E. Canellas, P. Vera, E. Garcia-Calvo, J.L. Luque-Garcia, C. Cámara, R. Ausejo, J. Miguel, and N. Mendoza, A common surfactant used in food packaging found to be toxic for reproduction in mammals, *Food and Chemical Toxicology*, Vol. 113, p. 115, 2018.
48. F. Wan, S. Liu, L. Wang, and S. Si, A novel salting-out extraction system for determination of ethylene glycol and diethylene glycol in liquid samples followed by GC-FID, *Microchemical Journal*, Vol. 179, p. 107491, 2022.

49. K.C.A. Francisco, P.F. Brandão, R.M. Ramos, L.M. Gonçalves, A.A. Cardoso, and J.A. Rodrigues, Salting-out assisted liquid–liquid extraction with dansyl chloride for the determination of biogenic amines in food, *International Journal of Food Science & Technology*, Vol. 55, p. 248, 2020.
50. A. Rózanska, M. Fabjanowicz, K. Kalinowska, Z. Polkowska, and J. Plotka-Wasylka, Green, simple analytical method for biogenic amines determination in fruit juice samples using salting-out assisted liquid-liquid microextraction andgas chromatography-mass spectrometry, *Food Chemistry*, Vol. 384, p. 132557, 2022.
51. M. Fabjanowicz, A. Rózanska, K. Kalinowska, and J. Plotka-Wasylka, Miniaturized, green salting-out liquid-liquid microextraction coupled with GC-MS used to evaluate biogenic amines in wine samples, *Microchemical Journal*, Vol. 180, p. 107616, 2022.
52. B.S. Szabó, P.P. Jakab, J. Hegedus, C. Kirchkeszner, N. Petrovics, Z. Nyiri, Z. Bodai, T. Rikker, and Z. Eke, Determination of 24 primary aromatic amines in aqueous food simulants by combining solid phase extraction and salting-out assisted liquid-liquid extraction with liquid chromatography tandem mass spectrometry, *Microchemical Journal*, Vol. 164, p. 105927, 2021.
53. A. Trakoli, IARC monographs on the evaluation of carcinogenic risks to humans. Volume 99: Some aromatic amines, organic dyes, and related exposures. International Agency for Research on Cancer, 2012.
54. R.-J. Luo, Q.-B. Lin, L. Zhu, J.-W. Yan, and Z. Li, Detection of primary aromatic amines content in food packaging ink and migration from printed plastic bags, *Food Packaging and Shelf Life*, Vol. 32, p. 100820, 2022.
55. Food contact materials for export. Polymers. Determination of 4,4-methylenediamine in food simulants, SNT Standard 2892, International Organization for Standardization, China, 2011.

7

Applications

7.1 Glaze-Like Coatings

Glazes work by forming a nonporous glass-like coating that seals any toxic materials in the clay and reduces the porosity (1). The nonporous, smooth surface avoids creating an environment where potentially harmful microorganisms can grow and reduces the potential for liquids to erode the underlying clay.

Coatings for oven-bake clay have been described that can make the surface of the resulting object food safe (1). Among the many different possibilities contemplated, the coating may contain one or more food-safe plastics that may be in particle form and may form suspensions or colloids when mixed with water, food-safe oil or other food-safe liquids.

Food-safe coatings for use with oven-bake clay can be produced by mixing food-safe plastic particles with a food-safe liquid, and optionally also adding a surfactant. The plastic particles should be less than 200 μm in mean size, less than 100 μm in mean size, and most preferable, less than 50 μm in mean size. The particles should be mixed in a ratio with less than 6 g of plastic per 10 ml of food-safe liquid, preferably 4 g of plastic or less per 10 ml of food-safe liquid. Optionally, less than 10 g of surfactant can be added per 10 ml of food-safe liquid. In a preferred form, less than 5 g of surfactant is added per 10 ml of the food-safe liquid, or more preferably less than 3 g of surfactant is added per 10 ml of food-safe liquid. Also, optionally, less than 2 g of dye can be added per 10 ml of food-safe liquid. In a preferred form, less than 1 g of dye is added per 10 ml of the food-safe liquid. The ingredients should be vigorously agitated, stirred or sonicated (or mixed using a combination of these methods) in order to form a uniform suspension or colloid. The resulting formulation can be stored in a container such as a plastic or glass bottle or coated carton. Just before use, if settling has occurred, the formulation can be agitated, for example, by hand for about 5 to 10 *s* or up to 30 *s*. Larger containers of more than 100 *ml* may require more agitation time to achieve a uniform suspension.

In particular, such a food-safe object is composed of (1):

1. An item used for eating and drinking consisting of oven-bake clay, and
2. A food-safe coating containing:
   A: One or more food-safe liquid, selected from ethanol, acetic acid, glycerin, cooking oil, and water, and
   B: One or more food-safe plastic selected from high-density poly(ethylene) (HDPE), low-density poly(ethylene) (LDPE), poly(propylene), poly(ethylene terephthalate), PS, and poly(carbonate).

In order to allow for easy application of the coating on an object, the plastic in the coating can form a uniform and stable suspension or a colloid. Also, the plastic particles are mixed with one or more food-safe liquid, such as glycerine, acetic acid, ethanol, cooking oil or water.

The plastic particles may be added most preferably in a ratio of less than 4 $g\,ml^{-1}$ of food-safe liquid.

Also, the coating may contain a surfactant to improve the formation of the suspension or the colloid. A dye may be added such that there is less than 1 4 $g\,ml^{-1}$ of food-safe liquid.

Figure 7.1 shows the results of coating from a mixture $g\,ml^{-1}$ of HDPE in water; and Figure 7.2 shows the results of coating from a mixture $g\,ml^{-1}$ of HDPE $g\,ml^{-1}$ of 1 HDPE of cottonseed oil.

Sculpey is the brand name for a type of polymer clay that can be molded and put into a conventional oven to harden (2). The primary ingredient in Sculpey is polyvinyl chloride, augmented with fillers, plasticizers and colorants.

The Sculpey was pre-baked for 15 *min* at 135°C and then the coating was baked with the Sculpey for 15 *min* at 135°C (1).

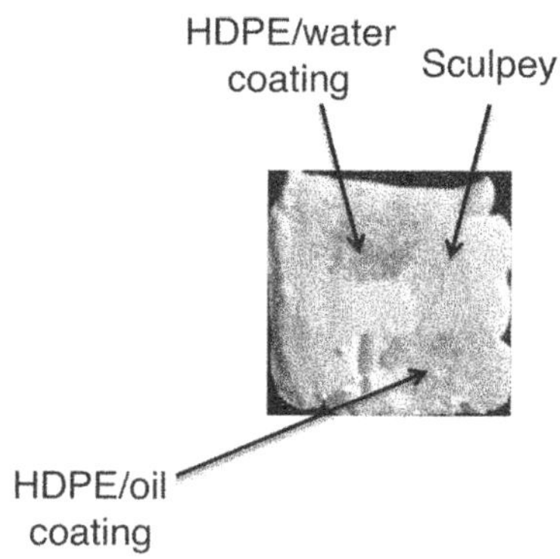

Figure 7.1 Coating from HDPE in water (1).

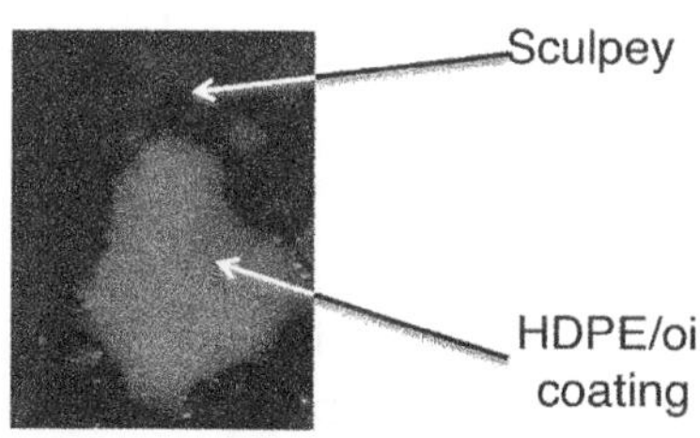

Figure 7.2 Coating from HDPE in cottonseed oil (1).

7.2 Functional Food Applications

A functional food is a food that should have an additional function (3) by adding some new ingredients. The term "functional food" may also apply to traits purposely bred into existing edible plants, such as purple or gold potatoes having increased anthocyanin or carotenoid contents, respectively.

Functional foods may be designed to have physiological benefits or to reduce the risk of chronic disease beyond basic nutritional functions, and may be similar in appearance to conventional food and consumed as part of a regular diet (3).

7.2.1 *Fructan*

A method has been described for preparing a functional food (4), in which the following steps are used:

1. Production of an aqueous solution of a malvaceous plant extract,
2. Addition of fructan to the solution, and
3. Adjusting the solution to a determined temperature to obtain a solution containing fructose and/or oligofructose hydrolyzed from the fructan in the presence of the malvaceous plant extract.

A functional food can be obtained from this method, which can be used to treat pathologies relating to calcium deficiency (4).

7.2.2 *Zogale*

Moringa oleifera, also called zogale, is grown and harvested as a tiny and middle-sized tree in different parts of the world (5). It has various industrial applications such as medicine, food, and cosmetics. This contains essential nutrients that could play a significant role in human nutrition.

Based on its chemical composition and nutritional properties, it can be used to overcome malnutrition, particularly in breastfeeding mothers and infants. Various parts of the Moringa tree, such as leaves, peels, seeds, and flowers, are good sources of phenolic compounds, essential amino acids, vitamins, proteins, and β-carotenes, which can play a regulatory and functional role (5).

The proportion of bioactive components is dependent on the source and variety of the plant. The flowers, seeds, and leaves of this plant have also been used in different food applications.

In a nut shell, the leaves, peels, seeds, and flowers of the Moringa tree are potential sources of different phenolic compounds, which act as anti-inflammatory, antidiabetic, neuro-enhancer, and anticancer agents (5).

7.2.3 *Garcinia indica*

Garcinia indica Choisy (kokum), a plant from the Clusiaceae family, is a fruit tree in the Western Ghats region. Kokum has been studied for its health benefits associated with numerous bioactive compounds, including phenolic acids, flavonoids, citric acids, and others (6).

Here, garcinol, hydroxycitric acid, and anthocyanins (cyanidin-3-glucoside and cyanidin-3-sambubioside) are the major bioactive compounds (6). These compounds are shown in Figure 7.3.

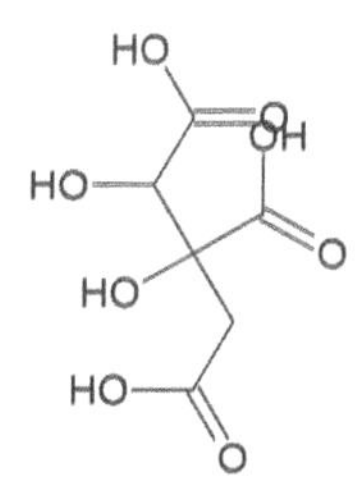

Hydroxycitric acid

Figure 7.3 Bioactive compounds in Kokum.

Garcinia indica fruits and fruit rinds have been reported to possess numerous therapeutic applications in various health conditions such as cancer, inflammation, diabetes, obesity, cardiovascular disease, and neurologic disorders.

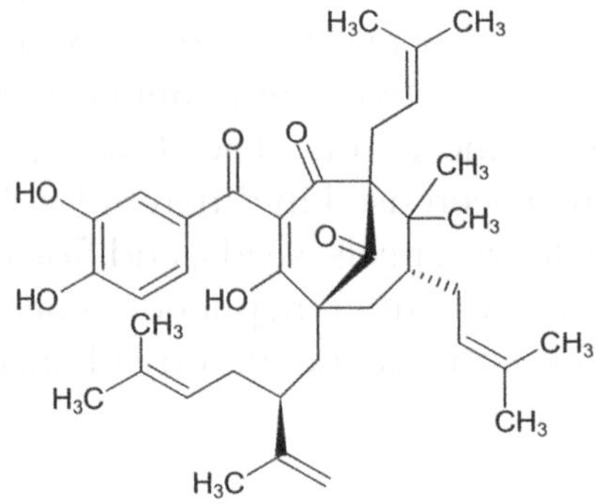

Garcinol

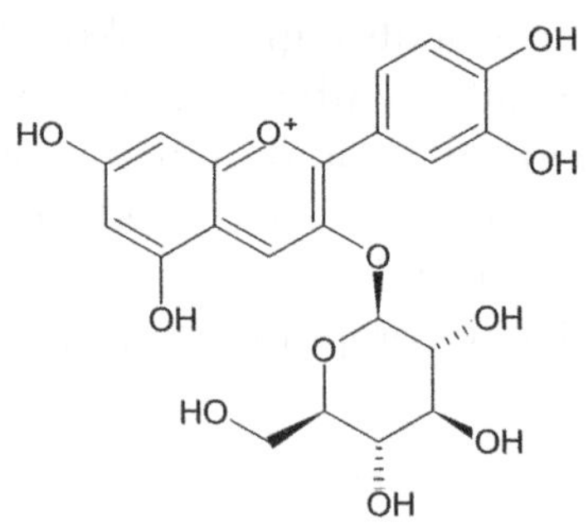

Cyanidin-3-glucoside

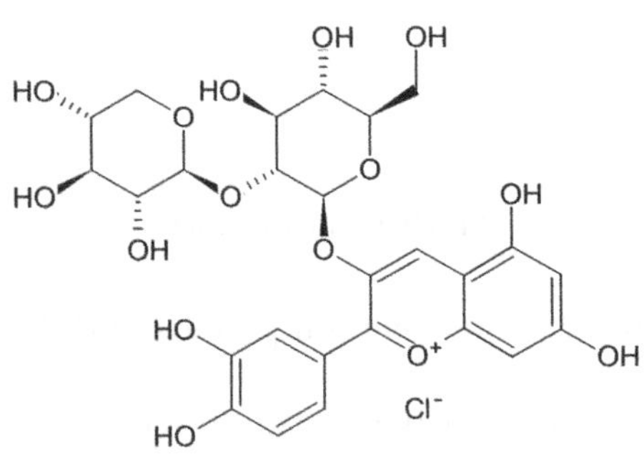

Cyanidin-3-sambubioside

Figure 7.3 (cont) Bioactive compounds in Kokum.

This plant can be used as follows (6):

1. Bioactive compounds present in *Garcinia indica* can be utilized for nutraceutical preparations.
2. *Garcinia indica* can be added to food products to convert them into functional foods.
3. Extraction of bioactive compounds can be done on an industrial scale.
4. Bioactive compounds can be extracted and used to commercialize life-saving drugs.

7.2.4 Mulberry Plants

Medicinal plants from the family Moraceae have diverse applications in agriculture, cosmetics, food, and in the pharmaceutical industry (7). Their extensive spectrum of pharmacological activity for treating numerous inflammatory illnesses, cancer, cardiovascular diseases, and gastrointestinal problems reflects their biological and therapeutic value.

The molecular mechanisms of action have been detailed (7). These are related to the biological implications of mulberry extracts, fractions, and isolated bioactive compounds from different parts in various health-related ailments. In addition, the food industry and animal nutrition applications were summarized.

This medicinal plant has phytochemicals such as steroids, saponins, alkaloids, glycosides, poly(saccharide)s, and phenolic compounds, including terpenoids, flavonoids, anthocyanins, and tannins. The aqueous, ethanolic, and methanolic extracts, as well as bioactive compounds, have antioxidative, hypoglycemic, nephroprotective, antimicrobial, neuroprotective, antimutagenic, hepatoprotective, anthelmintic, immune-modulatory, cardioprotective, and skin protecting activities.

Mulberry supplementation in food products improves the stability of phenolics, sensory properties, antioxidant activity, and antimicrobial properties. Mulberry leaves in animal feed increase the nutrient digestibility, growth parameters, antimicrobial, and antioxidant properties (7).

7.2.5 *Citrus Seeds*

Citrus fruits provide considerable non-edible residues, which reach 80% in juice production (8). They are considered as agri-wastes composed of peel, pulp and seeds.

Previous investigations have focused on peel and pulp to recover value-added products. Fatty acids, phytosterols and tocopherols are the major bioactives in citrus seeds, in addition to limonoids, dietary fibers and flavonoids.

Citrus fruits contain a substantial amount of flavanone-7-O-glycosides, flavones, polymethoxylated flavones, flavonols, and anthocyanin compounds. Some bioactive compounds of citrus fruits are shown in Figure 7.4.

Besides their nutritional values, these chemicals have promising applications, including in the production of biodiesel, food enhancers and antioxidants, especially from mandarin and grapefruit seeds.

The optimum conditions of the valorization of different citrus seeds were discussed to improve extraction yield and lessen environmental hazards of solvent extraction (8).

7.2.6 *Vine Tea*

Herbal teas like vine tea (*Ampelopsis grossedentata*) have been traditionally consumed worldwide because of their health-promoting effects and pleasant taste (9). Vine tea and its main bioactive component, dihydromyricetin, c.f. Figure 4.15, have gained attention because of their potential applications in food, material, and pharmaceutical sciences.

Vine tea and dihydromyricetin have been suggested as potential natural antioxidants to extend the shelf life of foods. Studies have also suggested a potential application in packaging and food safety.

In addition, dietary supplementation with vine tea extract has shown great potential for preventing metabolic diseases, which can justify its application in novel functional foods.

The flavonoid compounds that have been identified in vine tea extract are shown in Table 7.1 and some are shown in Figure 7.5.

Dihydromyricetin is the main bioactive component in vine tea. The potential health benefits of this flavonoid have been described

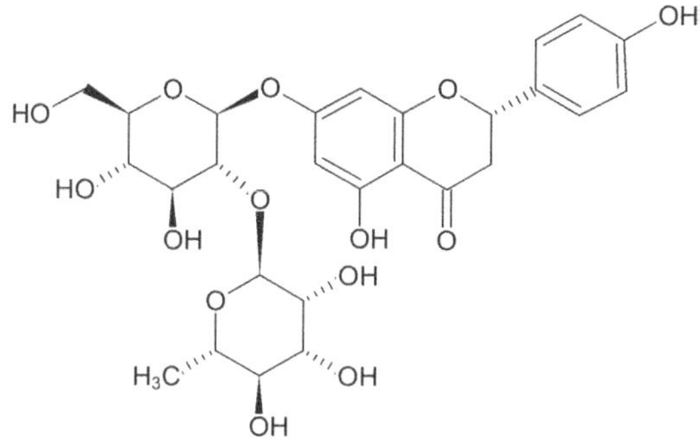

Naringin

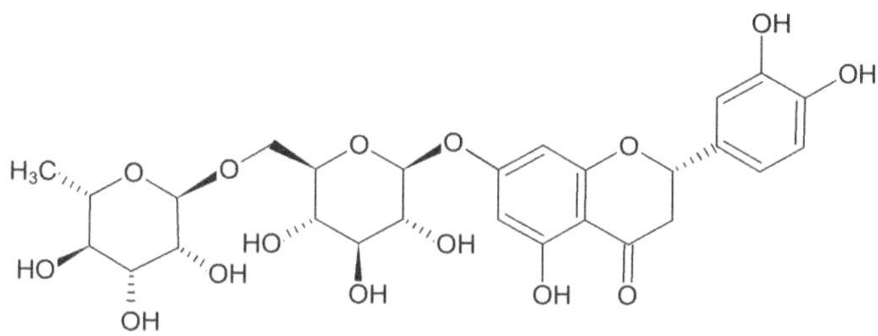

Eriocitrin

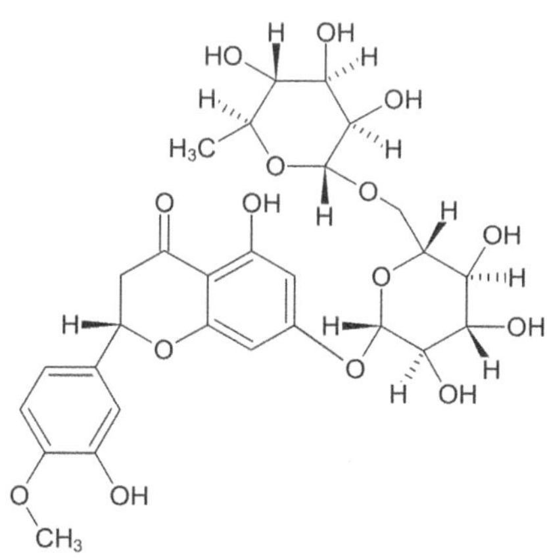

Hesperidin

Figure 7.4 Compounds in citrus fruits.

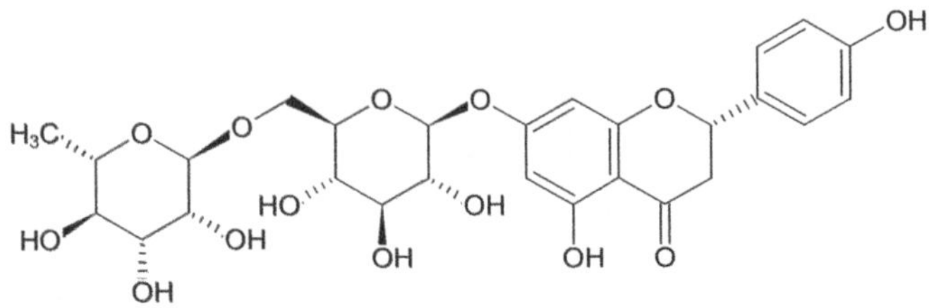

Narirutin

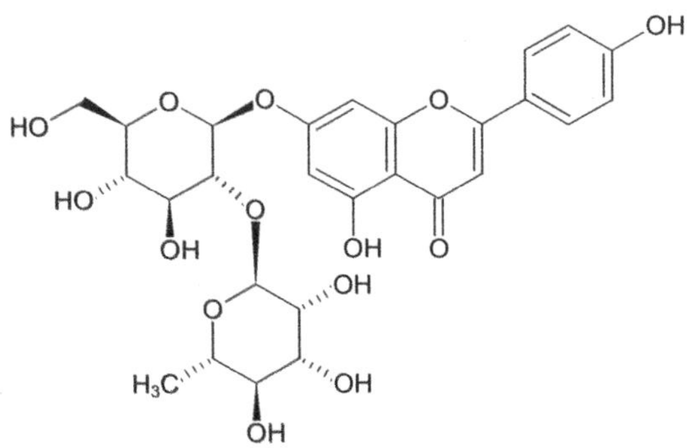

Rhoifolin

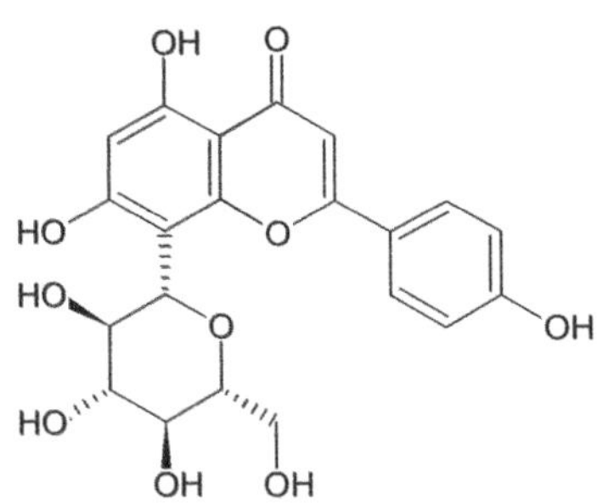

Vitexin

Figure 7.4 (cont) Compounds in citrus fruits.

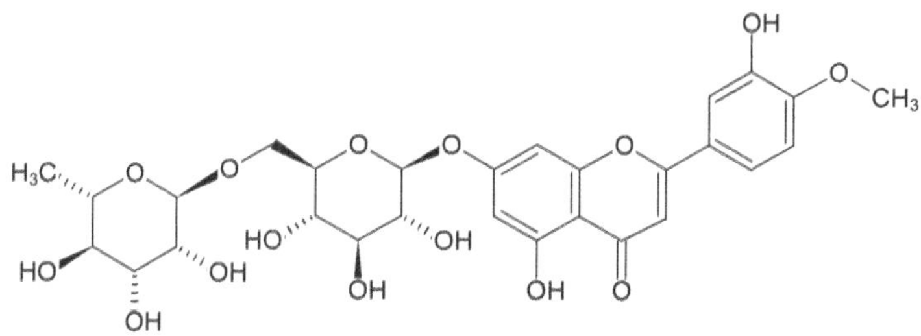

Diosmin

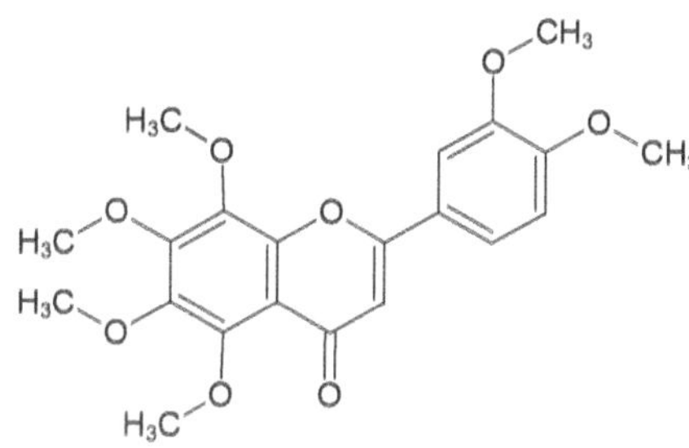

Nobiletin

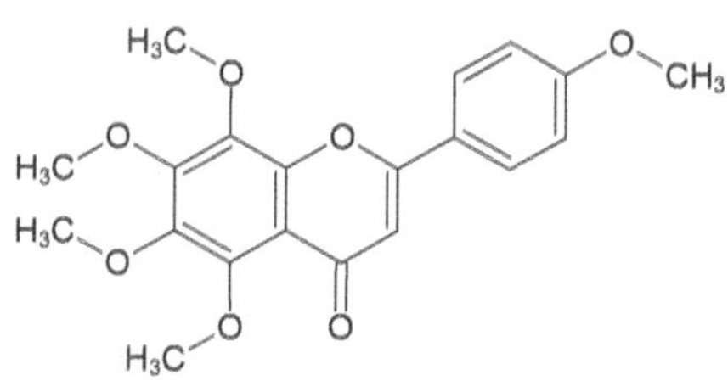

Tangeritin

Figure 7.4 (cont) Compounds in citrus fruits.

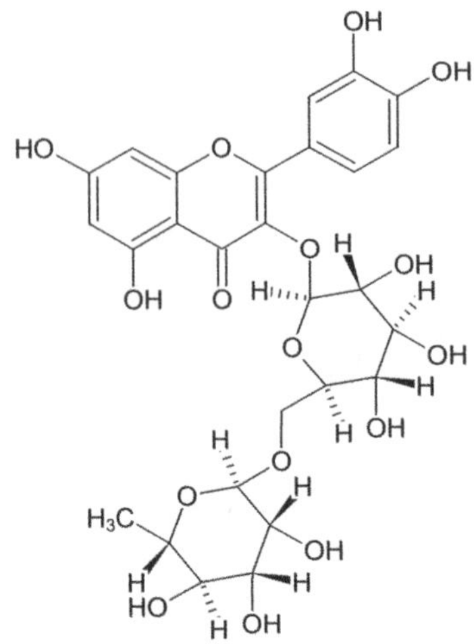

Rutin

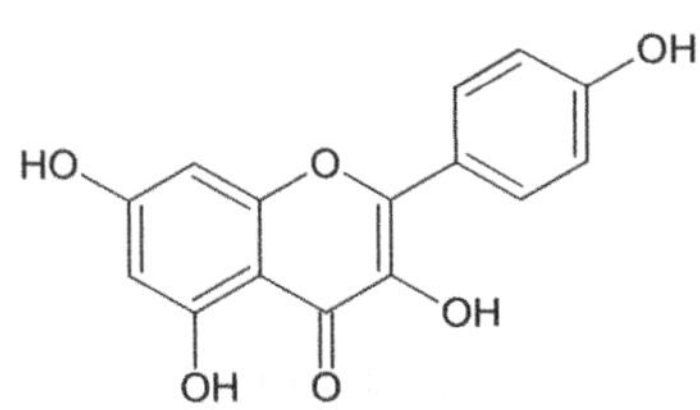

Kaempferol

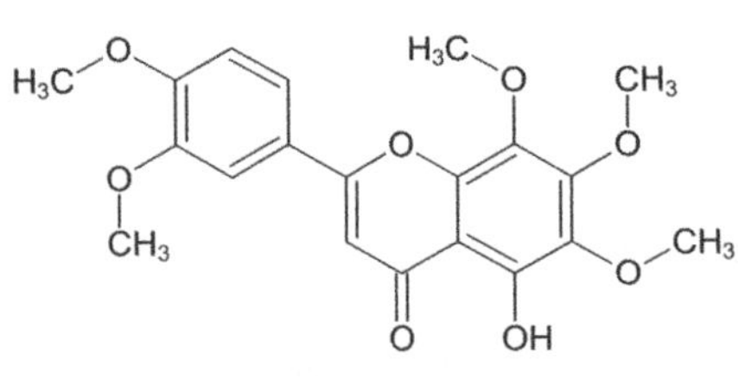

5-Demethyl nobiletin

Figure 7.4 (cont) Compounds in citrus fruits.

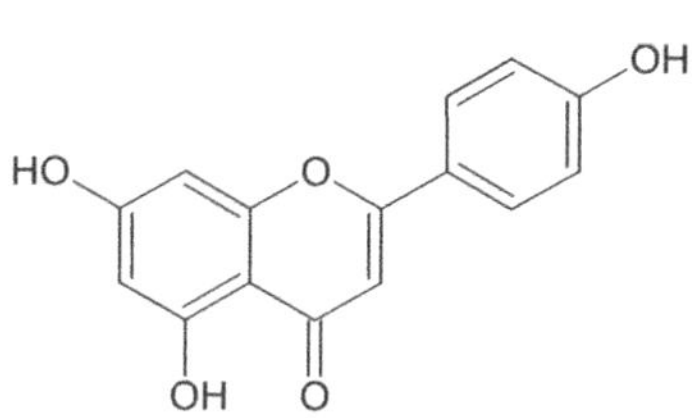

Apigenin

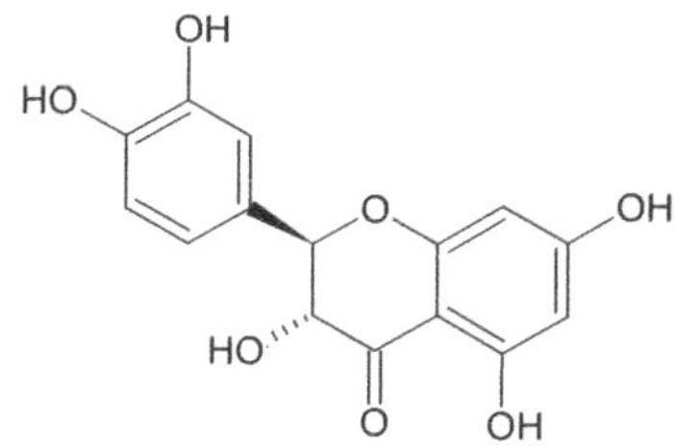

Dihydroquercetin

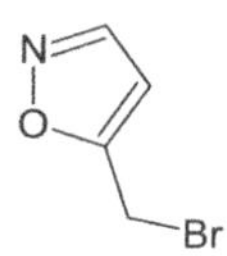

Hesperitin

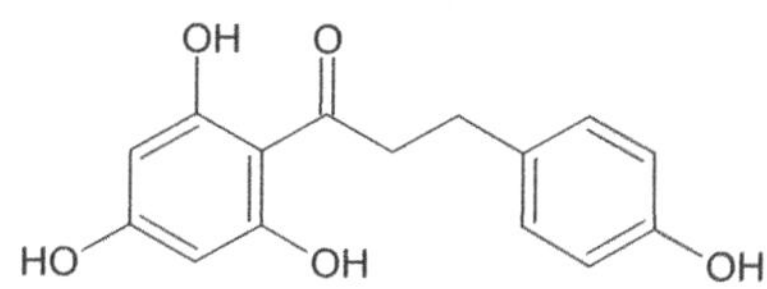

Phloretin

H_2O H_2O

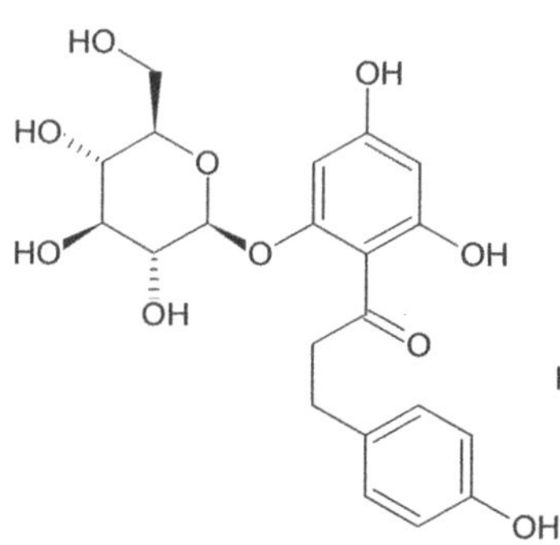

Phloridzin

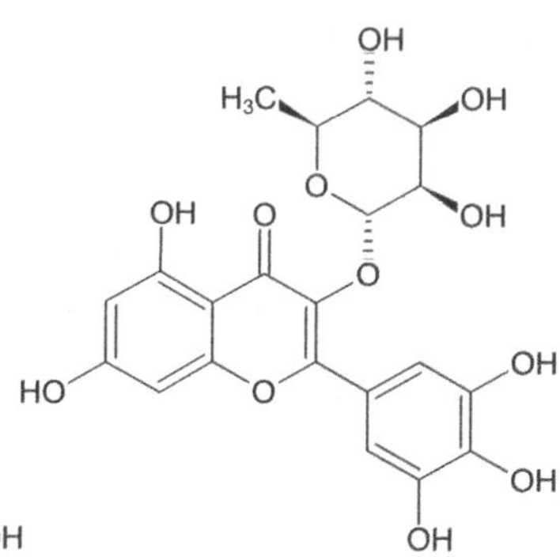

Myricitrin

Figure 7.5 Flavonoid compounds in vine tea extract.

Table 7.1 Flavonoid compounds in vine tea extract (10, 11).

Compound	Compound
Apigenin	Dihydromyricetin
Dihydroquercetin	3-Dihydroxyquercetin
Hesperitin	*i*-Dihydromyricetin
Kaempferol	Kaempferol-3-O-a-L-rhamnoside
Luteolin	Myricetin
Myricetin-3-O-β-D-glucoside	Myricitrin
Phloretin	Phloridzin
Quercetin	Quercetin-3-O-a-L-rhamnoside
Rutin	Vitexin

(12–15), such as antioxidant, anticancer, (e.g., inhibition of apoptosis, regulation of proliferation, metastatic inhibition), anti-inflammatory, anti-hypertension, hepatoprotective and neuroprotective activities; as well as its promising use in preventing diseases such as diabetes mellitus, atherosclerosis, and osteoporosis.

Besides flavonoids, four isoflavones were isolated from a methanol extract of vine tea (whole plants of *A. grossedentata*). These materials are shown in Table 7.2 and some are shown in Figure 7.6.

Table 7.2 Isoflavone compounds in vine tea extract (16).

Compound
6,7-Dihydroxy-3′-methoxy-4′,5′-methylenedioxy-isoflavone
6-O-β-D-Glucopyranoside
6-O-a-L-Rhamnopyranoside
6-O-β-D-Xylopyranosyl-(1-6)-β-D-glucopyranoside

7.2.7 Anti-Stress Drugs

Mental stress has a great influence on the circulatory system (17). For example, when one undergoes stress, angiotensins II and/or vasopressin increase, and due to sodium reabsorbancy an excess of intracorporeal sodium is created, which causes a rise in blood pressure. However, suffering stress not only causes a rise in blood pressure, but also influences various factors, such that it is believed

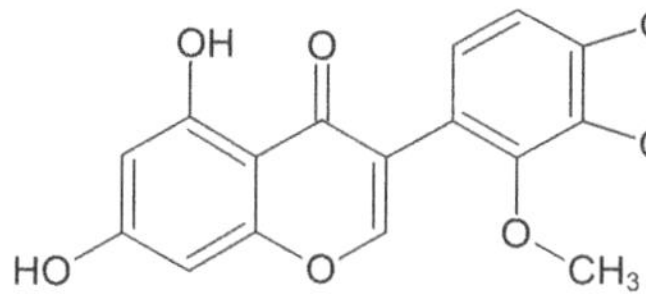

6,7-Dihydroxy-3′-methoxy-4′,5′-methylenedioxyisoflavone

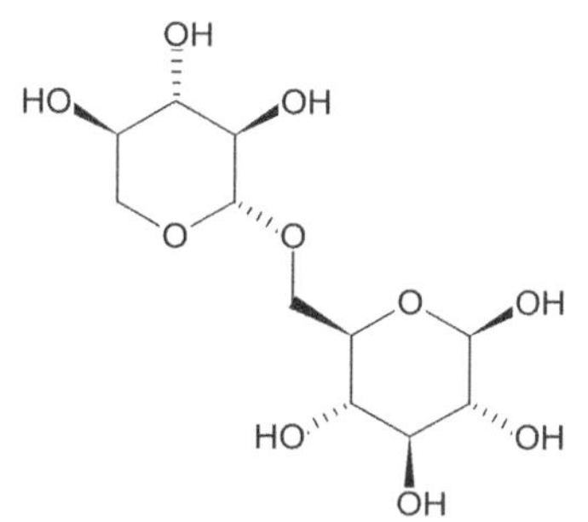

6-O-β-D-Xylopyranosyl-(1-6)-β-D-glucopyranoside

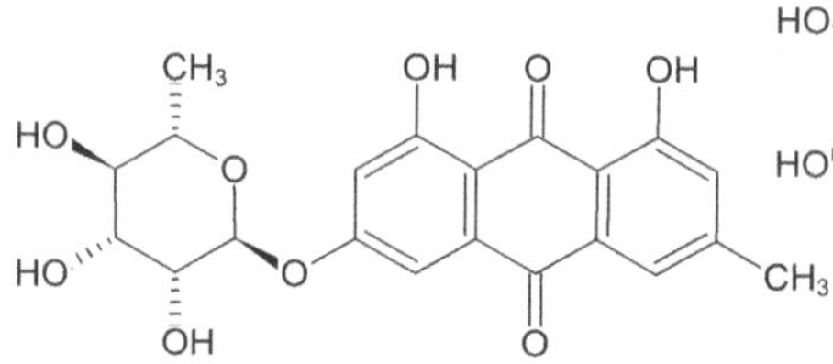

6-O-a-L-Rhamnopyranoside

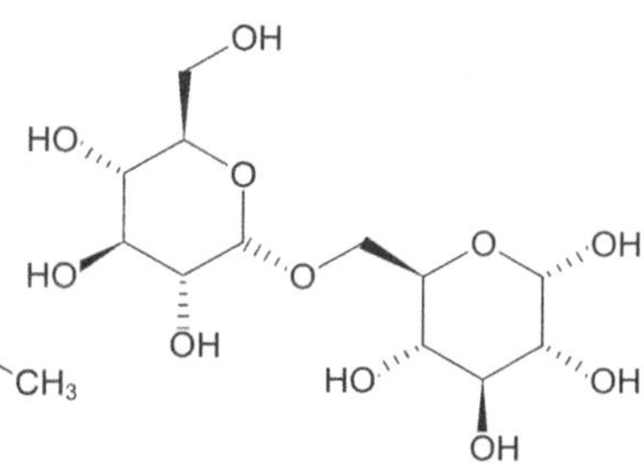

6-O-β-D-Glucopyranoside

Figure 7.6 Isoflavone compounds in vine tea extract.

to cause stomach ulcers, ischemic cardiac diseases, cerebrovascular diseases, hypertension, hyperlipemia, or the like.

An anti-stress agent has been developed that can be taken repeatedly and daily without any safety concerns and which can mitigate and prevent mental and physical symptoms caused by stress.

The anti-stress agent contains fermented sour milk as effective agent having an anti-stress effect. The anti-stress effect of this anti-stress agent can be confirmed, employing the rise in blood pressure, the rise in heart rate, by determining the suppressing effects of the agent before and after the intake of the agent.

The functional food may suitably contain sugars, proteins, lipids, vitamins, minerals, flavors, or coloring agents in addition to the anti-stress agent, depending on the type of food (17),

7.2.8 *Microalgae*

Microalgae are ancestral living organisms that constitute the basis of aquatic food chains (18). They are a phylogenetically diverse group, encompassing a number of different phyla and classes of organisms, in some cases, cyanobacteria are also included (19, 20). Microalgae grow well not only in freshwater, seawater, and hypersaline environments, but also in moist soils and rocks (21).

Microalgae may be regarded as a promising food or feed ingredient, owing to their nutritional features (22). A number of studies have claimed benefits of such microalgal compounds upon human health due to their anticancer, anti-inflammatory, antioxidant, antimicrobial, and anti-obesity capacities, in addition to their hypocholesterolemic character. Hence, they may serve as nutraceuticals (23), and their market value in food is expected to rise in the near future (24).

7.2.9 *Phenolics*

Phenolics are a class of plant-derived natural secondary metabolites with good anticancer, anti-inflammatory, antimicrobial and antioxidant activities (25). Phenolic compounds may also have the potential to be ideal food additives and preservatives.

Unlike the traditional plant extraction and chemical biosynthesis methods, metabolic engineering of microbes provides a possible way to sustainably produce phenolic compounds on a large scale, with high commercial availability and low environmental impact. In the last decade, many phenolic compounds have been successfully synthesized and optimized in the model organisms, especially in *Escherichia coli* and *Saccharomyces cerevisiae*. The recent efforts concerning their sustainable biosynthesis of phenolic compounds for functional food applications have been discussed (25).

7.2.10 Liposomes

Liposomes play a significant role in the encapsulation of various bioactive compounds, including functional food ingredients to improve the stability of the core (26). This technology can be used for promoting an effective application in functional food and nutraceuticals. The incorporation of traditional and emerging methods for the developments of liposome for loading bioactive compounds resulted in viable and stable liposome formulations for industrial applications. Thus, advanced technologies such as supercritical fluidic methods, microfluidization, and ultrasonication with traditional methods have been discussed (26).

Liposomes loaded with plant and animal bioactive compounds have been introduced for functional food and nutraceutical applications. In general, the application of liposome systems improves the stability, delivery, and bioavailability of bioactive compounds in functional food systems and nutraceuticals (26).

7.2.11 Diacylglycerol Functional Oil

Diacylglycerol (DAG) is a world leading anti-obesity functional cooking oil. It is synthesized via the structural modification of conventional fats and oils. DAG exists in three stereoisomers, namely sn-1,2-DAG, sn-1,3-DAG, and sn-2,3-DAG. 1,2-Diacylglycerol is shown in Figure 7.7.

It has been demonstrated that an sn-1,3-DAG has potential in suppressing body fat accumulation and lowering the postprandial serum triacylglycerol, cholesterol and the glucose level. DAG also was shown to improve bone health. This is attributed to the structure of DAG that causes it to absorb and digest via a different metabolic pathway than conventional fats and oils.

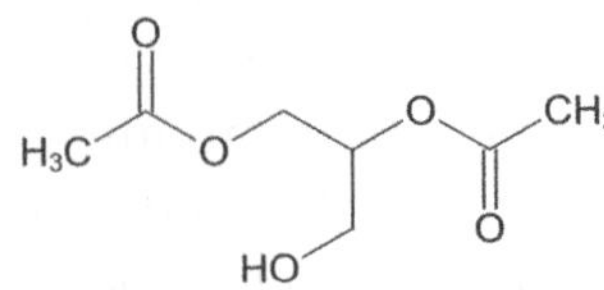

Figure 7.7 1,2-Diacylglycerol.

DAG has received much attention as a low calorie fat substitute and has been incorporated into various food matrixes.

However, DAG was banned from sale as it was found to contain probable carcinogen glycidol fatty acid esters (27).

7.3 Prevention of Oxidation

Milk, chocolate, butter, and other foods, when exposed to light, such as sunlight or fluorescent light, may develop a characteristic off-flavor caused by photo-oxidation or air oxidation (28). This tendency to develop an off-flavor significantly reduces the shelf life of foods susceptible to photo-oxidation or air oxidation. Photo-oxidation or air oxidation can also cause plastics to develop undesired characteristics over time, and reduce the stability of pharmaceuticals.

7.3.1 Photo-oxidation

A method of reducing photo-oxidation or air oxidation in susceptible materials such as food, plastics or pharmaceuticals has been developed (28).

This can be done by adding an antioxidation composition comprising a carboxylic acid, a metal ion, and a compound comprising an amino acid or an amino acid moiety.

Lysine and aspartic acid were used as amino acids. In addition, malic acid and citric acid were used. These compounds are shown in Figure 7.8.

An example of an antioxidation composition that may be used to protect food against photo-oxidation or air oxidation is shown in Table 7.3.

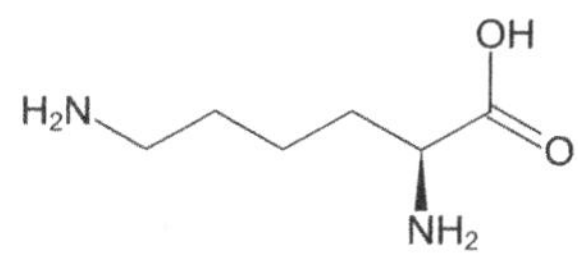

Lysine

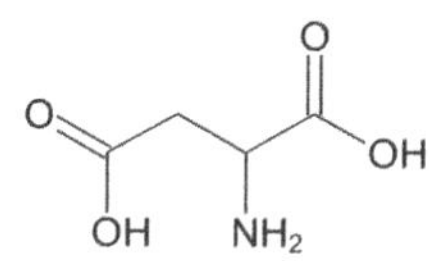

Aspartic acid

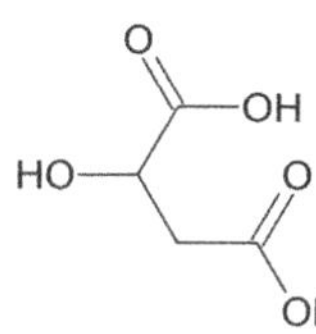

Malic acid

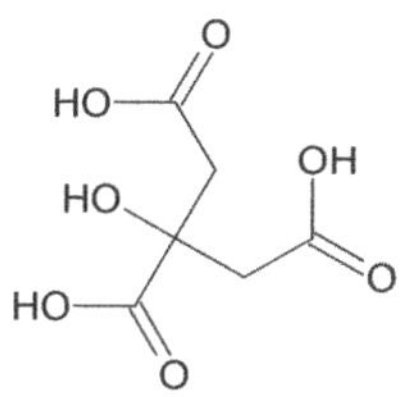

Citric acid

Figure 7.8 Acids for antioxidation composition (28).

Table 7.3 Antioxidation composition (28).

Compound	Amount/[g]
Water	160
Lysine	110
Magnesium oxide	40
Malic acid	130
Citric acid	62

The pH of this solution varies from 4.4 to 4.8 and is approximately 65% solid solution (28).

7.3.2 *Lipid Oxidation*

Lipid oxidation is the main cause of spoilage, color change and degradation in foodstuffs, which significantly limits the preservation and shelf life of packaged food (29). Non-natural antioxidants have been used in the polymer industry to prevent thermal degradation of polymers during processing. However, the potential toxicity from such antioxidants in food products makes them unsuitable for use in food packaging. Furthermore, consumers have a strong prejudice against the presence of non-natural substances in food.

7.3.3 *Natural Antioxidants*

Synthetic antioxidants, such as butylated hydroxytoluene and butylated hydroxyanisole, are the compounds usually used to maintain the quality of ready-to-eat food products (30). However, because of safety concerns the food industry is interested in natural alternatives.

Phenolic antioxidants can inhibit free radical formation and/or interrupt propagation of autoxidation. Fat-soluble vitamin E (α-tocopherol) and water-soluble vitamin C (*L*-ascorbic acid) are both effective in the appropriate matrix. Plant extracts, generally used for their flavoring characteristics, often have strong H-donating activity, thus making them extremely effective antioxidants. This antioxidant activity often occurs due to phenolic acids, phenolic diterpenes, flavonoids, and volatile oils. Some plant pigments can chelate metals and donate hydrogen to oxygen radicals, thus slowing the oxidation (30).

A food packaging material has been developed that contains a polymeric material and a natural antioxidant (29). The natural antioxidant may be extracted, isolated or derived from a plant material.

The natural antioxidant may be selected from compounds shown in Table 7.4. Some of these compounds are also shown in Figure 7.9.

The preparation of antioxidant-containing films has been described as follows (29):

Table 7.4 Natural antioxidants (29).

Compound	Compound	Compound
Rutin	Capsaicin	Pinocarveol
Lupane	Carvacrol	Rofficerone
Lutein	Eupafolin	Cirsimaritin
Thymol	Genkwanin	Hydroquinone
Borneol	Oleanolic acid	Ursolic acid
Botulin	Quercetin	*β*-Amyrin
Camphor	Rhamnetin	Acetyleugenol
Carotol	Rosmadial	Atuntzensin A
Cuminal	Verbenone	Linoleic acid
Diosmin	Zingerone	Rosmriquinone
Eugenol	1,8-Cineole	Safficinolide
Myrcene	Eriocitrin	Terpinen-4-ol
Sageone	Eucalyptol	*β*-Pinene
Shogaol	Hesperidin	Betulinic acid
Tannins	Hispidulin	Bornyl acetate
Rosmanol	Isoeugenol	Rosmariquinone
Apigenin	Kaempherol	*α*-Amyrin
Catechin	Salvigenin	*α*-Pinene
Gingerol	Anthocyanin	Homoplantaginin
Linalool	Capsaicinol	Rosmaridiphenol
Luteolin	Epirosmanol	*β*-Carotene
Oleanane	Eriodictyol	Dihydroquercetin
Physcion	Gallic acid	Methyl jasmonate
Isorosmanol	*p*-Cymene	
Dihydrokaempherol		
β-Sitosterol		
γ-Terpinene		
o-Quinone		
p-Quinone		
Ascorbyl palmitate		
Dimethoxy-rosmanol		
7-Methylepirosmanol		
Apigenin-7-O-glucoside		
Luteolin-7-O-glucoside		
Rosmanol-7-ethylether		
p-Cumene-2,3-diol		
4-Hydroxycinnamoylmethane		
Isoscutellarein 7-O-glucoside		
5-Hydroxy-7,4′-dimethoxy flavone		
5,4′-Dihydroxy-6,7,8-trimethoxy flavone		
1-Methyl-2-(1-methylethyl)benzene		

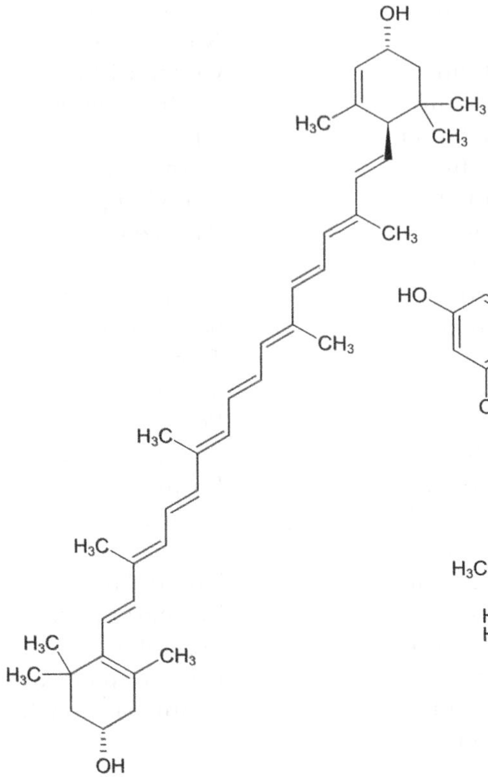

Lutein

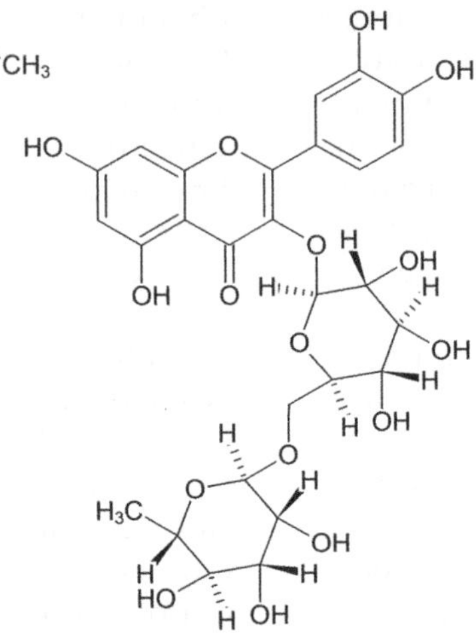

Rutin

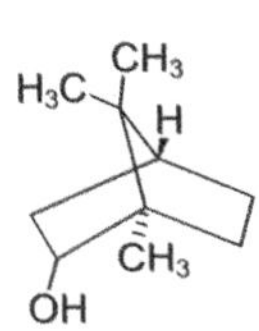

Borneol

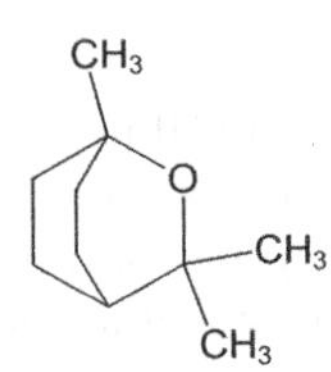

1,8-Cineole

Figure 7.9 Natural antioxidants.

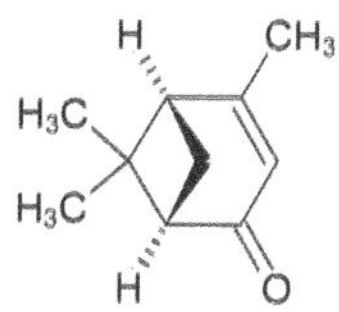

Verbenone

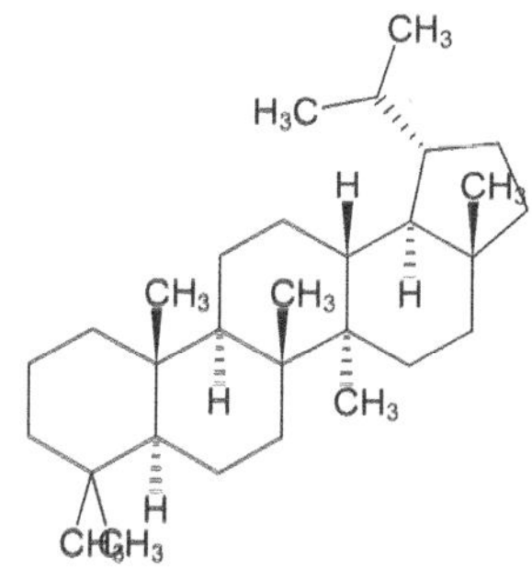

Lupane

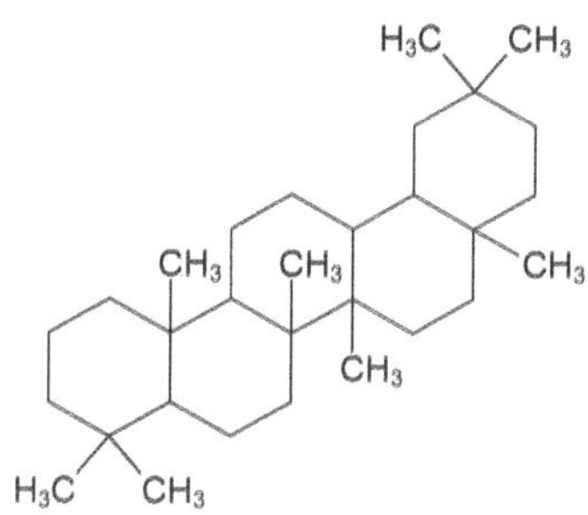

Oleanane

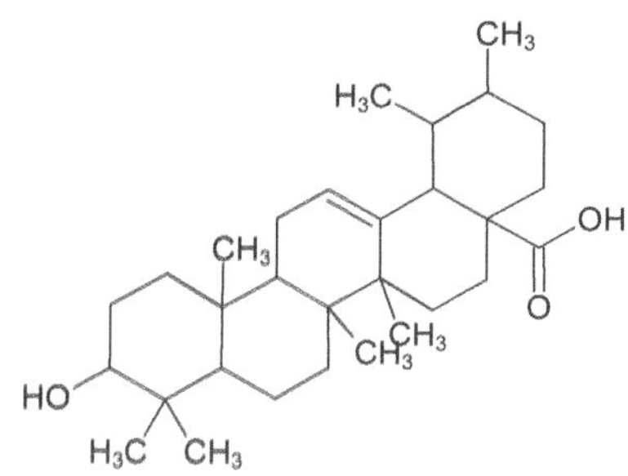

Ursolic acid

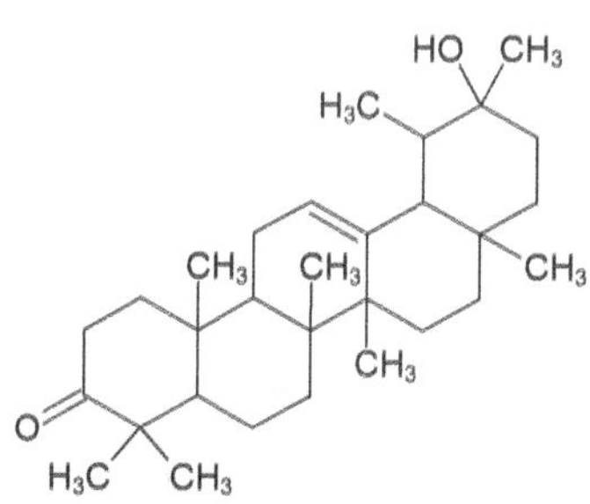

Rofficerone

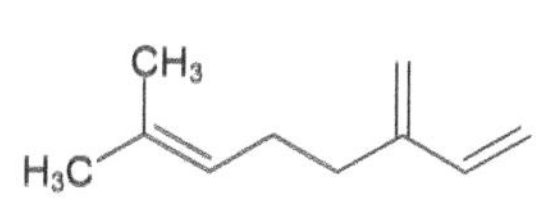

Myrcene

Figure 7.9 (cont) Natural antioxidants.

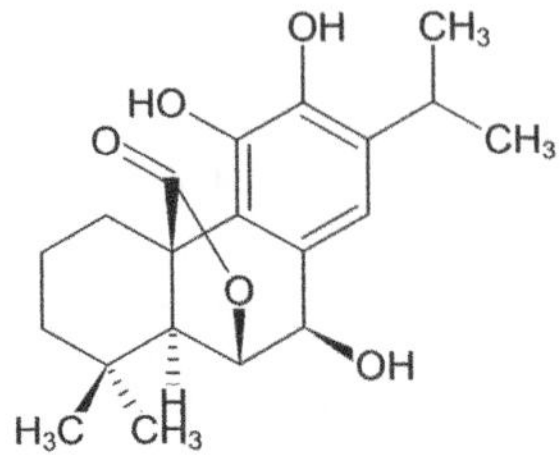

Epirosmanol

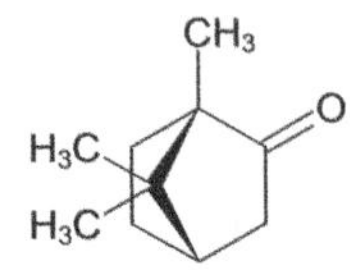

Camphor

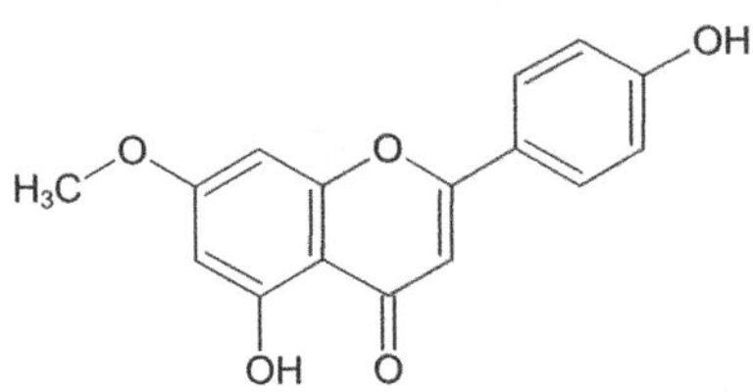

Genkwanin

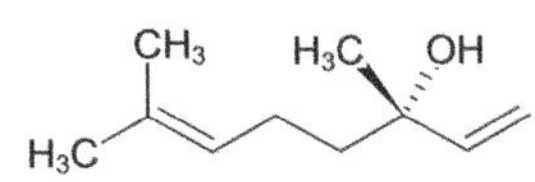

Linalool

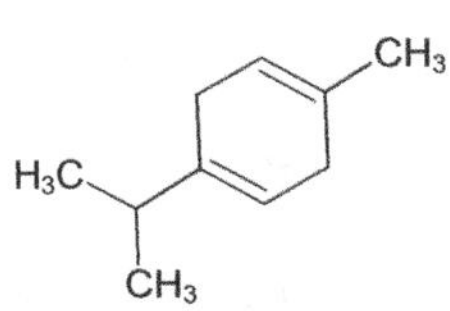

γ-Terpinene

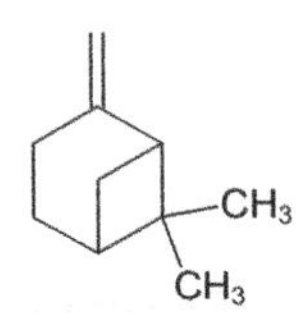

β-Pinene

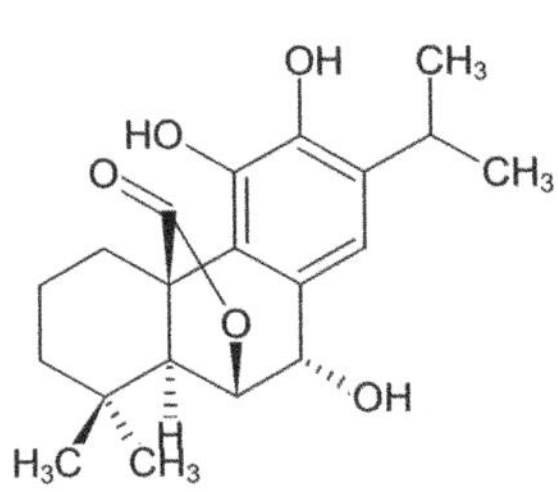

Rosmanol

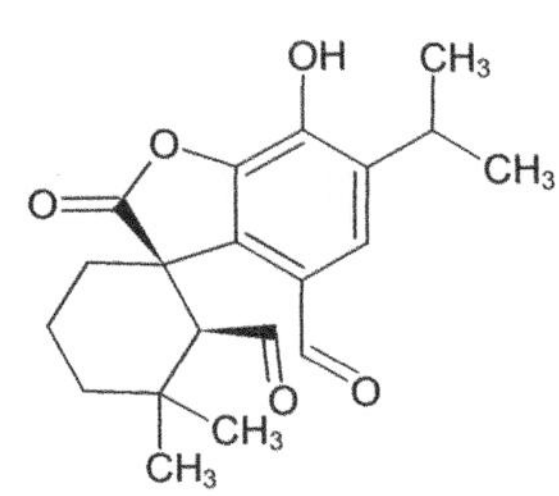

Rosmadial

Figure 7.9 (cont) Natural antioxidants.

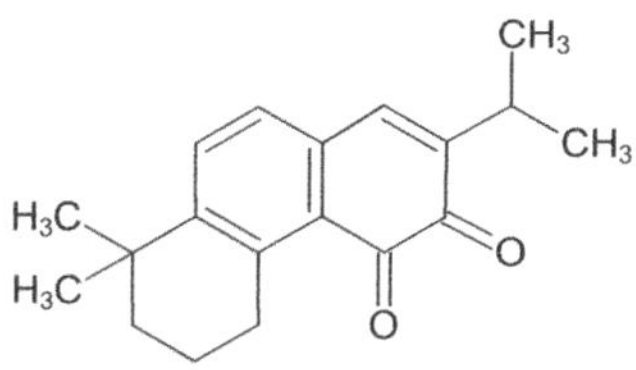

Rosmariquinone

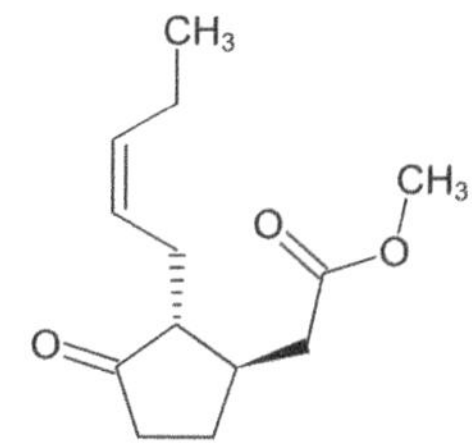

Methyl jasmonate

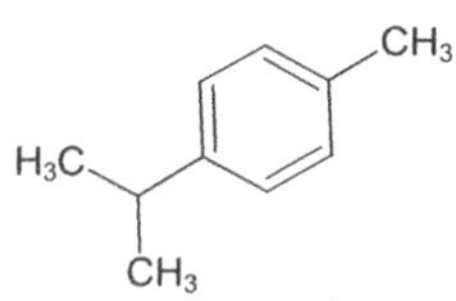

p-Cymene

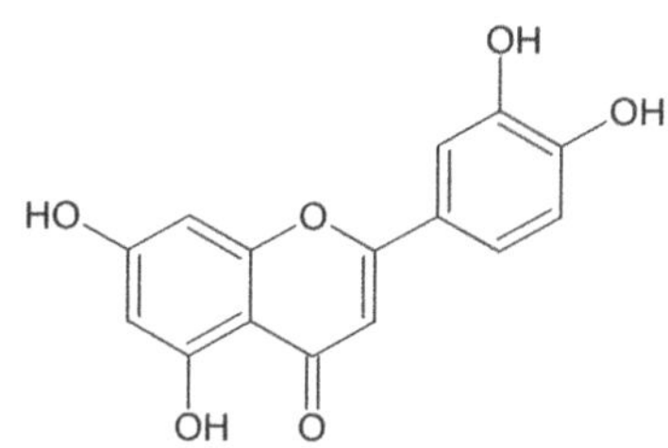

Luteolin

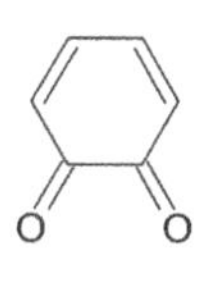

o-Quinone

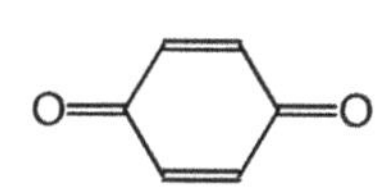

p-Quinone

Figure 7.9 (cont) Natural antioxidants.

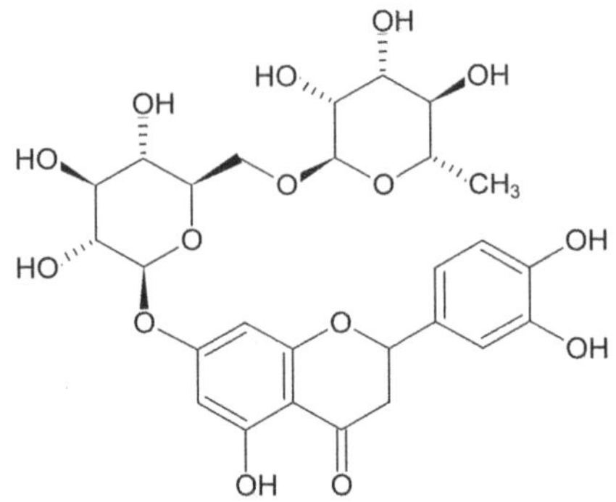

Eriocitrin

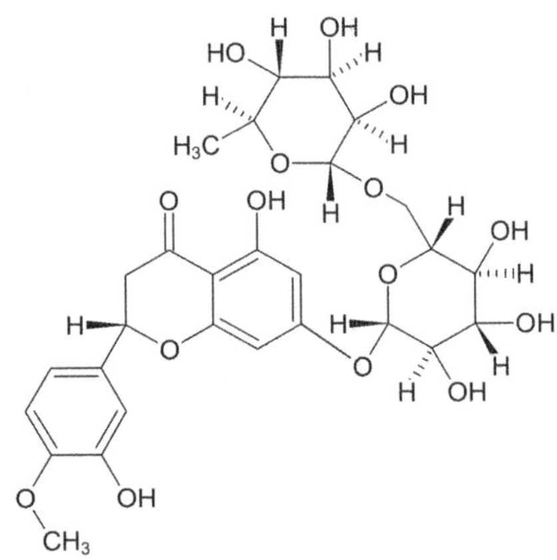

Hesperidin

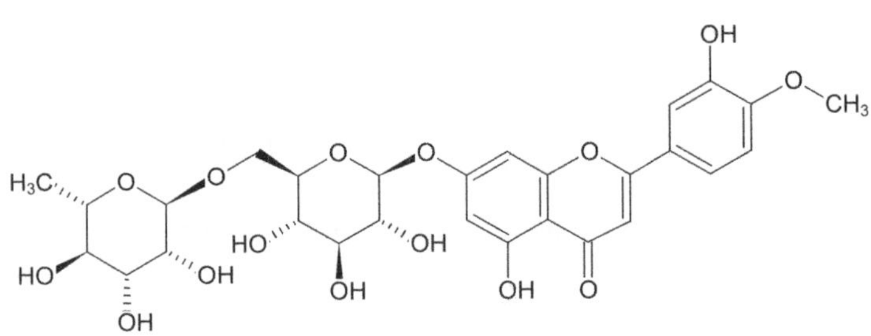

Diosmin

Figure 7.9 (cont) Natural antioxidants.

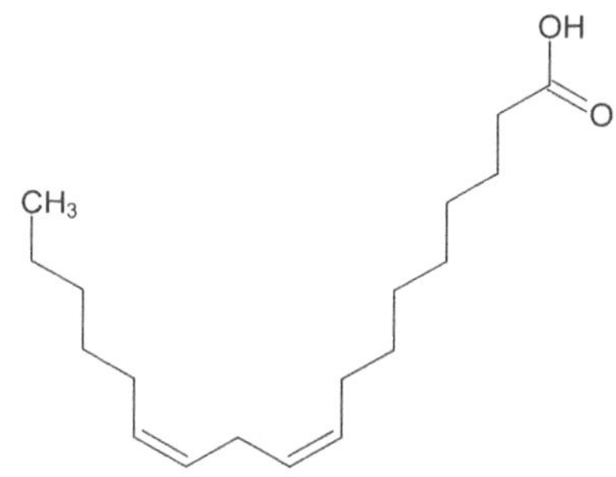

Linoleic acid

Figure 7.9 (cont) Natural antioxidants.

Preparation 7–1: A masterbatch consisting of a food grade rosemary extract and a poly(ethylene) (PE) resin was prepared. The rosemary extract may contain 24-26% total antioxidative phenolic diterpenes with >16% of carnosic acid, 1-4% essential oil, <1% water, less than 2% alcohol, sunflower oil (organic) and cuticular waxes. The masterbatch was added to a polymer resin and then extruded to form a thin film.

7.3.3.1 Food Ingredients from Plants

Interest in the content of natural antioxidants in plant-based foods can be from the human health perspective, in terms of how these compounds might help promote one's health and wellness, or from the storage point-of-view, as the endogenous antioxidant constituents aid in extending a foodstuff's shelf life.

In a chapter of a monograph, essential information is given about the mechanism of the action of antioxidants and the methods employed for determination of their activity (31). Furthermore, the classes of phenolic compounds, sources of plant antioxidants, and the extraction strategies of phenolic compounds from a plant material are detailed. Also, the influence of processing and storage on the content of natural antioxidants in foods and their antioxidant activity are described.

7.3.4 *Extraction Methods for Natural Antioxidants*

Antioxidants can be extracted from different parts in plants, such as leaves, roots, stems, fruits, seeds and peels (32, 33).

The quality of natural extracts and their antioxidant power depends not only on the quality of the original source (e.g., geographic origin, nutritional aspects and storage) but also on the technologies applied for their extraction. So far, extraction processes have been mainly performed on a laboratory scale. Scale-up is not direct because it strongly depends, for example, on complex transport phenomena. However, some authors have already reported applicability at higher scales.

The extraction of polyphenols from lettuce was investigated by a solvent-free microwave extraction method (34).

Solvent extraction for phenolic compounds from pomegranate waste was used, which was also successfully implemented on a pilot plant scale (35).

Furthermore, the scaling-up of the supercritical fluid extraction of phenolic and glucosinolate from rocket salad was studied (36).

There are no standard procedures for performing the extraction of all natural antioxidants, since each compound has its own chemical and physical properties, and they are present in quite different solid matrixes (33).

However, the extraction with organic solvents is the most used common process (37, 38)

Along with conventional solvent extraction procedures, other strategies may be used, such as extraction with supercritical fluids, high hydrostatic pressure, microwaves and ultrasound (39). However, these more recent technologies usually present higher investment costs, but often lower environmental impact (34).

Some examples of methods used for the extractions of natural antioxidant compounds from different sources are shown in Table 7.5.

7.3.4.1 *Solvent Extraction*

Solvent extraction, i.e., solid-liquid extraction or liquid-liquid extraction, involves the choice of solvents and the use of heat and/or stirring (40).

Table 7.5 Methods for the extraction of natural antioxidants (33).

Extraction Process	Source	Material extracted
Organic Solvents		
Ethanol, dichloromethane, hexane	Coffee leaves	Chlorophylls and carotenoids
Ethanol, acetone and water	*Baccharides* species	Phenolic content
	Sweet potato	Polyphenols and anthocyanins
Methanol, ethanol and acetone	Spent grain	Phenolic content
	Spice herb	Phenolic content
	Peel of eggplant	Total phenolics, flavonoids, tannins and anthocyanins
Methanol, ethanol, acetone and water	Peach fruit Flavonoids and phenolic compounds	
Ethanol	Jussara fruit	Phenolic compounds
Acetone, acetic acid, water	Basil leaves	Phenolic compounds
Water, acetone and ethanol	Ginkgo biloba leaves Flavonols	
N-Hexane and methanol	Pineapple peel	Polyphenols
Hexane, ethyl acetate, chloroform, butanol, methanol and water	Ripe bananas Phenols and flavonoids	
Methanol, water and ethanol	Chestnut byproducts	Polyphenols
	Kumquat peel	Phenolic and flavonoid content
Water and ethanol	*Withnaia somnifera* herb	Phenolic compounds
	Propolis	Phenolic compounds

Table 7.5 (cont) Methods for the extraction of antioxidants (33).

Extraction Process	Source	Material extracted
Supercritical fluid	Mango peel	Carotenoids
Supercritical fluid	Apple pomace	Phenolic compounds
Supercritical fluid	Myrtle leaves and berries	Phenolic acids, flavonoids and anthocyanins
Supercritical fluid	Green algae	Carotenoids and phenolic compounds
Supercritical fluid	Cape gooseberry	Phenolic compounds and β-carotene
High Hydrostatic Pressure	Red macroalgae	Proteins, polyphenols and poly(saccharide)s
High Hydrostatic Pressure	Tomato pulp	Flavonoids and lycopene
High Hydrostatic Pressure	Watercress	Phenolic acids and flavonoids from watercress
High Hydrostatic Pressure	Papaya seeds	Phenolic content
Pressurized liquid extraction	Peppermint	Phenolic compounds and essential oils
Pressurized liquid extraction	Carrot byproducts	Carotenoids
Pressurized liquid extraction	Goldenberry	Polyphenols
Pressurized liquid extraction	Goji berry	Phenolic compounds
Pressurized liquid extraction	Feijoa peel	Phenolic compounds
Pressurized liquid extraction	Grape marc	Anthocyanins and phenolic compounds
Pressurized liquid extraction	Spent coffee grounds	Polyphenols
Pressurized liquid extraction	Olive pomace	Phenolic compounds

Table 7.5 (cont) Methods for the extraction of antioxidants (33).

Extraction Process	Source	Material extracted
Ultrasound-assisted extraction	Green propolis	Phenolic compounds
Ultrasound-assisted extraction	Strawberry trees	Anthocyanins
Ultrasound-assisted extraction	Carotenoids	Pomegranate wastes
Ultrasound-assisted extraction	Blueberry pomace	Phenolic compounds
Ultrasound-assisted extraction	Mango peel	Pectin and phenolic compounds
Microwave-assisted extraction	Pomegranate peels	Phenolic compounds
Microwave-assisted extraction	*Phaleria macrocarpa* fruit peel	Phenolic compounds
Microwave-assisted extraction	Gac peel	Carotenoids

Surface active agents, also called surfactants, are amphiphilic molecules composed of a polar *head* group connected to hydrophobic *tail* group. These compounds may be classified as:

1. Anionic surfactants, containing anionic functional groups at the head,
2. Cationic surfactants, containing cationic functional groups at the head,
3. Zwitterionic surfactants, that have both anionic and cationic moieties at the same molecule, and
4. Nonionic surfactants.

A substance goes into supercritical state when its forced to a pressure and temperature above the critical point. Physicochemical properties of supercritical fluid are placed between the liquid and the gas, so this state has been defined as a state when liquid and gas cannot be distinguished. A substance that becomes supercritical fluid has a density similar to liquid, viscosity of a gas and diffusivity between these two states (41). Compared with conventional extraction techniques, such as solid liquid extraction, Soxhlet extraction or classic solvent extraction, besides decreased consumption of toxic solvents and better selectivity, supercritical fluid extraction is also characterized by some additional advantages (40):

1. Better mass transport resulting from low viscosity and high diffusivity is desired, especially in the case of extraction of target compounds from solid samples,
2. A supercritical fluid can easily diffuse through a solid sample matrix, so a faster extraction is achieved,
3. A better efficiency is obtained by increasing extraction yields and use of smaller amounts of solvents,
4. The possibility of direct coupling of supercritical fluid extraction with analytical instrumental techniques like gas chromatography (41),
5. The possibility of modifying physical conditions (like density) to extract the target compounds (42), and
6. It produces no toxic residues (43).

A solid-liquid extraction process is normally executed in a Soxhlet apparatus. The advantages of using a Soxhlet extractor include, among others, the repetitiveness of placing fresh solvent in contact with the solid matrix, and that no filtration is required at the end of the process.

However, it also has disadvantages, such as the need for large quantities of solvents, and consequently, an evaporation/concentration process and a possible thermal degradation of the compounds, as the process is usually carried out at the boiling point of the solvents for a long period of time (38, 44, 45).

7.3.4.2 Supercritical Fluid Extraction

Supercritical fluid extraction is an environmentally friendly alternative process to conventional organic solvent extraction, as it uses fluids in their supercritical states as solvents, avoiding the use of large quantities of toxic solvents that are undesirable in the food industry. In addition to this, it is carried out in the absence of light and oxygen, which reduces the degradation of the compounds (46, 47).

Carbon dioxide is the most commonly used solvent to obtain supercritical conditions, mainly due to its low toxicity, low cost, easy separation from extracted solutes, compatibility with foodstuffs and possibility to be used when low temperatures are required (48).

Supercritical fluid extraction with carbon dioxide is an efficient alternative process to conventional solvent extraction methods, especially for extracting lipophilic plant materials such as lipids, essential oils and aroma compounds. The same does not happen with more hydrophilic substances, such as phenolics, alkaloids and glycosidic compounds, which are poorly soluble in supercritical carbon dioxide (45).

This process has been applied in the extractions of carotenoids from mango peel (49) and phenolic compounds from apple pomace (50).

7.3.4.3 High Hydrostatic Pressure and Pressurized Liquid Extraction

In both high hydrostatic pressure and pressurized liquid extraction processes, the pressure is applied in order to increase the mass transfer rate between solid matrices and extraction solvent, resulting in fast and effective alternative methods for active compounds' extractions. However, high pressure can enhance the extraction efficiency, but with higher investment costs. When applying high hydrostatic pressure, a raw material/solvent mixture is hermetically sealed in a package and introduced to a vessel containing a pressure-transmitting medium (e.g., water, hydrophilic and lipophilic organic solvents at different concentrations) (51, 52).

Pressure is applied, normally in the range of 100-1000 *MPa,* at a controlled temperature, including ambient temperature, which is an advantage when heat-sensitive compounds are involved (53)

This method has been used successfully, for example, in the extraction of flavonoids and lycopene from tomato pulp and phenolic acids and flavonoids from watercress (54).

In addition, it was demonstrated (53) that high hydrostatic pressure enabled a higher yield and shorter extraction time, compared to other extraction methods (ultrasounds and conventional extractions) when applied to recover phenolic compounds and oil from papaya seeds. In the case of the pressurized liquid extraction process, the material/solvent mixture is directly introduced in a closed vessel, after which high pressures (3.3-20.3 *MPa*) and temperatures (40–200°C) are applied, enabling rapid extraction periods (3–20 *min*). Pressurized solvents remain in liquid states above their boiling points, allowing for high-temperature extraction. There is an improvement of compounds' solubilities and the desorption kinetics from the matrices (55). This process has been used to extract several bioactive natural compounds.

7.3.4.4 *Ultrasound-Assisted Extraction*

Ultrasound-assisted extraction is a method in which ultrasound is applied to a mixture of a solvent with the solids containing the target molecules. It is a process whereby a combination of mixing effects with physical impacts of ultrasound on raw material (e.g., fragmentation, erosion, sonocapillary effect, sonoporation, local shear stress and disintegration of plant cell walls) increases the mass transfer rate, which may explain its enhanced extraction performance. Eventually, it allows one to reduce the time of extraction, the amount of solvent used and the energy consumption. In addition, ultrasound may also be applied in combination with other methods, such as Soxhlet extraction, microwave extraction, supercritical fluid extraction and extrusion (56).

Ultrasound-assisted extraction has been used in the extraction of bioactive compounds such as carotenoids from pomegranate wastes (57), phenolics from blueberry pomace (58) and edible oils from plant sources (45, 59). This technology has also been used as a supplementary technique to conventional ones.

An ultrasound technique was used for pectin recovery after the extraction of phenolic compounds from mango peel with a conventional ethanol-water extraction, with the purpose of generating value from the whole residue (60).

7.3.4.5 *Microwave-Assisted Extraction*

Using microwaves is another green extraction method, which is based on the direct impact on polar compounds. Microwave-assisted extraction offers a rapid delivery of energy to a total volume of solvent and solid matrix with subsequent efficient and homogeneous heating of both phases. It enables the reduction of the extraction time and solvent volumes, and can be performed in open or closed systems. In the latter case, the solvent and sample are contained in sealed vessels under a controlled temperature and pressure. The closed vessels allow the temperature of the solvent to rise above its boiling point, which decreases extraction time and subsequently increases extraction efficiency (61).

However, when solvents are nonpolar or volatile, the efficiency of this method can be very low. As such, polar solvents, such as ethanol, methanol and water should be used (45).

This process has been applied to recover antioxidants from a large number of plant materials, such as phenolics from pomegranate peels (62) and carotenoids from Gac peel (63). Even though microwave-assisted extraction is commonly applied to solid/solvent mixtures, studies have been reported in which microwaves were applied only to the plant material; namely, in the extraction of polyphenols from olive tree leaves (64).

7.3.4.6 *Bamboo*

Bamboo is a multipurpose plant known mostly for its industrial uses but is now being recognized as a potential source of bioactive compounds and natural antioxidants (65). All the parts of the bamboo plant, such as rhizome, culm shaving, leaves, roots, shoots and seeds, have clinical applications.

Studies have revealed that bamboo is a rich source of antioxidants. The main antioxidants in bamboo leaves and shoots are phenols, vitamins C and E, and mineral elements such as selenium, copper, zinc, iron and manganese. Natural antioxidants are in great demand as synthetic antioxidants being used in food and pharmaceuticals may be deleterious to health. Hence, bamboo, a fast growing plant with huge biomass, can serve as an alternative for the production of natural antioxidants (65).

7.3.4.7 *Bioactive Peptides*

Antioxidant peptides can be extracted from non-antioxidant precursor proteins from different origins by the activity of either proteolytic microorganisms or isolated enzymes (66). The main sources of bioactive peptides have been detailed. Also, the current strategies to obtain these compounds as well as their health benefits and *in vivo* biological effects were evaluated.

The utilization of fermentation processes and enzymes has been established to obtain antioxidant bioactive peptides from proteins, with isolated enzymes being the most commonly used method, due to their superior control over releasing and obtaining targeted peptides.

Antioxidant peptides have the ability to reduce the formation of oxidative products along with the induction of antioxidant enzymes *in vivo*.

In food technology, successful application in meat products strengthens the role of selected peptides as antioxidant additives, although there is a need to observe the effects of the isolated bioactive peptides in other food matrices along with studies to scale-up its production (66).

7.3.4.8 Use in Meat Products

The lipoperoxidation and the oxidation of pigments and proteins reduces the quality and nutritional value of meat products. The use of antioxidants slows down this reaction, thereby preserving the characteristics of the product during its storage and prolonging its useful life (67).

The use of synthetic antioxidants in food products has been the subject of several toxicological studies.

Investigations in this sense should be conducted considering the complex mechanism of reactive oxygen species and interactions with cellular constituents to elucidate the mechanism of action of synthetic antioxidants and natural sources. Although natural additives appear as an alternative to meet the various market niches and associate natural antioxidants to active packaging as they are progressively released into the product.

The research with respect to natural antioxidants that could be used satisfactorily in meat products have been presented, in addition to recent studies that use them in active packaging (67).

7.4 Heterocyclic Aromatic Amines in Prepared Food

Heterocyclic aromatic amines are a major class of poisonous compounds that are formed from proteinaceous foods during heat processing and flavor forming (68). These toxicants have detrimental effects on the human body and eventually bring about mutagenicity and carcinogenicity.

Owing to the presence of heterocyclic aromatic amines at a marginal level in the intricate tissue of food and their interactivity, an effective sample preparation should be employed to extract these controversial chemicals from the food matrix.

For separation and detection of heterocyclic aromatic amines, advanced extraction methods and instrumental techniques have been applied. According to the type of sample preparation, sensitivity of the experiment, laboratory conditions and available facilities, the choice of analytical equipment will be different. Various cooked food samples containing heterocyclic aromatic amines and heat processing of them have been listed. Also, sample pretreatment methods and analytical techniques that have been applied to determine heterocyclic aromatic amines have been detailed (68).

2-Amlno-1-methyl-6-phenylimidazo[4,5-b]pyrldine, c.f. Figure 7.10, is produced during cooking. It is mutagenic to bacteria and cultured mammalian cells. This material was found to induce high incidences of colon and mammary carcinomas in F344 rats when administered at a concentration of 400 *ppm* in the diet for 52 weeks. Since 2-amlno-1-methyl-6-phenylimidazo[4,5-b]pyrldine is the most abundant of the mutagenic heterocyclic amines in cooked meat and fish, the compound might be related to malignancies of the colon and breast in humans.

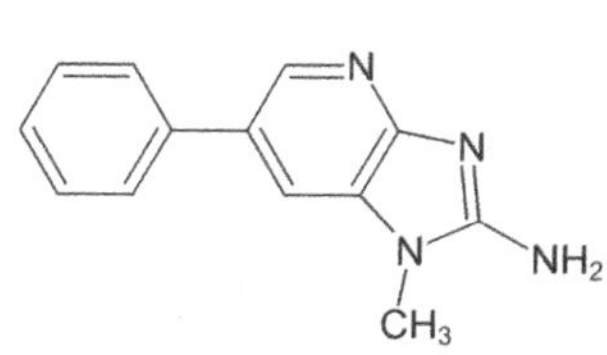

Figure 7.10 2-Amlno-1-methyl-6-phenylimidazo[4,5-b]pyrldine.

The association between prostate cancer risk and the estimated daily intake of heterocyclic amines from cooked meat and also the level of cooked-meat doneness were examined (69). Here, a population-based, case-control study involving 317 case patients with prostate cancer and 480 age-matched control subjects was carried out in Auckland, New Zealand.

The levels of meat doneness and daily intake of heterocyclic amines were determined from self-reported dietary data and experimentally measured heterocyclic amine levels in locally sourced meat samples cooked under controlled conditions to varying degrees of doneness. The heterocyclic amines found in the highest concentrations in the meat samples were 2-amino-1,6-dimethylfuro[3,2-e]imidazo[4,5-b]pyridine, c.f. Figure 7.11, and 2-amino-1-methyl-6-phenylimidazo[4,5-b]pyridine, c.f. Figure 7.10, from well-done chicken and pork and very well-done beefsteak (69).

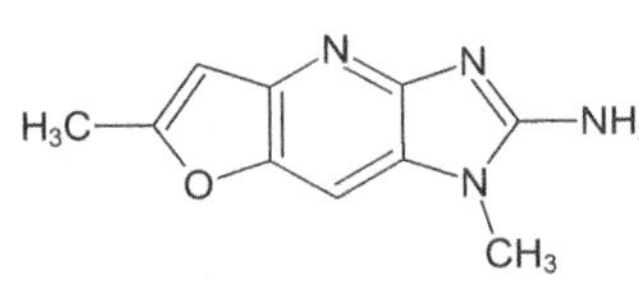

Figure 7.11 2-Amino-1,6-dimethylfuro[3,2-e]imidazo[4,5-b]pyridine.

The content and the profile of heterocyclic aromatic amines formed during the process of grilling beef under various conditions were determined. Also, different cuts of beef meat were tested (70).

The changes of the content of heterocyclic aromatic amines, depending on the chemical composition of beef meat, were investigated.

The analyzed compounds included free amino acids, glucose and protein as well as free nitrogenous bases and their nucleosides.

It was demonstrated that heterocyclic aromatic amines formation strongly correlates with the presence of various amino acids in raw beef as well as that of glucose and protein (correlation coefficient 0.84 – 0.93). It was also demonstrated that free purine bases, pyrimidine bases and their nucleosides, which had never previously been studied in terms of their role as substrates for heterocyclic aromatic amines synthesis, correlate very strongly with heterocyclic aromatic amine formation in grilled beef (correlation coefficient 0.78 to 0.99). The influence of thermal processing at high temperatures on the thermal decomposition of aminoimidazoazaarenes has not been corroborated by the present study.

2-Amino-3,4,8-trimethyl-3H-imidazo[4,5-f]quinoxaline was the most prevalent (4.50 to 7.55 $ng\,g^{-1}$), the next were 2-amino-38-dimethylimidazo[4,5-f]quinoxaline (1.10 to 7.20 $ng\,g^{-1}$), 2-amino-34-dimethyl-3H-imidazo[4,5-f]quinolone (1.70 to 5.30 $ng\,g^{-1}$), and 2-amino-4-methylimidazo[4,5-f]quinolone (0.38 to 2.00 $ng\,g^{-1}$). Some heterocyclic aromatic amines are shown in Figure 7.12.

7.5 Radiation-Curable Compositions

Migrateable residues in coatings used for packaging of foodstuffs present a health risk and consequently they should be kept to an absolute minimum (71). Known measures to reduce extractables of the photo-initiating system from cured ink layers include the use of polymeric or co-polymerizable photoinitiators and synergists instead of the usual low molecular weight compounds.

However, in many cases it was reported that the curing speed was significantly reduced and this was believed to be due to the reduced mobility of the photoinitiator or synergist.

It was found that low viscous liquid radiation-curable compositions and inkjet inks could be prepared exhibiting high shelf life stability and a very low amount of extractables in printing applications by using a diffusion hindered photoinitiator in combination with a specific polymerizable aromatic tertiary amine.

A liquid radiation-curable composition for inkjet printing has been developed (71). The composition contains a photoinitiating system consisting of one or more diffusion hindered photoinitiators.

The liquid radiation-curable composition can be advantageously used for reducing the amount of extractables in printing applications for food packaging.

The liquid radiation-curable composition contains a photoinitiating system consisting of one or more diffusion hindered photoinitiators and polymerizable co-initiators.

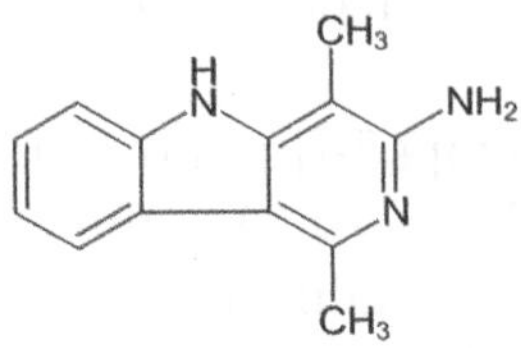

3-Amino-1,4-dimethyl-
5H-pyrido[4,3-b]indole

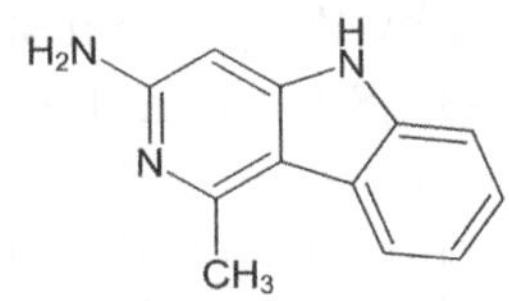

3-Amino-1-methyl-
5H-pyrido[4,3-b]indole

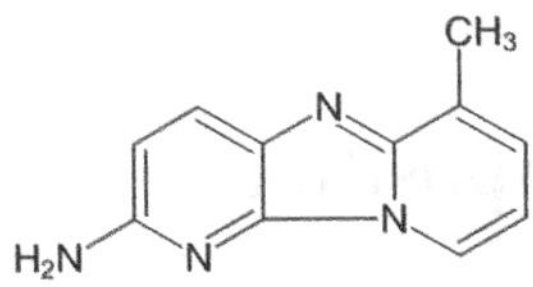

6-Methylimidazo[1,2-a:5,4-b′]
dipyridin-2-amine

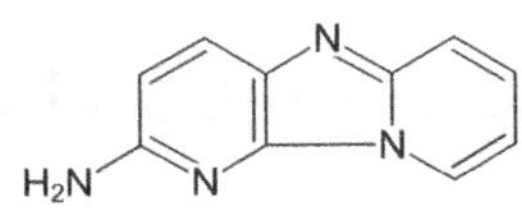

2-Aminodipyrido
[1,2-a:3,2′-d]imidazole

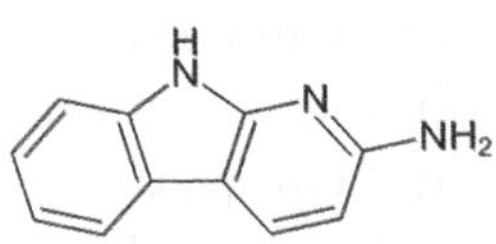

1,9-Dihydro-2H-pyrido
2,3-indol-2-imine

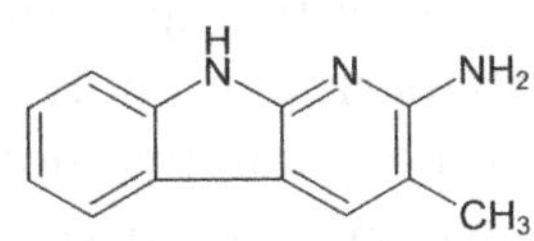

2-Amino-3-methyl-9H-
pyrido-2,3-indole

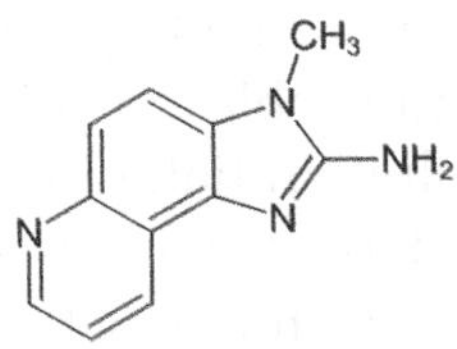

2-Amino-3-methylimidazo
4,5-f-quinolone

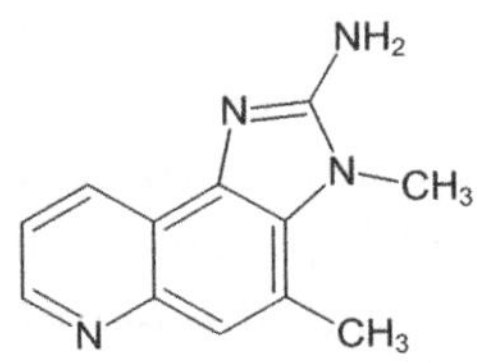

2-Amino-3,4-dimethyl-
3H-imidazo-4,5-f-quinolone

Figure 7.12 Heterocyclic aromatic amines.

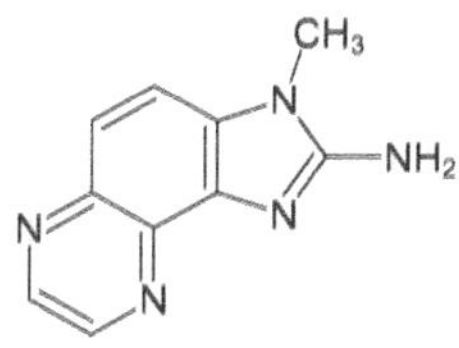

2-Amino-3-methylimidazo
(4,5-f)quinoxaline

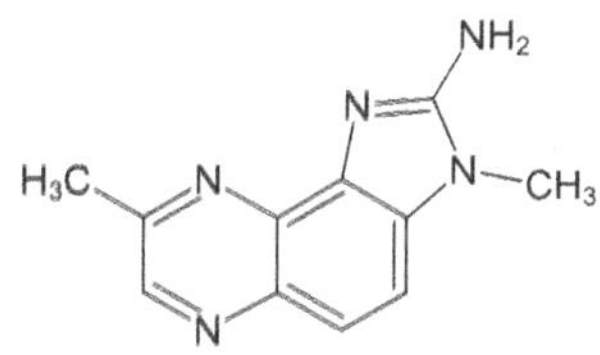

3,8-Dimethyl-3H-imidazo
(4,5-f)quinoxalin-2-amine

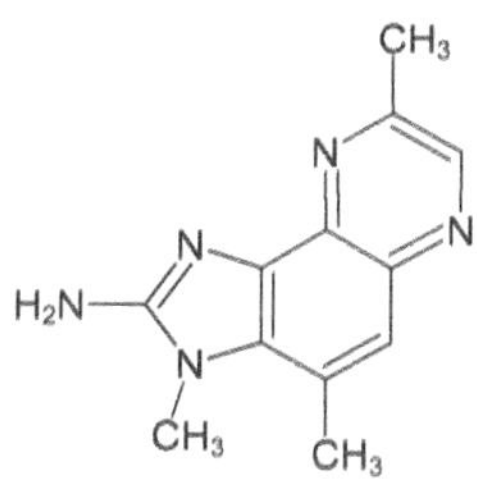

2-Amino-3,4,8-trimethyl-
3H-imidazo(4,5-f)quinoxaline

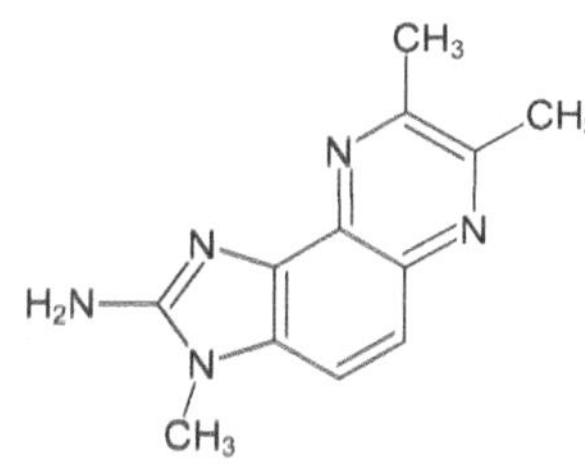

2-Amino-3,7,8-trimethyl-
3H-imidazo(4,5-f)quinoxaline

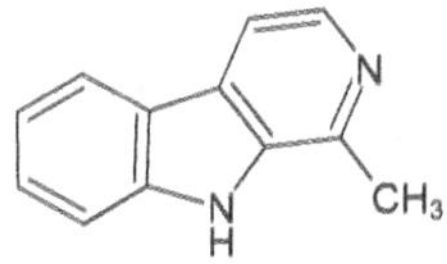

Harmane

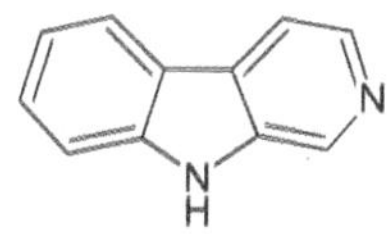

Norharmane

Figure 7.12 (cont) Heterocyclic aromatic amines.

A diffusion hindered photoinitiator is a photoinitiator which exhibits a lower mobility in a cured layer of the liquid radiation-curable composition than a monofunctional photoinitiator, such as benzophenone. Several methods can be used to lower the mobility of the photoinitiator. One way is to increase the molecular weight of the photoinitiator so that the diffusion speed is reduced, e.g., difunctional photoinitiators or polymeric photoiniators. Another way is to increase its reactivity so that it is built into the polymerizing network, e.g., multifunctional photoinitiators and polymerizable photoinitiators. The diffusion hindered photoinitiator is preferably selected from the group consisting of non-polymeric difunctional or multifunctional photoinitiators, oligomeric or polymeric photoinitiators and polymerizable photoinitiators. Non-polymeric difunctional or multifunctional photoinitiators are considered to have a molecular weight between 300 *D* and 900 *D*.

7.6 Antimycotics

The spoilage of food caused by foodborne pathogens and microorganisms is a serious problem (72). Antibacterial agents are essential to inhibit the growth of bacteria in food. To keep food fresh and prolong the shelf life, different kinds of antibacterial agents were used. Compared with natural antibacterial agents, synthetic antibacterial agents have the advantages of low cost and high activity, but their toxicity is usually higher than that of natural antibacterial agents (72).

Antimicrobial agents that have been incorporated into a polymeric matrix and their applications in food packaging are listed in Table 7.6. Some of the antimicrobial agents are shown in Figure 7.13.

Table 7.6 Antimicrobial agents for polymeric matrices (72).

Antimicrobial agents	Polymers	Targeted Microorganisms	References
Origanum vulgare and *Thymus vulgaris*	LDPE	*Salmonella typhimurium, Listeria monocytogenes, Escherichia coli*	(73)
Cinnamon essential oil	Active paper or sliced bread	*Rhizopus stolonifer*	(74)
Nisin bacteriocins	PE, ethylene vinyl acetate	*Listeria monocytogenes* V7	(75)
Nisin or pediocin	Corn starch	*Listeria monocytogenes C perfringens*	(76)
Plantaricin BM-1	PE, LDPE, HDPE	*Listeria monocytogenes*	(77)
Lysozyme	PP-*g*-PAA	*Listeria monocytogenes*	(78)
Sodium benzoate	Poly(butylene adipate-*co*-terephthalate)/organoclay nanocomposite	*Bacillus subtilis* and *Staphylococcus aureus*	(79)
Potassium sorbate and oregano essential oil	Thermoplastic starch, poly(butylene adipate-*co*-terephthalate)	Restructured chicken steaks *Escherichia coli*	(80)
Potassium sorbate and vanillin	Chitosan films	Butter cake mold	(81)
Citric acid and chitosan	Fish gelatin/chitosan composite films	*Escherichia coli*	(82)
Gallic acid grafted chitosan	Gallic acid grafted chitosan films	Agaricus bisporus	(83)

Table 7.6 (cont) Antimicrobial agents for polymeric matrices (72).

Antimicrobial agents	Polymers	Targeted Microorganisms	References
Chitosan	Acrylonitrile and acrylamide grafted chitosan	Apple and guava *Escherichia coli, Staphylococcus aureus* and *Pseudomonas aeruginosa*	(84)
Grapefruit seed extract	Carrageenan	*Escherichia coli, Staphylococcus aureus*, Bacillus cereus and Listeria monocytogenes	(85)
Grapefruit seed extract	Agar/alginate/collagen hydrogel films	Potatoes *Listeria monocytogenes, Escherichia coli*	(86)
Propyl paraben	Styrene-acrylate copolymers	*Saccharomyces cerevisiae*	(87)

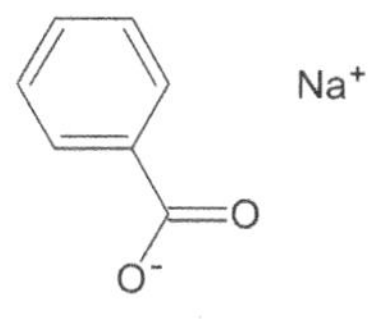

Sodium benzoate

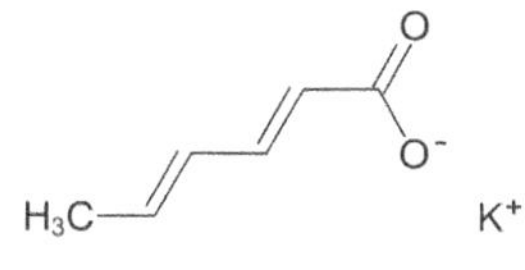

Potassium sorbate

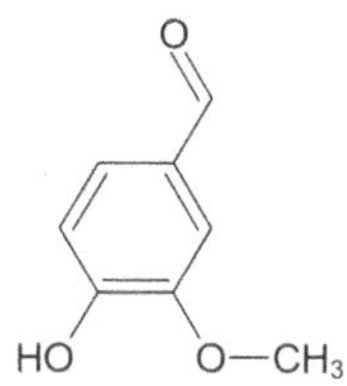

Vanillin

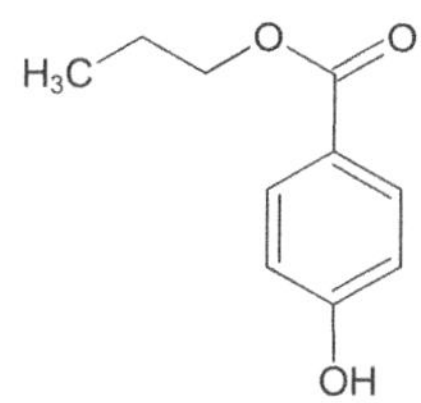

Propyl paraben

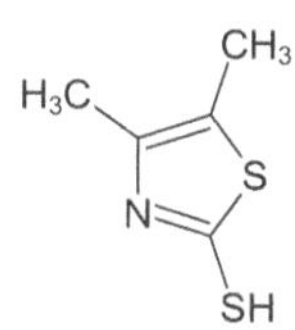

Chitosan

Figure 7.13 Antimicrobial agents.

References

1. N.L. Crane and A.L. Crane, Glaze-like coatings used to make oven bake clays food and drink safe, US Patent 10 053 391, assigned to Noah Lee Crane, Alan Laurence Crane, August 21, 2018.
2. Wikipedia contributors, Sculpey — Wikipedia, the free encyclopedia, https://en.wikipedia.org/w/index.php?title=Sculpey&oldid=1021255988, 2021. [Online; accessed 26-March-2022].
3. Wikipedia contributors, Functional food — Wikipedia, the free encyclopedia, https://en.wikipedia.org/w/index.php?title=Functional_food&oldid=1097449029, 2022. [Online; accessed 13-August-2022].

4. A. Koyazounda and J.-C. Millet, Functional food and method for preparation thereof, US Patent Application 20080057108, assigned to A'Gozz Nutrition Sarl, March 06, 2008.
5. R. Iqbal, A. Liaqat, F. Saeed, A. Khaliq, M.F.J. Chughtai, M. Afzaal, S. Tehseen, M. Aziz, M. Hussain, and F.M. Anjum, Zogale (Moringaolifera) as a functional ingredient: A review on its nutraceutical properties and food applications, *International Journal of Food Properties*, Vol. 24, p. 1202, 2021.
6. S. Desai, P. Sharma, P. Kashyap, B. Choudhary, and J. Kaur, Bioactive compounds, bio-functional properties, and food applications of Garcinia indica: A review, *Journal of Food Biochemistry*, p. e14344, 2022.
7. M. Maqsood, R. Anam Saeed, A. Sahar, and M.I. Khan, Mulberry plant as a source of functional food with therapeutic and nutritional applications: A review, *Journal of Food Biochemistry*, p. e14263, 2022.
8. A. Zayed, M.T. Badawy, and M.A. Farag, Valorization and extraction optimization of citrus seeds for food and functional food applications, *Food Chemistry*, Vol. 355, p. 129609, 2021.
9. R.C.V. Carneiro, L. Ye, N. Baek, G.H.A. Teixeira, and S.F. O'Keefe, Vine tea (Ampelopsis grossedentata): A review of chemical composition, functional properties, and potential food applications, *Journal of Functional Foods*, Vol. 76, p. 104317, 2021.
10. Y. Le, X. Ping, H. Shuangru, and W. Yuefei, Antioxidant activity of bioactive compounds extracted from Ampelopsis grossedentata leaves by optimized supercritical carbon dioxide, *Journal of Medicinal Plants Research*, Vol. 5, p. 4373, 2011.
11. Q. Gao, R. Ma, L. Chen, S. Shi, P. Cai, S. Zhang, and H. Xiang, Antioxidant profiling of vine tea (Ampelopsis grossedentata): Off-line coupling heart-cutting HSCCC with HPLC-DAD-QTOF-MS/MS, *Food Chemistry*, Vol. 225, p. 55, 2017.
12. X. Kou and N. Chen, Pharmacological potential of ampelopsin in rattan tea, *Food Science and Human Wellness*, Vol. 1, p. 14, 2012.
13. H. Li, Q. Li, Z. Liu, K. Yang, Z. Chen, Q. Cheng, and L. Wu, The versatile effects of dihydromyricetin in health, *Evidence-Based Complementary and Alternative Medicine*, Vol. 2017, 2017.
14. D. Liu, Y. Mao, L. Ding, and X.-A. Zeng, Dihydromyricetin: A review on identification and quantification methods, biological activities, chemical stability, metabolism and approaches to enhance its bioavailability, *Trends in Food Science & Technology*, Vol. 91, p. 586, 2019.
15. H. Tong, X. Zhang, L. Tan, R. Jin, S. Huang, and X. Li, Multitarget and promising role of dihydromyricetin in the treatment of metabolic diseases, *European Journal of Pharmacology*, Vol. 870, p. 172888, 2020.

16. D.-Y. Wang, Z.-Z. Zheng, S.-Y. Xu, and S.-Z. Zheng, Four new isoflavones from Ampelopsis grossedentata, *Journal of Asian Natural Products Research*, Vol. 4, p. 303, 2002.
17. A. Masuyama and M. Yasui, Anti-stress drugs and functional foods having anti-stress effects, US Patent 6 596 301, assigned to Groupe Danone SA, Calpis Co. Ltd., July 22, 2003.
18. F. Camacho, A. Macedo, and F. Malcata, Potential industrial applications and commercialization of microalgae in the functional food and feed industries: A short review, *Marine Drugs*, Vol. 17, p. 312, 2019.
19. R.A. Kay and L.L. Barton, Microalgae as food and supplement, *Critical Reviews in Food Science & Nutrition*, Vol. 30, p. 555, 1991.
20. M. Hayes, H. Skomedal, K. Skjøanes, H. Mazur-Marzec, A. Torunska-Sitarz, M. Catala, M. Isleten Hosoglu, and M. García-Vaquero, Microalgal proteins for feed, food and health in C. Gonzalez-Fernandez and R. Muñoz, (Eds.), *Microalgae-Based Biofuels and Bioproducts*, Woodhead Publishing Series in Energy, chapter 15, pp. 347–368. Woodhead Publishing, 2017.
21. S.M. Dittami, S. Heesch, J.L. Olsen, and J. Collén, Transitions between marine and freshwater environments provide new clues about the origins of multicellular plants and algae, *Journal of Phycology*, Vol. 53, p. 731, 2017.
22. A.P. Batista, A. Niccolai, P. Fradinho, S. Fragoso, I. Bursic, L. Rodolfi, N. Biondi, M.R. Tredici, I. Sousa, and A. Raymundo, Microalgae biomass as an alternative ingredient in cookies: Sensory, physical and chemical properties, antioxidant activity and in vitro digestibility, *Algal Research*, Vol. 26, p. 161, 2017.
23. J.K. Vidanarachchi, M.S. Kurukulasuriya, A. Malshani Samaraweera, and K.F.S.T. Silva, Applications of marine nutraceuticals in dairy products in S.-K. Kim, (Ed.), *Marine Medicinal Foods*, Vol. 65 of *Advances in Food and Nutrition Research*, chapter 30, pp. 457–478. Academic Press, 2012.
24. I. Priyadarshani and B. Rath, Commercial and industrial applications of micro algae – A review, *Journal of Algal Biomass Utilization*, Vol. 3, p. 89, 2012.
25. R. Mark, X. Lyu, J.J.L. Lee, R. Parra-Salívar, and W.N. Chen, Sustainable production of natural phenolics for functional food applications, *Journal of Functional Foods*, Vol. 57, p. 233, 2019.
26. K.K. Ajeeshkumar, P.A. Aneesh, N. Raju, M. Suseela, C.N. Ravishankar, and S. Benjakul, Advancements in liposome technology: Preparation techniques and applications in food, functional foods, and bioactive delivery: A review, *Comprehensive Reviews in Food Science and Food Safety*, Vol. 20, p. 1280, 2021.
27. Y.-Y. Lee, T.-K. Tang, E.-T. Phuah, C.-P. Tan, Y. Wang, Y. Li, L.-Z. Cheong, and O.-M. Lai, Production, safety, health effects and

applications of diacylglycerol functional oil in food systems: A review, *Critical Reviews in Food Science and Nutrition*, Vol. 60, p. 2509, 2020.

28. J.I. Dulebohn and R.J. Carlotti, Method for preventing photooxidation or air oxidation in food, pharmaceuticals and plastics, US Patent 6 730 518, assigned to Natura Inc., May 04, 2004.
29. P. Withers, Antioxidant active food packaging, US Patent 10 723 536, assigned to International Consolidated Business Pty Ltd., July 28, 2020.
30. M.S. Brewer, Natural antioxidants: Sources, compounds, mechanisms of action, and potential applications, *Comprehensive Reviews in Food Science and Food Safety*, Vol. 10, p. 221, 2011.
31. R. Amarowicz and R.B. Pegg, Natural antioxidants of plant origin in I.C.F.R. Ferreira and L. Barros, (Eds.), *Functional Food Ingredients from Plants*, Vol. 90 of *Advances in Food and Nutrition Research*, chapter 1, pp. 1–81. Academic Press, 2019.
32. M.A. Shah, S.J.D. Bosco, and S.A. Mir, Plant extracts as natural antioxidants in meat and meat products, *Meat Science*, Vol. 98, p. 21, 2014.
33. S.C. Lourenço, M. Moldão-Martins, and V.D. Alves, Antioxidants of natural plant origins: From sources to food industry applications, *Molecules*, Vol. 24, p. 4132, 2019.
34. S. Périno, J.T. Pierson, K. Ruiz, G. Cravotto, and F. Chemat, Laboratory to pilot scale: Microwave extraction for polyphenols lettuce, *Food Chemistry*, Vol. 204, p. 108, 2016.
35. S. Saffarzadeh-Matin and F.M. Khosrowshahi, Phenolic compounds extraction from Iranian pomegranate (Punica granatum) industrial waste applicable to pilot plant scale, *Industrial Crops and Products*, Vol. 108, p. 583, 2017.
36. M. Solana, S. Mirofci, and A. Bertucco, Production of phenolic and glucosinolate extracts from rocket salad by supercritical fluid extraction: Process design and cost benefits analysis, *Journal of Food Engineering*, Vol. 168, p. 35, 2016.
37. M. Naczk and F. Shahidi, Extraction and analysis of phenolics in food, *Journal of Chromatography A*, Vol. 1054, p. 95, 2004. Food Science.
38. J. Azmir, I.S.M. Zaidul, M.M. Rahman, K.M. Sharif, A. Mohamed, F. Sahena, M.H.A. Jahurul, K. Ghafoor, N.A.N. Norulaini, and A.K.M. Omar, Techniques for extraction of bioactive compounds from plant materials: A review, *Journal of Food Engineering*, Vol. 117, p. 426, 2013.
39. M. Oroian and I. Escriche, Antioxidants: Characterization, natural sources, extraction and analysis, *Food Research International*, Vol. 74, p. 10, 2015.

40. J. Płotka-Wasylka, M. Rutkowska, K. Owczarek, M. Tobiszewski, and J. Namieśnik, Extraction with environmentally friendly solvents, *Trends in Analytical Chemistry*, Vol. 91, p. 12, 2017.
41. M. Herrero, A. Cifuentes, and E. Ibañez, Sub- and supercritical fluid extraction of functional ingredients from different natural sources: Plants, food-by-products, algae and microalgae: A review, *Food Chemistry*, Vol. 98, p. 136, 2006.
42. C.G. Pereira and M.A.A. Meireles, Supercritical fluid extraction of bioactive compounds: Fundamentals, applications and economic perspectives, *Food and Bioprocess Technology*, Vol. 3, p. 340, 2010.
43. F. Sahena, I. Zaidul, S. Jinap, A. Karim, K. Abbas, N. Norulaini, and A. Omar, Application of supercritical CO_2 in lipid extraction – A review, *Journal of Food Engineering*, Vol. 95, p. 240, 2009.
44. B. Dimitrios, Sources of natural phenolic antioxidants, *Trends in Food Science & Technology*, Vol. 17, p. 505, 2006.
45. L. Wang and C.L. Weller, Recent advances in extraction of nutraceuticals from plants, *Trends in Food Science & Technology*, Vol. 17, p. 300, 2006.
46. M. Bleve, L. Ciurlia, E. Erroi, G. Lionetto, L. Longo, L. Rescio, T. Schettino, and G. Vasapollo, An innovative method for the purification of anthocyanins from grape skin extracts by using liquid and sub-critical carbon dioxide, *Separation and Purification Technology*, Vol. 64, p. 192, 2008.
47. M. Bimakr, R.A. Rahman, F.S. Taip, A. Ganjloo, L.M. Salleh, J. Selamat, A. Hamid, and I. Zaidul, Comparison of different extraction methods for the extraction of major bioactive flavonoid compounds from spearmint (Mentha spicata l.) leaves, *Food and Bioproducts Processing*, Vol. 89, p. 67, 2011.
48. I. Ignat, I. Volf, and V.I. Popa, A critical review of methods for characterisation of polyphenolic compounds in fruits and vegetables, *Food Chemistry*, Vol. 126, p. 1821, 2011.
49. A. del Pilar Sánchez-Camargo, L.-F. Gutiérrez, S.M. Vargas, H.A. Martinez-Correa, F. Parada-Alfonso, and C.-E. Narváez-Cuenca, Valorisation of mango peel: Proximate composition, supercritical fluid extraction of carotenoids, and application as an antioxidant additive for an edible oil, *The Journal of Supercritical Fluids*, Vol. 152, p. 104574, 2019.
50. G. Ferrentino, K. Morozova, O.K. Mosibo, M. Ramezani, and M. Scampicchio, Biorecovery of antioxidants from apple pomace by supercritical fluid extraction, *Journal of Cleaner Production*, Vol. 186, p. 253, 2018.
51. S.A. Socaci, A.C. Fărcaş, Z.M. Diaconeasa, D.C. Vodnar, B. Rusu, and M. Tofană, Influence of the extraction solvent on phenolic content,

antioxidant, antimicrobial and antimutagenic activities of brewers' spent grain, *Journal of Cereal Science*, Vol. 80, p. 180, 2018.
52. V. Briones-Labarca, C. Giovagnoli-Vicuña, and R. Cañas-Sarazúa, Optimization of extraction yield, flavonoids and lycopene from tomato pulp by high hydrostatic pressure-assisted extraction, *Food Chemistry*, Vol. 278, p. 751, 2019.
53. V. Briones-Labarca, M. Plaza-Morales, C. Giovagnoli-Vicuña, and F. Jamett, High hydrostatic pressure and ultrasound extractions of antioxidant compounds, sulforaphane and fatty acids from Chilean papaya (Vasconcellea pubescens) seeds: Effects of extraction conditions and methods, *LWT - Food Science and Technology*, Vol. 60, p. 525, 2015.
54. J. Pinela, M. Prieto, L. Barros, A.M. Carvalho, M.B.P.P. Oliveira, J.A. Saraiva, and I.C.F.R. Ferreira, Cold extraction of phenolic compounds from watercress by high hydrostatic pressure: Process modelling and optimization, *Separation and Purification Technology*, Vol. 192, p. 501, 2018.
55. Z.Y. Ju and L.R. Howard, Effects of solvent and temperature on pressurized liquid extraction of anthocyanins and total phenolics from dried red grape skin, *Journal of Agricultural and Food Chemistry*, Vol. 51, p. 5207, 2003.
56. F. Chemat, N. Rombaut, A.-G. Sicaire, A. Meullemiestre, A.-S. Fabiano-Tixier, and M. Abert-Vian, Ultrasound assisted extraction of food and natural products. mechanisms, techniques, combinations, protocols and applications. a review, *Ultrasonics Sonochemistry*, Vol. 34, p. 540, 2017.
57. A.M. Goula, M. Ververi, A. Adamopoulou, and K. Kaderides, Green ultrasound-assisted extraction of carotenoids from pomegranate wastes using vegetable oils, *Ultrasonics Sonochemistry*, Vol. 34, p. 821, 2017.
58. B.S.B. Bamba, J. Shi, C.C. Tranchant, S.J. Xue, C.F. Forney, and L.-T. Lim, Influence of extraction conditions on ultrasound-assisted recovery of bioactive phenolics from blueberry pomace and their antioxidant activity, *Molecules*, Vol. 23, p. 1685, 2018.
59. T. Li, X.-Y. Qu, Q.-A. Zhang, and Z.-Z. Wang, Ultrasound-assisted extraction and profile characteristics of seed oil from isatis indigotica fort, *Industrial Crops and Products*, Vol. 35, p. 98, 2012.
60. B.B.V. Guandalini, N.P. Rodrigues, and L.D.F. Marczak, Sequential extraction of phenolics and pectin from mango peel assisted by ultrasound, *Food Research International*, Vol. 119, p. 455, 2019.
61. T.S. Ballard, P. Mallikarjunan, K. Zhou, and S. O'Keefe, Microwave-assisted extraction of phenolic antioxidant compounds from peanut skins, *Food Chemistry*, Vol. 120, p. 1185, 2010.

62. K. Kaderides, L. Papaoikonomou, M. Serafim, and A.M. Goula, Microwave-assisted extraction of phenolics from pomegranate peels: Optimization, kinetics, and comparison with ultrasounds extraction, *Chemical Engineering and Processing - Process Intensification*, Vol. 137, p. 1, 2019.
63. H.V. Chuyen, M.H. Nguyen, P.D. Roach, J.B. Golding, and S.E. Parks, Microwave-assisted extraction and ultrasound-assisted extraction for recovering carotenoids from Gac peel and their effects on antioxidant capacity of the extracts, *Food Science & Nutrition*, Vol. 6, p. 189, 2018.
64. S. Šahin, R. Samli, A.S.B. Tan, F.J. Barba, F. Chemat, G. Cravotto, and J.M. Lorenzo, Solvent-free microwave-assisted extraction of polyphenols from olive tree leaves: Antioxidant and antimicrobial properties, *Molecules*, Vol. 22, 2017.
65. C. Nirmala, M.S. Bisht, H.K. Bajwa, and O. Santosh, Bamboo: A rich source of natural antioxidants and its applications in the food and pharmaceutical industry, *Trends in Food Science & Technology*, Vol. 77, p. 91, 2018.
66. J.M. Lorenzo, P.E.S. Munekata, B. Gómez, F.J. Barba, L. Mora, C. Pérez-Santaescolástica, and F. Toldrá, Bioactive peptides as natural antioxidants in food products — A review, *Trends in Food Science & Technology*, Vol. 79, p. 136, 2018.
67. J.S. Ribeiro, M.J.M. Cordeiro Santos, L.K.R. Silva, L.C.L. Pereira, I.A. Santos, S.C. da Silva Lannes, and M.V. da Silva, Natural antioxidants used in meat products: A brief review, *Meat Science*, Vol. 148, p. 181, 2019.
68. F. Barzegar, M. Kamankesh, and A. Mohammadi, Heterocyclic aromatic amines in cooked food: A review on formation, health risk-toxicology and their analytical techniques, *Food Chemistry*, Vol. 280, p. 240, 2019.
69. A.E. Norrish, L.R. Ferguson, M.G. Knize, J.S. Felton, S.J. Sharpe, and R.T. Jackson, Heterocyclic amine content of cooked meat and risk of prostate cancer, *Journal of the National Cancer Institute*, Vol. 91, p. 2038, 1999.
70. A. Szterk, Heterocyclic aromatic amines in grilled beef: The influence of free amino acids, nitrogenous bases, nucleosides, protein and glucose on HAAs content, *Journal of Food Composition and Analysis*, Vol. 40, p. 39, 2015.
71. J. Loccufier and R. Claes, Radiation curable compositions for food applications, US Patent 8 344 038, assigned to Agfa Graphics NV, January 01, 2013.
72. T. Huang, Y. Qian, J. Wei, and C. Zhou, Polymeric antimicrobial food packaging and its applications, *Polymers*, Vol. 11, 2019.

73. A.C. Valderrama Solano and C. de Rojas Gante, Two different processes to obtain antimicrobial packaging containing natural oils, *Food and Bioprocess Technology*, Vol. 5, p. 2522, 2012.
74. A. Rodriguez, C. Nerín, and R. Batlle, New cinnamon-based active paper packaging against rhizopusstolonifer food spoilage, *Journal of Agricultural and Food Chemistry*, Vol. 56, p. 6364, 2008.
75. A. La Storia, D. Ercolini, F. Marinello, and G. Mauriello, Characterization of bacteriocin-coated antimicrobial polyethylene films by atomic force microscopy, *Journal of Food Science*, Vol. 73, p. T48, 2008.
76. S.M.M. Meira, G. Zehetmeyer, J.O. Werner, and A. Brandelli, A novel active packaging material based on starch-halloysite nanocomposites incorporating antimicrobial peptides, *Food Hydrocolloids*, Vol. 63, p. 561, 2017.
77. M. Zhang, X. Gao, H. Zhang, H. Liu, J. Jin, W. Yang, and Y. Xie, Development and antilisterial activity of PE-based biological preservative films incorporating plantaricin BM-1, *FEMS Microbiology Letters*, Vol. 364, p. fnw283, 2017.
78. M.J. Roman, E.A. Decker, and J.M. Goddard, Metal-chelating active packaging film enhances lysozyme inhibition of listeria monocytogenes, *Journal of Food Protection*, Vol. 77, p. 1153, 2014.
79. D. Mondal, B. Bhowmick, D. Maity, M.M.R. Mollick, D. Rana, V. Rangarajan, R. Sen, and D. Chattopadhyay, Investigation on sodium benzoate release from poly (butylene adipate-co-terephthalate)/organoclay/sodium benzoate based nanocomposite film and their antimicrobial activity, *Journal of Food Science*, Vol. 80, p. E602, 2015.
80. L. Cestari, R.C. Gaiotto, J.L. Antigo, M.R.S. Scapim, G.S. Madrona, F. Yamashita, M.S.S. Pozza, and I.N. Prado, Effect of active packaging on low-sodium restructured chicken steaks, *Journal of Food Science and Technology*, Vol. 52, p. 3376, 2015.
81. J. Sangsuwan, N. Rattanapanone, and I. Pongsirikul, Development of active chitosan films incorporating potassium sorbate or vanillin to extend the shelf life of butter cake, *International Journal of Food Science & Technology*, Vol. 50, p. 323, 2015.
82. J. Uranga, A.I. Puertas, A. Etxabide, M.T. Dueñas, P. Guerrero, and K. De La Caba, Citric acid-incorporated fish gelatin/chitosan composite films, *Food Hydrocolloids*, Vol. 86, p. 95, 2019.
83. J. Liu, S. Liu, X. Zhang, J. Kan, and C. Jin, Effect of gallic acid grafted chitosan film packaging on the postharvest quality of white button mushroom (Agaricus bisporus), *Postharvest Biology and Technology*, Vol. 147, p. 39, 2019.
84. D. Kumar, P. Kumar, and J. Pandey, Binary grafted chitosan film: Synthesis, characterization, antibacterial activity and prospects for

food packaging, *International Journal of Biological Macromolecules*, Vol. 115, p. 341, 2018.

85. P. Kanmani and J.-W. Rhim, Development and characterization of carrageenan/grapefruit seed extract composite films for active packaging, *International Journal of Biological Macromolecules*, Vol. 68, p. 258, 2014.
86. L.-F. Wang and J.-W. Rhim, Preparation and application of agar/alginate/collagen ternary blend functional food packaging films, *International Journal of Biological Macromolecules*, Vol. 80, p. 460, 2015.
87. D. Chung, M.L. Chikindas, and K.L. Yam, Inhibition of Saccharomyces cerevisiae by slow release of propyl paraben from a polymer coating, *Journal of Food Protection*, Vol. 64, p. 1420, 2001.

8

Recycling

Globally, there are only low recycling rates for single-use plastic packaging materials with only 14% of plastic packaging being collected for recycling and only 5% of it successfully recycled into new plastic materials (1, 2).

Common single-use plastic packaging waste materials polluting the environment in order of quantity include drinking bottles, bottle caps, food wrappers, grocery bags, lids, straws, stirrers and foam take-out containers (3, 4).

The food industry is encouraged to adopt the reduce, reuse and recycle waste control options rather than adopt the easy direct disposal method. This waste that is disposed in landfills results in methane gas creation, a greenhouse gas that surpasses carbon dioxide by more than 2000% in causing global warming (5).

It has been observed that plastic plays a vital role in maintaining food safety, extending shelf life and minimizing food waste. However, it is currently not possible to recycle the multilayer plastic packaging widely used throughout the food industry, and there are currently no viable alternatives offering the same level of protection. Unless possible substitutes for multilayer plastics offering the same level of food protection can be developed, there will be detrimental effects on food quality, safety and shelf life, which will lead to increased food waste, additional food costs and a reduction in the variety and availability of certain foods.

The principal conventional polymers and bioplastics used in food packaging in terms of their resources or raw materials, properties, uses and applications have been described in a chapter of a monograph (6).

Food packaging plastics represent an important part of total plastic waste that ends up in landfills or accumulates in the environment when they are not properly managed. For counteracting this contamination problem, plastic recycling is the best alternative. One of the challenges to guaranteeing recycling efficiency is the adequate polymer separation of plastic waste streams because of contaminants or the presence of polymers different than the target recycled plastic, which can cause processability problems or variation of its original properties (6).

The principal polymer identification techniques have been detailed that are applied to post-consumer plastic streams in order to improve the sorting and separation of plastics before recycling. Plastic waste recycling methods associated with single and multilayer polymer systems and their stages are described. Also, a comparative analysis of the properties variation of polymers used in food packaging with mechanical recycling is presented based on recent research studies, which shows the principal degradation reactions taking place for each polymer (6).

In 2018, the recycling rate of plastic packaging was only 40% (7). However, this is an overestimation since in practice this 40% accounts for the plastic waste which is collected for recycling. The actual net recycling rate of plastic packaging was in fact lower. A study on the post-consumer plastic packaging recycling network in the Netherlands, for instance, has indicated an effective recycling yield of only 26% instead of 50% as was reported by the official authorities (8).

Besides improving recycling rates, the European Commission also stated that it is important for recyclers to work closely together with the chemical industry to help identify wider and high-value applications for their output. This together with the idea of a more circular economy implies the preference of the EU for closed-loop recycling scenarios, in which, for instance, food packaging is recycled to become a food contact material once more (7).

The recycling of food packaging materials can be estimated to be only 5% in a closed-loop process within the EU (9). This recycling process is mainly done by means of a mechanical recycling.

Here, washed and sorted plastic waste is remelted and processed into new food packaging. However, the recycling process may cause several issues such as the safe use of the recycled plastics as food packaging due to an increase in possible contamination sources, which might migrate into the food, thus imposing a risk to human health (7).

8.1 Identification of the Materials

8.1.1 *Spectroscopic Methods*

In the polymer recycling process, at least two steps might require an automatic identification tool:

1. The sorting of the plastics, and
2. The qualification of the regenerated product.

The efficiencies of reflectance FTMIR and FT-Raman were tested with respect to these identification applications. Although it was possible with both techniques to recognize most usual polymers, severe limitations of FTMIR were evidenced, such as a high sensitivity to the surface state and a weak (noisy) reflectance signal for CH bonds. On the other hand, the FT-Raman technique proved to be a rapid (1 *s* per object) and highly selective method, giving information even on the mineral fillers present in plastics (10).

8.1.2 *Real Time Mass Spectrometry*

Distinguishing poly(amide) (PA) polymers is useful in directing monomer residual testing, product testing, and reverse engineering, but is analytically challenging. The applicability of direct analysis in real time (DART) mass spectrometry for screening PA polymers was tested (11).

A DART ion source coupled to a quadrupole Orbitrap high-resolution mass spectrometer was employed for this study. Ten types of PA polymers and four retail samples were evaluated. The so obtained data for these samples and their repeatability was assessed across days/calibrations, operators, and equipment.

Ions corresponding to the cyclic or linear monomers and oligomers of each PA polymer were detected in each mass spectrum. Although similar mass spectra were obtained for PA6, PA66, and PA6/PA66 (polymer blends of PA6 and PA66), their DART tandem mass spectra were completely different.

The analysis was repeatable, and nearly identical DART tandem mass spectra were obtained on different days, by different operators, and with different equipment.

So, this technique could be successfully applied to commercially available samples (11).

8.1.3 Optical Identification

A technology for automatic sorting of plastics, based upon optical identification of fluorescence signatures of dyes incorporated in such materials in trace concentrations prior to product manufacturing, has been described (12).

Three commercial tracers were selected primarily on the basis of their good absorbency in the 310 *nm* to 370 *nm* spectral band and their identifiable narrow-band fluorescence signatures in the visible band of the spectrum when present in binary combinations. This absorption band was selected because of the availability of strong emission lines in this band from a commercial mercury (Hg) arc lamp and high fluorescence quantum yields of the tracers at this excitation wavelength band.

The plastics chosen for tracing and identification are high-density poly(ethylene) (HDPE), low-density poly(ethylene) (LDPE), poly(propylene) (PP), ethylene vinyl acetate, poly(vinyl chloride) (PVC) and poly(ethylene terephthalate) (PET), and the tracers were compatible and chemically nonreactive with the host matrices and did not affect the transparency of the plastics. The design of a monochromatic and collimated excitation source by a sensor system is described and its performances in identifying and sorting plastics doped with tracers at a few parts per million concentration levels were evaluated. In an industrial sorting system, the sensor was able to sort 300 *mm* long plastic bottles at a conveyor belt speed of 3.5 m, s^{-1} with a sorting purity of around 95%. The limitation was imposed due to mechanical singulation irregularities at high speed and the limited processing speed of the computer used (12).

8.1.4 *Chemical Safety Aspects*

Food packaging facilitates the storage, handling, transport, and preservation of food (13). Despite these beneficial aspects, food packaging raises environmental concerns due to its high production volume, often short usage time, and problems related to waste management and littering.

Chemical safety aspects of recycled food packaging were investigated. Recycling may increase the levels of potentially hazardous chemicals in the packaging and after migration in the food. Since the exposure to certain chemicals migrating from food packaging has been associated with chronic diseases, it is of high importance to assess the safety of recycled packaging products.

Typical migrants from all types of recycled food packaging materials have been listed. Also, some approaches to reduce chemical contamination are described (13).

General properties of food packaging materials with respect to their recyclability are listed in Table 8.1.

Table 8.1 General properties of food packaging materials (13).

Materials	Recycling feasibility	Comments
Thermoplastics	Yes	PET beverage bottles
Thermosets	No	
Plastic multilayers	Yes	
Paper & board	Yes	No contact with food
Metals	Yes	
Glass	Yes	

8.1.4.1 *Phthalates*

Phthalates are a group of diesters of *o*-phthalic acid, i.e., dialkyl or alkyl aryl esters of 1,2-benzenedicarboxylic acid (14).

High-molecular-weight phthalates, such as di-2-ethylhexyl phthalate, are primarily used as plasticizers to soften polyvinyl chloride (PVC) products, while the low-molecular-weight phthalates, such as diethyl phthalate, di-*n*-butyl phthalate, and butyl benzyl phthalate, are widely used as solvents to hold color and scent in various consumer and personal care products. Phthalates have become ubiquitous environmental contaminants due to volatilization and leaching from their widespread applications, and thus contamination of the environment has become another important source for phthalates in food, in addition to migration from packaging materials (14).

Some phthalates and phthalic acids are shown in Figure 8.1.

Human exposure to phthalates is increasingly becoming a concern due to the findings from toxicology studies in animals (14). Di-2-ethylhexyl phthalate, one of the important widely used phthalates, is a rodent liver carcinogen. Di-2-ethylhexyl phthalate, di-*n*-butyl phthalate, butyl benzyl phthalate, and several phthalate metabolites, such as monobutyl phthalate, monobenzyl phthalate, and mono-(2-ethylhexyl) phthalate, are teratogenic in animals. Since foods are the major source of exposure to phthalates, information on levels of phthalates in foods is important for human exposure assessment.

Identifying the knowledge gaps for future investigations by reviewing levels of a wide range of phthalates in a variety of foods, such as bottled water, soft drinks, infant formula, human milk, total diet foods, and others, migration of phthalates from various food-packaging materials, and traditional and new methodologies for the determination of phthalates in foods is the objective of a review (14).

For more than two decades, PET beverage bottles have been recycled and safely reprocessed into new food contact packaging applications (15). During the recollection of post-consumer PET beverage bottles, PET containers from non-food products are inevitably co-collected and thereby enter the PET recycling feed stream. To explore the impact of this mixing on the safety-in-use of recycled PET bottles, the concentrations of post-consumer substances in PET containers were determined that are used for a range of non-food product applications taken from the market.

Based on the chemical nature and amounts of these post-consumer substances, their potential carry-over into beverages filled in recycled PET bottles was evaluated, starting from different fractions of non-food PET in the recollection systems and taking worst-case cleaning efficiencies of super-clean recycling processes into account (15).

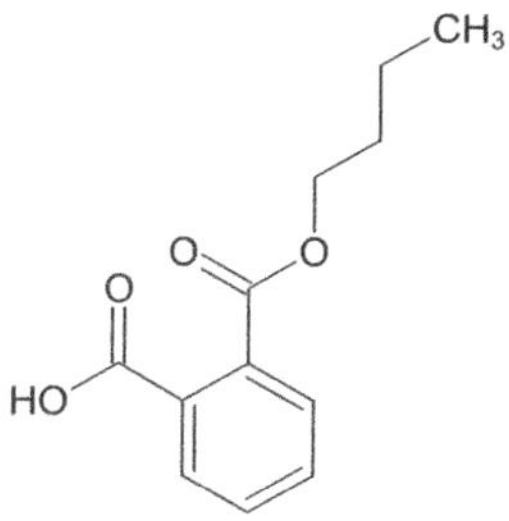

Monobutyl phthalate

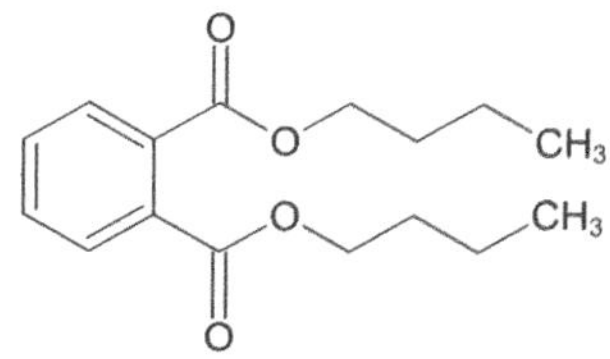

Di-*n*-butyl phthalate

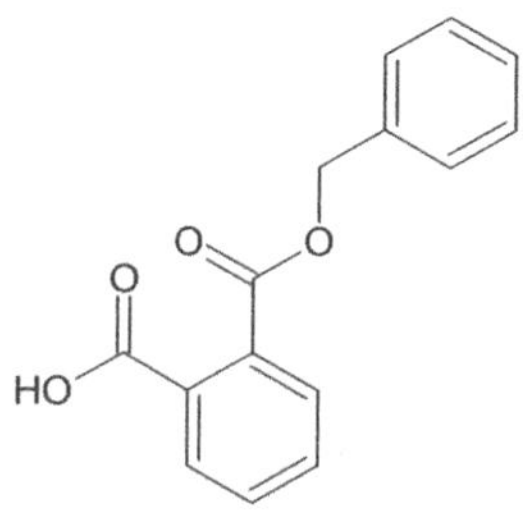

Monobenzyl phthalate

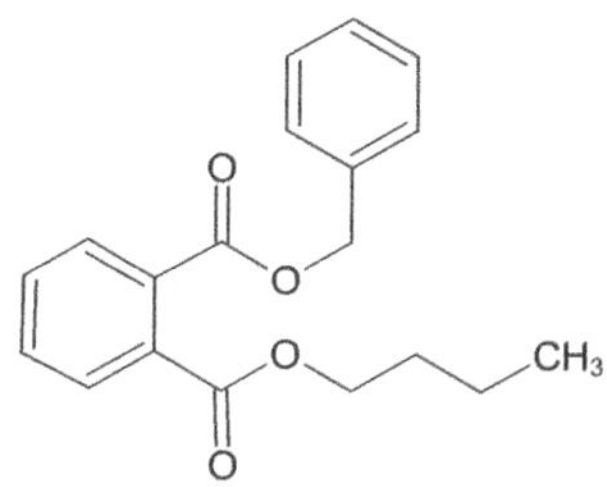

Butyl benzyl phthalate

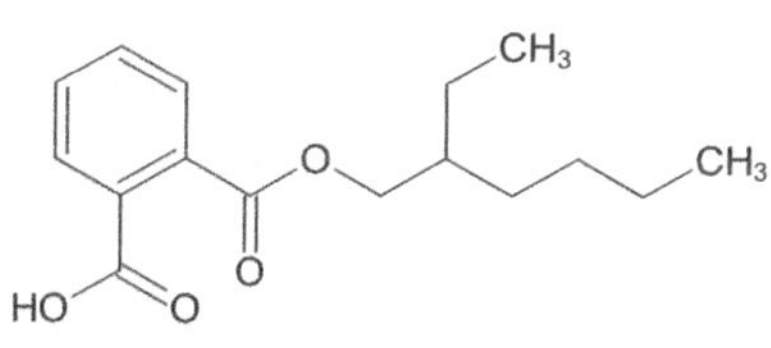

Mono(2-ethylhexyl) phthalate

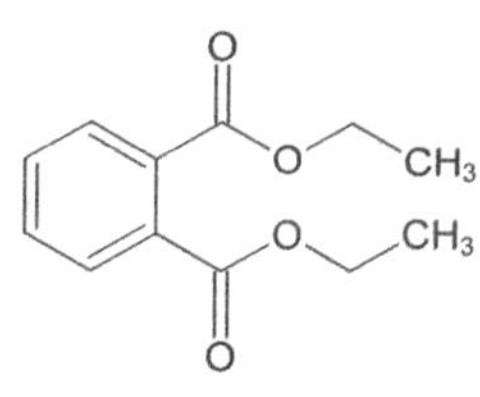

Diethyl phthalate

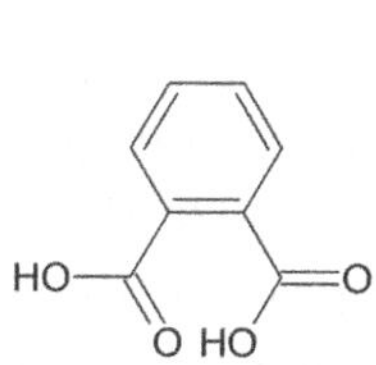

1,2-Benzenedicarboxylic acid

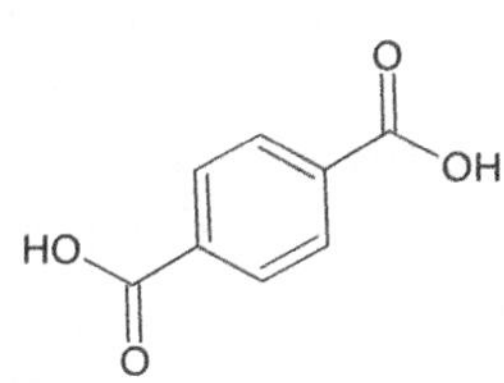

1,4-Benzenedicarboxylic acid

Figure 8.1 Phthalates and phthalic acids.

On the basis of the Threshold of Toxicological Concern concept and Cramer classification tools, a risk assessment was presented for potential exposure of the consumer to the identified contaminants as well as unidentified, potentially genotoxic substances in beverages. As a result, a fraction of 5% non-food PET in the recycling feed stream, which is very likely to occur in the usual recollection systems, does not pose any risk to the consumer. The data showed that fractions of up to 20%, which may sporadically be contained in certain, local recollection systems, would also not raise a safety concern (15).

8.1.5 Migration of Contaminants

8.1.5.1 Model Contaminants

The migration of several model contaminants from PET bottles into an Iranian yogurt drink has been investigated (16).

PET is produced from ethylene glycol and either terephthalic acid or dimethyl terephthalate. It is commonly used as packaging material for drinking water, mineral water, carbonated beverages and edible oils.

The strength and permeability properties of PET, its resistance to chemicals and its high degree of transparency are the main factors that make it more applicable. Hazardous compounds should be avoided as the bottles are manufactured by a commercial plant that produces food contact bottles. However, PET is known to contain small amounts of low-molecular-weight compounds ranging from monomers to pentamers.

The level of these compounds depends on the type of PET. Migration from packaging materials into product generally is concerned with minor constituents that influence the quality of the contained product by sensory or toxicological hazards.

There has been extensive research performed on the use of food contact PET materials. Investigations have been carried out in order to establish the type and concentration of compounds related to post-consumer PET recycling (17–21).

The amount obtained by Soxhlet extraction represents the maximum level of migration of contaminants.

Such an amount may never be attained even under long-term storage conditions. The results of gas chromatography analysis of an extract solution showed that among the three selected contaminants, di(2-ethylhexyl) phthalate had the highest concentration in a PET bottle. The results are shown in Table 8.2.

Di(2-ethylhexyl) adipate and di(2-ethylhexyl) phthalate are shown in Figure 8.2.

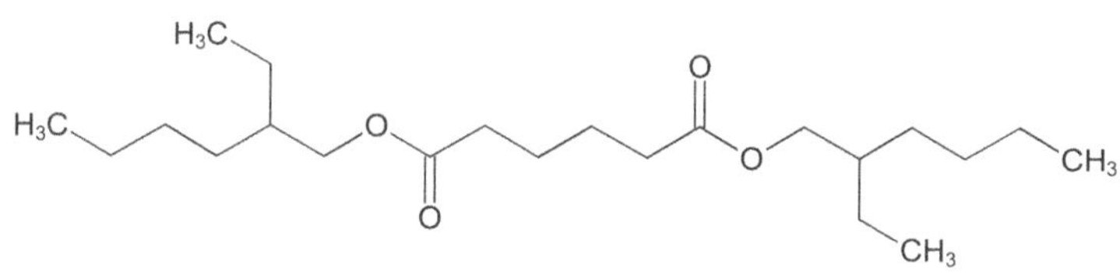

Di(2-ethylhexyl) adipate

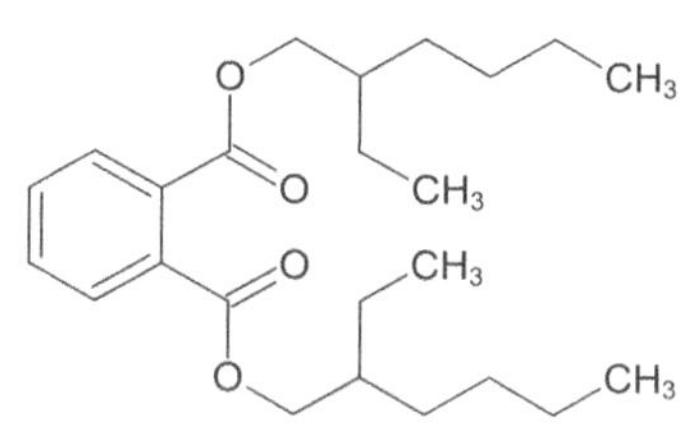

Di(2-ethylhexyl) phthalate

Figure 8.2 Di(2-ethylhexyl) adipate and di(2-ethylhexyl) phthalate.

Table 8.2 Amount of migrants (16).

Compound	Amount/[$\mu g/g$ PET]
Ethylene glycol	21
Di(2-ethylhexyl) phthalate	670
Di(2-ethylhexyl) adipate	510

Furthermore, the migration of di(2-ethylhexyl) adipate, di(2-ethylhexyl) phthalate and ethylene glycol from PET bottles into the stimulant at 25±1°C as a function of time is shown in Table 8.3.

Table 8.3 Migration as function of time (16).

	Compounds		
	Di(2-ethylhexyl) adipate	Di(2-ethylhexyl) phthalate	Ethylene glycol
Time/[d]	Amount of migration/[$mg\,l^{-1}$]		
0	0.0	0.0	0.0
9	1.1	1.4	1.4
16	1.7	1.8	2.2
25	2.0	2.1	2.5
52	2.1	2.3	2.8
66	2.3	2.5	2.9
115	2.3	2.5	3.0

The equilibrium level of migration is not just related to the primary concentration of the monomers in the PET bottles, but the solubility of these compounds in solution is an inevitable factor affecting this phenomenon. Ethylene glycol was found in the highest concentration in food simulant at both storage temperatures of 25 and 45°C. Possibly the higher solubility of ethylene glycol in 3% acetic acid is the main reason for a higher migration of this monomer compared to the other compounds.

8.1.6 *Migration of Plasticizers*

The migration of di(2-ethylhexyl) adipate and acetyltributyl citrate plasticizers from a food-grade PVC film into a sweetened sesame paste (*Halawa tehineh*) was investigated (22).

A loss of 54.7% di(2-ethylhexyl) adipate from the PVC film was observed. This value is slightly higher than the limit of 3 $mg\,dm^{-2}$ of a film surface set by the European Union for di(2-ethylhexyl) adipate.

8.1.7 *Migration of Aluminum and Silicon*

Samples of PET/clay nanocomposite were prepared by melt blending PET and Cloisite 20A nanoparticles (23, 24). A stretch blow-molding machine was used to produce bottles from neat PET and

PET nanocomposite. Tests were performed on the migration of aluminum and silicon from PET nanocomposite bottles into acidic food simulant.

The samples were stored at room temperature of 25°C and 45°C for time durations ranging from 7 *d* to 90 *d*. A specific surface of sheets (prepared from PET/clay nanocomposite) immersed in acidic food simulant, and two-sided migration of Al and Si was investigated. According to X-ray diffraction analysis, the nanoclays show intercalated structure in the PET matrix.

Transmission electron microscopy and atomic force microscopy micrographs displayed both intercalation and exfoliation morphology for PET/clay nanocomposites. Inductively coupled plasma was used to quantify amounts of Al and Si that had migrated into the acidic food simulant. It was observed that the migration process is dependent on storage time and temperature, and the molar ratio of aluminum and silicon in the acidic aqueous solution (Al/Si)aq to the ratio in the solid phase of prepared nanocomposites (Al/Si)solid was about 23% higher in the samples stored at 45°C (24).

8.2 Recycling Methods

The recycling of multi-material multilayer plastic packaging (MMPP) is complex since the recycling industry is not able to identify, sort and separate the diverse layers with current standard technologies (25).

Therefore, MMPP is usually sorted as mixed plastic waste in post-consumer streams (also called other types of plastics) and frequently incinerated with energy recovery in European countries, which does not prioritize recycling as a circular economy practice (26).

In low-income countries, these packaging materials are usually disposed of at dumpsites or in landfills and a fraction also leaks into the environment due to the common lightweight and short lifespan (27).

8.2.1 Delamination

Multi-material multilayers may include adhesives for bonding interlayers. So, a recycling process may include delamination steps (25).

A post-consumer liquid packaging board delamination through bonding damage was investigated (28).

Acetone was the solvent used to delaminate food packaging containing a PE/Aluminum/PET composition (29).

The application of physical processes to delaminate multilayers is unusual due to the bonding's strength. However, basic structures could be separated by shredding and washing (30).

An innovative technology has been developed to establish a low-energy material recycling process to delaminate MMPP (31). This process includes a shredding step followed by a microemulsion with a surfactant that promotes the separation of the layers. Figure 8.3 is a diagram of a method for separating multilayer systems.

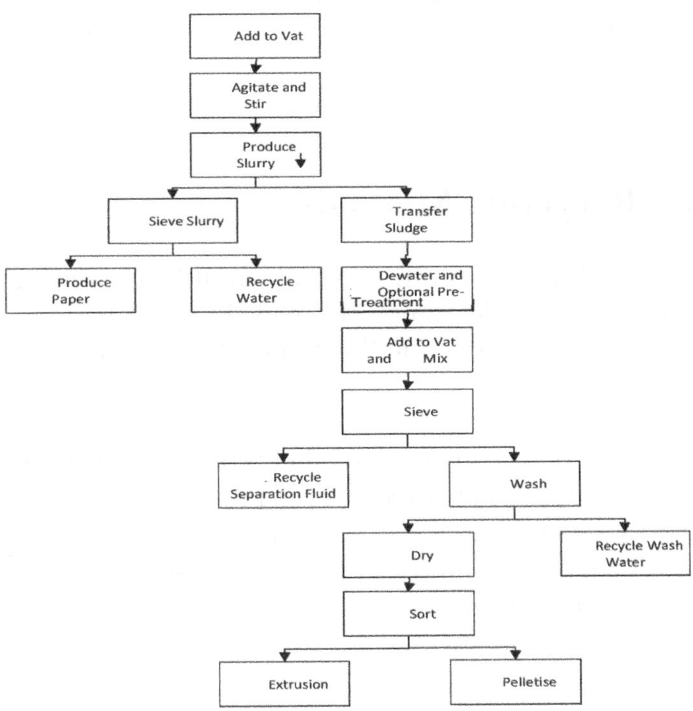

Figure 8.3 Method for separating multilayer systems (31).

This technique can delaminate structures such as PE/Aluminum, PP/Aluminum, and PE/PET by minimizing interlayers forces.

8.2.2 *Separation*

A raw mixture of plastics usually includes various kinds of waste plastics. This makes the separation an important process that should be carried out prior to recycling. However, the separation of mixed plastics encounters many problems due to the characteristics of plastics. So, this is one of the most problematic processes in the management system of plastic waste (32).

8.2.2.1 *Wet Separation Techniques*

Flotation Separation. As wetting agents for wet separation, sodium ligninsulfonate, tannic acid, diethylhexyl sodium sulfosuccinate and saponin can be used (33). These wetting agents are shown in Figure 8.4. As plastics for testing, PVC, poly(carbonate) (PC), poly(oxymethylene) (POM), and poly(phenylene ether) (PPE) were used.

The efficient flotation separation amongst these naturally hydrophobic polymers could be attributed to the selective reduction in their hydrophobicity as a consequence of the adsorption of the surfactants. The relative order of floatability was found to be PPE > POM > PC > PVC.

This order is consistent with the critical surface tension values determined for all these solids except PPE (33).

8.2.3 *Mechanical Recycling*

During mechanical recycling, polymer backbones can become partially degraded because heating can lead to breakage of intramolecular bonds resulting in a lower molecular weight distribution and changed mechanical and optical properties of the recycled polymer, e.g., increased brittleness and opacity (34, 35).

Additionally, plastic polymers may already have been damaged by external influences during use or waste management, e.g., contact with acidic substances, UV radiation, and extreme temperatures. These inherent properties of plastics have led to their classification as nonpermanent materials, although they are highly resistant to complete degradation in the environment.

Some types of plastic additives, e.g., antioxidants and stabilizers, form intended reaction products, thereby losing their original

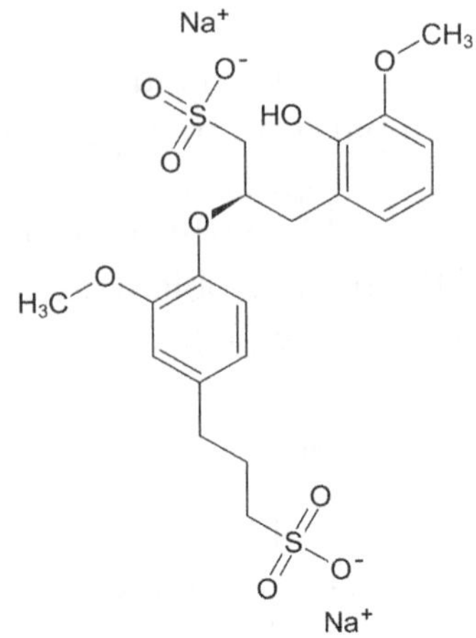

Sodium ligninsulfonate

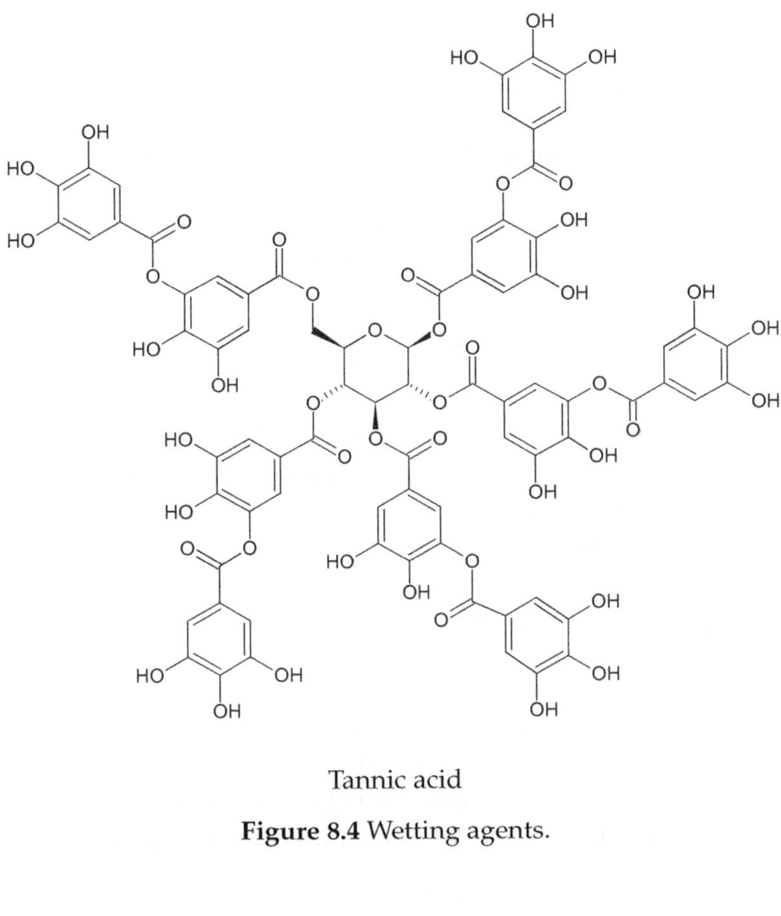

Tannic acid

Figure 8.4 Wetting agents.

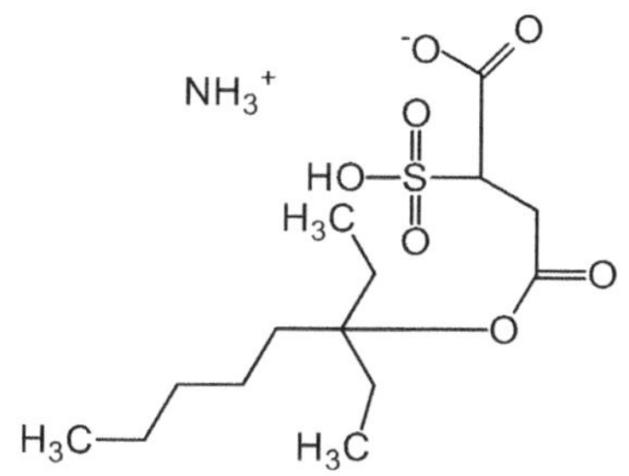

Diethylhexyl sodium sulfosuccinate

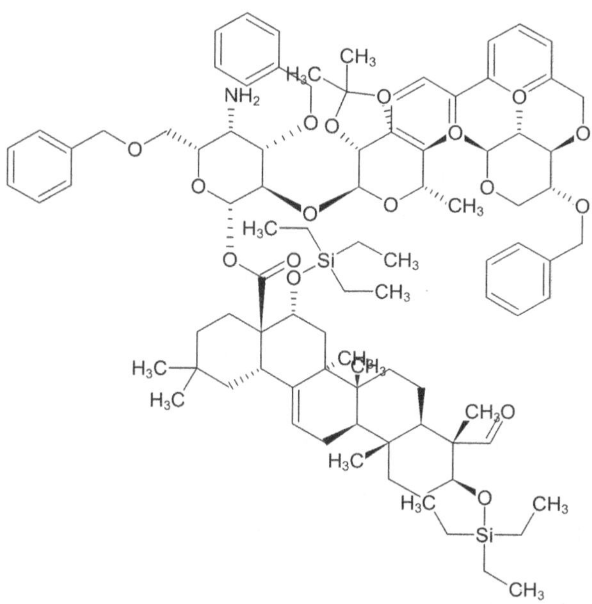

Saponin

Figure 8.4 (cont) Wetting agents.

function. To compensate for such unwanted changes, mechanical recycling processes require the addition of virgin polymers and/or further additives (36).

8.2.4 *Selective Dissolution*

Selective dissolution techniques are solvent-based methods to separate layers of MMPP. These methods take advantage of the different solubilities of polymers, where a specific solvent dissolves a targeted component and removes materials in steps (37).

For thermoplastic polymers, such as poly(ethylene) (PE), PP, poly(styrene) (PS), PET, and PVC, the method of dissolution is an important technique for recycling systems (38). The choice of the solvent is an essential step of this practice, while a range of solvents may allow the recovery of diverse layers (39).

Along the process, the target polymer is dissolved, and other polymers remain at a solid phase. Then, the solvent is used to target another polymer, or else, another solvent can be used (40).

Also, specific components of the multilayer structure can be selectively-dissolved for further recovery, with low contamination rates (41). The recycling of beverage cartons using solvent-based techniques was also evaluated (42).

In summary, these methods produce high-quality end products comparable to virgin materials (41) and show that solvent-based technologies are a realistic recycling approach (39).

8.2.5 *Dissolution-Reprecipitation Technique*

The application of the dissolution-reprecipitation technique in the recycling of polymers from waste plastic packaging materials has been investigated (43).

Here, the type of polymer in each packaging was identified with Fourier transform infrared (FTIR). Also, the experimental conditions of the recycling process, i.e., type of solvent/non-solvent, initial polymer concentration and dissolution temperature, were optimized using model polymers.

The dissolution/reprecipitation technique was applied in the recycling of a number of plastic materials based on PE, PP, PS, PET, and PVC.

The recovery of the polymer was measured and possible structural changes during the recycling procedure were assessed by FTIR.

Furthermore, the potential recycling-based degradation of the polymer was investigated by measuring the thermal properties, such as melting point, crystallinity and glass transition temperature, of the polymer before and after recycling, using differential scanning calorimetry, their molecular properties (average molecular weight) using viscosimetry, as well as their mechanical tensile properties (43).

High recoveries were found in most samples. The properties of the recycled materials were not substantially different from the original materials. However, a slight degradation was observed in a few samples (43).

A method for separating polymers from a physically commingled solid mixture containing a plurality of polymers can be done by the following steps (44):

1. Dissolving at least one first polymer of the mixture in a selected solvent at a first temperature to form a first solution and a first remaining solid component which contains at least two polymers of the mixture which are insoluble in the selected solvent at the first temperature but soluble in the selected solvent at higher temperatures;
2. Separating the first solution from the first remaining solid component;
3. Separating the at least one first polymer from the selected solvent of the first solution;
4. Dissolving at least one second polymer from the first remaining solid component in the selected solvent at a second temperature to form a second solution and a second remaining solid component which contains at least one third polymer of the mixture which is insoluble in the selected solvent at the second temperature but soluble in the selected solvent at a third higher temperature;
5. Separating the second solution from the second remaining solid component;
6. Separating the at least one second polymer from the selected solvent of the second solution;

7. Dissolving at least one third polymer from the second remaining solid component in the selected solvent at the third higher temperature to form a third solution; and
8. Separating the at least one third polymer from the selected solvent of the third solution.

8.2.6 *Compatibilization*

Multi-material multilayers usually include polymer blends that are mechanically unstable (45).

For this reason, the compatibilization technique uses chemical substances to increase the mechanical stability of the multilayers and further recycle all layers in a single stream without separation (46).

These methods make polymer blends less immiscible by adding another component to the structures (47).

8.2.7 *Feedstock Recycling*

An alternative emerging treatment for difficult to recycle plastic waste might be producing chemicals and fuels such as monomer or hydrocarbon feedstocks.

Processes known to perform this are gasification, pyrolysis, fluid-catalyzed cracking, and hydrocracking (47). Pyrolysis may become an active technology for plastic waste, such as MMPP, which is complex to depolymerize (48). The process occurs at high temperatures and in the absence of oxygen when polymers are broken down and, with further processing, results in petrochemical feedstock such as naphtha or diesel (47). This process, known as chemical recycling, includes thermolysis through catalysts to selectively generate gases, liquids, or waxes, which are developments that go beyond mechanical recycling (49).

At the moment, this is not a widespread recycling practice due to energy costs (50). However, the technology might evolve as a future feasible option for plastics or plastics-containing waste that is difficult to be materially recycled from an economical and systematic perspective, opening new routes and opportunities.

A study developed by Solis and Silveira (50) assessed eight technologies regarding feedstock recycling for mixed plastic waste and their technology readiness level (TRL).

The most developed techniques are pyrolysis, catalytic cracking, and conventional gasification, whereas other emerging technologies noticeably demand more investments to be at a commercial stage (50).

8.2.8 Closed-Loop Recycling

Poly(ethylene terephthalate) (PET) is a favored candidate for closed-loop recycling for new packaging applications (19). A PET-recycling process was investigated with respect to its cleaning efficiency and the compliance of the PET recyclate with food law.

The key technology of the investigated PET-recycling process to remove contaminants consists of a predecontamination-extruder combination. At the end of the recycling process, there is either a pelletizing system or downstream equipment to produce preforms or flat sheets.

Therefore, the process has two process options, an inline production of PET preforms and a batch option producing PET pellets. In the case of possible misuse of PET bottles by the consumer, the inline process produces higher concentrations in the bottle wall of the recyclate containing preforms (19).

Owing to the dilution of the PET output material by large amounts of uncontaminated PET, the batch option is the less critical process in terms of consumer protection. Regarding an appropriate testing procedure for the evaluation of a bottle-to-bottle recycling process, both process options have their own specific requirements with respect to the design of a challenge test (19).

A closed-loop control system for recycling products, such as food service ware, has been presented (51). It uses the following steps:

1. Organizing a number of entities in a group that together use sizable quantities of disposable PS food service ware;
2. Collecting the used service ware from the group, while avoiding mixing the used service ware with other service ware PS products that did not emanate from the PS service ware of the group;
3. Establishing a facility for washing, grinding and pelletizing the used PS food service ware; and

4. Transferring the pelletized used PS to an entity that combines the pelletized used PS with virgin PS in proportions allowed under government regulations and arranges for the conversion of the mixture to recycled PS food service ware, and for the sale of the service ware to the entities in the group, and continuing the cycle over again.

Figure 8.5 illustrates a closed-loop continuous system for recycling food service ware, such as expanded PS dishes and trays used in institutions such as public schools, colleges, universities, industrial catering, etc.

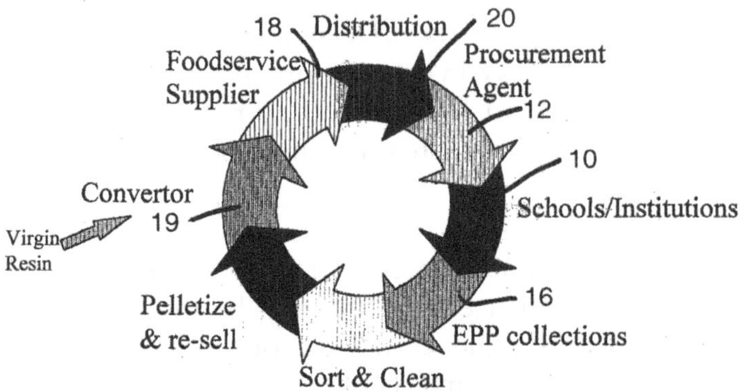

Figure 8.5 Closed-loop continuous system for recycling food service ware (51).

In this illustration, a number of participants are identified, such as schools and other institutions that employ disposable food service ware 10; a system coordinator or procurement agent 12 may oversee the collection and conversion of the waste products, and the converter 19 utilizes the washed and pelletized product and mixes it with additional designated virgin raw material to manufacture the new food service ware. Various functions are performed by each of the participants and some of their respective tasks may be carried out by one or more entities. From the converter 19, newly manufactured food service ware is ultimately supplied back to the schools or institutions 10 where the cycle is repeated over and over again. During each cycle of the system, additional virgin material is combined with the converted used service ware. In one embodiment, the newly manufactured food service ware is transported from the converter 19 to a distributor 20, and then to the procurement agent 12, to meet the supply demands of the school or institution 10. In some embodiments, the recycled material may be tested to detect any extractives or microbial contamination. This testing may occur once the finished product is produced and/or may also occur after the used food service ware is cleaned (51).

8.2.9 *Supercritical Ethanol*

Different properties, such as barriers to gas, light, flavor and water vapor, as well as flexibility, are necessary for the production of food packages. These properties may be obtained by combining different polymers. In spite of the advantages achieved with the use of multilayer films, the recycling process of such material is a challenging task (29). An alternative for recycling of a quite common type of food packaging was presented. Here PE, aluminum and PET were used as a multilayer film. The multilayer packages were delaminated with acetone.

The depolymerization of PET took place by the reaction with ethanol under supercritical conditions. The obtained products in this reaction are diethyl terephthalate and ethylene glycol. The diethyl terephthalate is easily obtained through precipitation in water. On the other hand, ethylene glycol could be purified from ethanol and water by distillation.

Diethyl terephthalate was obtained as the main product, presenting high purity and yield of 80%. Also, metallic aluminum was obtained. The optimal reaction time was 120 *min* (29).

8.3 Post-Consumer Polyolefins

8.3.1 *Predictive Models*

The use of post-consumer polyolefins is increasing due to the pledges of multiple organizations to use higher concentrations of recycled materials in packaging (52). However, direct food contact applications are challenging with various regulations around the world mandating minimum safety requirements, which is often determined by the condition of use (microwave, cold storage) and food type (fatty, aqueous, dry).

Various predictive models that have been developed to study contaminant migration from packaging to food contact articles have been described.

Also, multiple techniques and methods were examined that have been employed to quantify and assess the safety of recycled materials (e.g., United States and Europe), which serve as a base for the regulatory framework for many other countries. Also, the presence of different contaminants during polymer processing is discussed due to potential problems from the presence of harmful compounds affecting human health or degradation of material properties (52).

8.3.2 *Safety Concerns*

Plastic recycling is one of the important ways to mitigate plastic pollution. However, chemicals present in recycled plastics are one of the key qualities affecting their potential uses and deserves more attention.

The application of 475 migrants coming from 15 post-consumer recycled polyolefins was confirmed by direct-immersion solid-phase microextraction coupled with gas chromatography mass spectrometry, and atmospheric pressure gas chromatography coupled with quadrupole time-of-flight mass spectrometry. About 60% of them might not be a risk to humans because they were food additives/components or saturated hydrocarbons, fatty acyls, or prenol lipids. Most of them had a molecular weight between 150 *Da* and 210 *Da*, though high concern substances with high molecular weight (e.g., octocrylene) implied that high molecular weight surrogates are required to study the efficiency of recycling processes for polyolefins. The mean predicted octanol/water partition coefficient (XLogP) was about 6.5 and 3.5 for 95% ethanol and 3% acetic acid food simulants, respectively.

Octocrylene, 1-tetradecene, 1-dodecene, dodecyl acrylate, 2,4-di-*tert*-butylphenol, 1,4-benzenedicarboxylic acid, diethyl ester, benzenamine, 2,4-dichloro- and diethyl phthalate were of high concern depending on the potential food contact use of the materials. These compounds are collected in Figure 8.6.

The here presented results are informative and can be of great help to recyclers and lawmakers for recycling polyolefins for safe food contact use (53).

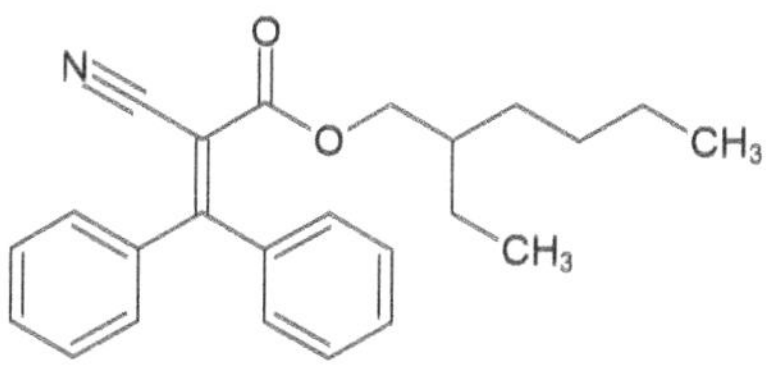

Octocrylene

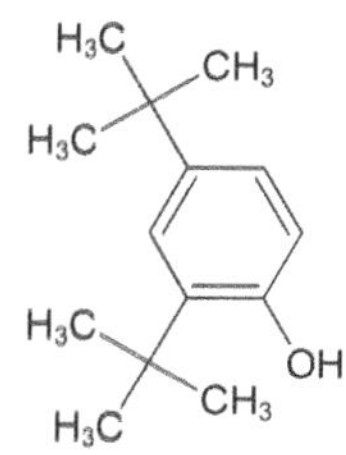

2,4-Di-*tert*-butylphenol

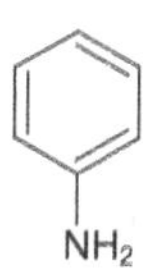

Benzenamine

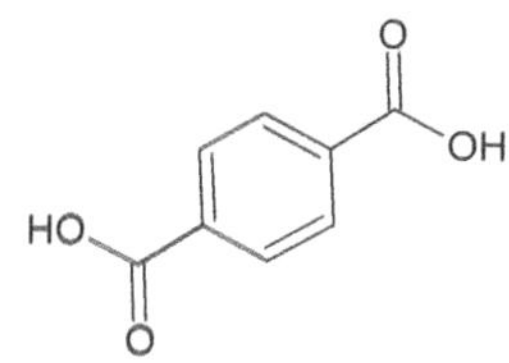

1,4-Benzenedicarboxylic acid

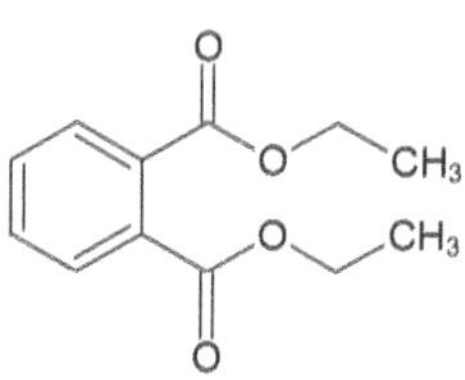

Diethyl phthalate

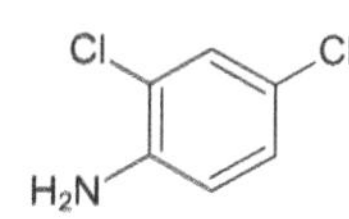

2,4-Dichloro benzenamine

Figure 8.6 Contaminants.

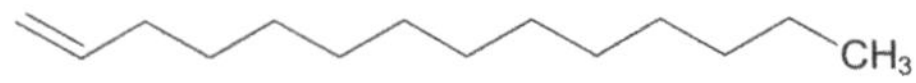

1-Tetradecene

1-Dodecene

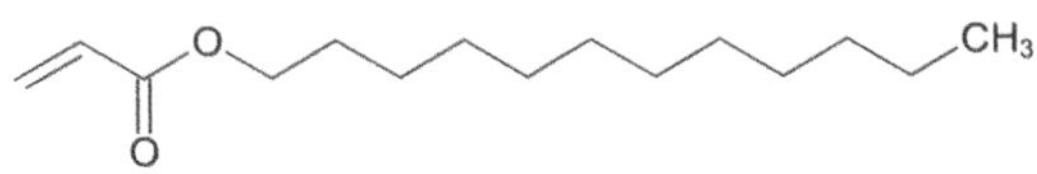

Dodecyl acrylate

Figure 8.6 (cont) Contaminants.

8.3.3 Multilayer Composites

Multi-material multilayer plastic packaging (MMPP) is widely applied in fast-moving consumer goods combining functionalities of distinct materials (54). These packaging structures can enhance properties, such as resource-use efficiency and barrier performance, leading to consequential benefits like a prolonged shelf life. Nevertheless, they represent a challenge for existing recycling systems, confronting circular economy principles.

In a study, the future of recycling technologies for MMPP was assessed. Future scenarios were identified, including (54):

1. High-performance material recycling,
2. Recycling into hydrocarbons,
3. Business as usual, and
4. Downcycling.

In-depth interviews and a feedback survey were methods used to validate the scenario matrix while defining some expectations towards the future.

Circular economy models have been studied as solutions to current and future environmental challenges (55).

The analysis showed that distinct technologies will develop unevenly in different parts of the world (54).

Packaging for fresh food products may consist of four to seven layers of different components (56). A wide range of substances with different physical and chemical characteristics form multi-material multilayers (37).

Various polymers are applied, such as polyolefins PE and PP and chemical variants (HDPE, LDPE, linear low-density poly(ethylene)), or polyesters such as PET and poly(lactic acid). Other materials are ethylene vinyl alcohol, poly(urethane), and poly(vinylidene chloride). These multilayer multi-material films and sheets are developed by co-extrusion or lamination techniques (57).

An overview of the main materials, their functionalities and applications are shown in Table 8.4.

8.3.3.1 HDPE Multilayer Composites

A method for recycling post-consumer multilayer wastes containing PET, PE and aluminum has been described (29).

Composite materials were prepared by using ground multilayer films as reinforcement phase in recycled HDPE matrix. The PET and aluminum worked as reinforcing fillers to establish better mechanical properties, while PE improved the adhesion between matrix and fillers. The composites were characterized by scanning electron microscopy, thermal analysis, water uptake, tensile, flexural and Izod impact tests.

The addition of a multilayer matter in the PE matrix provided a material with excellent mechanical properties such as tensile impact strength and high elasticity. The presence of different components in the multilayer packaging, which was the main obstacle preventing its recycling, became the key feature to develop a material with superior properties compared to those presented by commercial products. The composites presented a low water uptake and high thermal stability. One of the greatest advantages in the preparation of such composites was the possibility to range the formulation PE/filler up to 50% of filler, providing final materials with tailored mechanical properties according to their composition.

The approach presented offers new possibilities for process and

Table 8.4 Materials applied for multi-material multilayer plastic packaging (54).

Material	Layer Functionality	Application Types
PE	- heat-sealable - moisture barrier - combination with gas/aroma barriers (e.g., PA, EVOH)	Beverage cartons (usually contains LDPE layer)
PP	- moisture barrier - coated with heat seal coatings (PVDC, acrylate) for mechanical strength -combined with gas/aroma barriers	Toiletries, cosmetics, microwavable food packaging
PET	-gas/aroma barrier -moisture barrier -mechanical strength -heat-sealable	Food packaging such as bottles, meat, or cheese
PS	-gas permeability -printability -combined with gas/aroma barriers	Printable outside layer for fresh food packaging
PVC	-gas/aroma barrier -mechanical strength	Stretch wrap
PA	-gas/aroma barrier -mechanical strength -heat-sealable	Thermoformed food packaging
EVOH	-oxygen barrier	Oxygen-sensitive food packaging

Compound	Abbreviation
Poly(ethylene)	PE
Poly(propylene)	PP
Poly(ethylene terephthalate)	PET
Poly(styrene)	PS
Poly(vinyl chloride)	PVC
Poly(amide)	PA
Ethylene vinyl alcohol	EVOH

Table 8.4 (cont) Materials applied for multi-material multilayer plastic packaging (54).

Material	Layer Functionality	Application Types
EVA	-adhesive -heat-sealable	Interlayer to bond PE to PVC
PUR	-adhesive (dry bond lamination) -heat- sealable	Cosmetics, pharmaceutical packaging
PVDC	-gas/aroma barrier -moisture barrier -abrasion protection	Film wraps
PLA	-gas/aroma barrier -moisture barrier -mechanical strength	Cups, containers, take-out packaging
Bio-PE	-heat-sealable -resistant to hydrolysis/oxidation -flexibility	Similar uses as fossil fuel PE (uses the same recycling stream)
PHA	-gas/aroma barrier -moisture barrier -mechanical strength	Applications in development (may replace PET)
PEF	-gas/aroma barrier -moisture barrier -mechanical strength	Applications in development (may replace PET)

Compound	Abbreviation
Ethylene vinyl acetate	EVA
Poly(urethane)	PU
Poly(vinylidene chloride)	PVDC
Poly(lactic acid)	PLA
Bio-Poly(ethylene)	Bio-PE
Poly(hydroxyalkanoate)s	PHA
Poly(ethylene furanoate)	PEF

product development from flexible multilayer wastes contributing to the mitigation of future environmental issues (29).

8.3.4 *Recyclability Enhancement of Food Containers*

Pizza boxes and take-out food containers are particular culprits when it comes to recycling (58). Proper recycling of these containers relies in part on proper curbside preparations. In the case of take-out food containers, proper washing with soap and hot water is required to remove grease. Pizza boxes cannot be washed, and generally have the bottom portion torn off and discarded separately.

To solve these problems, a special food container was fabricated. The food container includes a first interior face having disposed thereon a first protective layer for protecting the first interior face from exposure to contaminants that would interfere with recycling of the food container. The first protective layer is affixed to the first interior face by an adhesive interface so as to be peelable from the first interior face.

The protective layer offers numerous advantages, including protecting the surface of the container from contaminants, and in some cases increasing the shelf life of the content of the container (58).

8.4 Plastic Waste Generation

The packaging industry requires materials that are lightweight so as to reduce the amount used to package a product, thus reducing product weight for transportation cost savings as well as reducing the amount of end-of-life packaging waste material (59). Plastics have managed to fulfill this role very well and have remained unchallenged.

For example, a study revealed that the replacement of glass bottles with plastic bottles for beverage packaging in airlines resulted in savings of over USD 1 million in fuel costs as a result of the weight loss (60).

For recycling of plastics, the most important steps are collection and sorting. Sorting involves the selection of waste which is suitable for recycling using manual or automated means. This can be done at the source of the waste and by using separate collection systems or at facilities by mechanical processing and sorting of mixed waste (61).

8.5 Recycled Plastics in Food Contact

Sustainability in packaging is definitely needed. One way for consumer product companies to reach their sustainability goals is to have their packages made from recycled plastic. There is a wide variety of legislation covering recycled plastics for use in food contact applications. Some jurisdictions allow it, and some have denied it (62).

8.5.1 *Improving Safety and Quality*

The use of recycled plastics for food packaging will likely increase due to the high price of crude oil, a feedstock for the petrochemicals used in the manufacture of plastics, as well as the recent advances in plastic recycling technologies that allow improved quality and safety of post-consumer recycled plastics intended for food packaging applications.

Some efforts at improving the safety and quality of recycled plastics for food contact applications have been presented (63).

Several types of plastics are used in packaging. To provide a consistent national system that facilitates the collection and recycling of post-consumer plastics through the normal channels for collecting recyclable materials from household waste, the Society of the Plastics Industry (SPI) developed a resin identification code system for common plastics used in the manufacture of containers. The SPI code system applies to six homogeneous plastics and one heterogeneous material category.

To improve the recyclability of plastic packaging for food contact use, controls on the source of post-consumer plastics as well as limitations on the use of the specific recycled article are needed to ensure the quality and safety of the recycled materials for the proposed food contact application (63).

References

1. P. Dauvergne, Why is the global governance of plastic failing the oceans?, *Global Environmental Change*, Vol. 51, p. 22, 2018.

2. J.N. Hahladakis and E. Iacovidou, Closing the loop on plastic packaging materials: What is quality and how does it affect their circularity?, *Science of the Total Environment*, Vol. 630, p. 1394, 2018.
3. C. Giacovelli, *Single-Use Plastics: A Roadmap for Sustainability*, International Environmental Technology Centre, 2018.
4. T.Y. Chitaka, V. Russo, and H. von Blottnitz, In pursuit of environmentally friendly straws: A comparative life cycle assessment of five straw material options in South Africa, *The International Journal of Life Cycle Assessment*, Vol. 25, p. 1818, 2020.
5. N. Singh, D. Hui, R. Singh, I.P.S. Ahuja, L. Feo, and F. Fraternali, Recycling of plastic solid waste: A state of art review and future applications, *Composites Part B: Engineering*, Vol. 115, p. 409, 2017.
6. E. Velásquez, M. Guerrero Correa, L. Garrido, A. Guarda, M.J. Galotto, and C. López de Dicastillo, Food packaging plastics: Identification and recycling in *Recent Developments in Plastic Recycling*, pp. 311–343. Springer, 2021.
7. E. De Tandt, C. Demuytere, E. Van Asbroeck, H. Moerman, N. Mys, G. Vyncke, L. Delva, A. Vermeulen, P. Ragaert, S. De Meester, and K. Ragaert, A recycler's perspective on the implications of REACH and food contact material (FCM) regulations for the mechanical recycling of FCM plastics, *Waste Management*, Vol. 119, p. 315, 2021.
8. M. Brouwer, C. Picuno, E.U.T. van Velzen, K. Kuchta, S. De Meester, and K. Ragaert, The impact of collection portfolio expansion on key performance indicators of the Dutch recycling system for post-consumer plastic packaging waste, a comparison between 2014 and 2017, *Waste Management*, Vol. 100, p. 112, 2019.
9. L. Lugal, A. Grant, M. Cordle, and D. Fletcher, PET market in Europe: State of play, *Eunomia Reports*, 2020.
10. J. Florestan, A. Lachambre, and N. Mermilliod, Recycling of plastics: Automatic identification of polymers by spectroscopic methods, *Resources, Conservation and Recycling*, Vol. 10, 1994.
11. Y. Abe, L.K. Ackerman, and M. Mutsuga, Rapid identification of polyamides using direct analysis in real time mass spectrometry, *Rapid Commun Mass Spectrom*, Vol. 34, 2020.
12. S.R. Ahmad, A new technology for automatic identification and sorting of plastics for recycling, *Environmental Technology*, Vol. 25, 2004.
13. B. Geueke, K. Groh, and J. Muncke, Food packaging in the circular economy: Overview of chemical safety aspects for commonly used materials, *Journal of Cleaner Production*, Vol. 193, 2018.
14. X.-L. Cao, Phthalate esters in foods: Sources, occurrence, and analytical methods, *Comprehensive Reviews in Food Science and Food Safety*, Vol. 9, p. 21, 2010.

15. R. Franz and F. Welle, Contamination levels in recollected PET bottles from non-food applications and their impact on the safety of recycled PET for food contact, *Molecules*, Vol. 25, p. 4998, 2020.
16. M. Farhoodi, Z. Emam-Djomeh, M.R. Ehsani, and A. Oromiehie, Migration of model contaminants (ethylene glycol, DEHA and DEHP) from PET bottles into Iranian yogurt drink, *e-Polymers*, Vol. 8, 2008.
17. C.P. Rader, S.D. Baldwin, D.D. Cornell, G.D. Sadler, and R.F. Stockel, *Plastics, Rubber, and Paper Recycling: A Pragmatic Approach*, ACS Publications, 1995.
18. M. Huber and R. Franz, Studies on contamination of post consumer plastics from controlled resources for recycling into food packaging applications, *Deutsche Lebensmittel-Rundschau (Germany)*, Vol. 93, p. 328, 1997.
19. R. Franz and F. Welle, Recycled poly(ethylene terephthalate) for direct food contact applications: Challenge test of an inline recycling process, *Food Additives & Contaminants*, Vol. 19, 2002.
20. F.L. Bayer, Polyethylene terephthalate recycling for food-contact applications: Testing, safety and technologies: A global perspective, *Food Additives & Contaminants*, Vol. 19, p. 111, 2002.
21. C. Nerin, J. Albinana, M.R. Philo, L. Castle, B. Raffael, and C. Simoneau, Evaluation of some screening methods for the analysis of contaminants in recycled polyethylene terephthalate flakes, *Food Additives & Contaminants*, Vol. 20, p. 668, 2003.
22. A.E. Goulas, P. Zygoura, and A. Karatapanis, Migration of di(2-ethylhexyl) adipate and acetyltributyl citrate plasticizers from food-grade PVC film into sweetened sesame paste (halawa tehineh): Kinetic and penetration study, *Food Chem. Toxicol.*, Vol. 45, 2007.
23. M. Farhoodi, S.M.A. Mousavi, R. Sotudeh Gharebagh, Z. Emam-Djomeh, and A. Oromiehie, Migration of silicon from nanocomposite packaging materials into acidic food simulant, in *Advanced Materials Research*, Vol. 622, pp. 873–877. Trans Tech Publ., 2013.
24. M. Farhoodi, S.M. Mousavi, R. Sotudeh-Gharebagh, Z. Emam-Djomeh, and A. Oromiehie, Migration of aluminum and silicon from PET/clay nanocomposite bottles into acidic food simulant, *Packaging Technology and Science*, Vol. 27, p. 161, 2014.
25. O. Horodytska, F. Valdés, and A. Fullana, Plastic flexible films waste management – a state of art review, *Waste Management*, Vol. 77, p. 413, 2018.
26. K. Kaiser, M. Schmid, and M. Schlummer, Recycling of polymer-based multilayer packaging: A review, *Recycling*, Vol. 3, p. 1, 2017.
27. J.N. Hahladakis and E. Iacovidou, An overview of the challenges and trade-offs in closing the loop of post-consumer plastic waste (PCPW):

Focus on recycling, *Journal of Hazardous Materials*, Vol. 380, p. 120887, 2019.
28. S. Zhang, K. Luo, L. Zhang, X. Mei, S. Cao, and B. Wang, Interfacial separation and characterization of Al–PE composites during delamination of post-consumer Tetra Pak materials, *Journal of Chemical Technology & Biotechnology*, Vol. 90, p. 1152, 2015.
29. S.L. Fávaro, A.R. Freitas, T.A. Ganzerli, A.G.B. Pereira, A.L. Cardozo, O. Baron, E.C. Muniz, E.M. Girotto, and E. Radovanovic, PET and aluminum recycling from multilayer food packaging using supercritical ethanol, *Journal of Supercritical Fluids*, Vol. 75, 2013.
30. M. Perick, Method for recycling plastics from a film packaging bag, EP Patent 3 059 061, assigned to Mondi AG, July 19, 2017.
31. F. Lovis, H. Seibt, and S. Kernbaum, Method and apparatus for recycling packaging material, US Patent 10 682 788, assigned to Saperatec GmbH, June 16, 2020.
32. G. Dodbiba and T. Fujita, Progress in separating plastic materials for recycling, *Physical Separation in Science and Engineering*, Vol. 13, 2004.
33. J. Shibata, S. Matsumoto, H. Yamamoto, E. Kusaka, and Pradip, Flotation separation of plastics using selective depressants, *International Journal of Mineral Processing*, Vol. 48, p. 127, 1996.
34. F. De Santis and R. Pantani, Optical properties of polypropylene upon recycling, *The Scientific World Journal*, Vol. 2013, 2013.
35. I.A. Ignatyev, W. Thielemans, and B. Vander Beke, Recycling of polymers: A review, *ChemSusChem*, Vol. 7, p. 1579, 2014.
36. G. Curtzwiler, K. Vorst, J.E. Danes, R. Auras, and J. Singh, Effect of recycled poly(ethylene terephthalate) content on properties of extruded poly(ethylene terephthalate) sheets, *Journal of Plastic Film & Sheeting*, Vol. 27, p. 65, 2011.
37. A. Mieth, E. Hoekstra, and C. Simoneau, Guidance for the identification of polymers in multilayer films used in food contact materials, *European Commission JRC Technical Reports*, pp. 1–70, 2016.
38. A.J. Hadi, G.F. Najmuldeen, and I. Ahmed, Polyolefins waste materials reconditioning using dissolution/reprecipitation method, *APCBEE Procedia*, Vol. 3, p. 281, 2012. 2nd International Conference on Chemistry and Chemical Process (ICCCP 2012) May 5-6, 2012.
39. T.W. Walker, N. Frelka, Z. Shen, A.K. Chew, J. Banick, S. Grey, M.S. Kim, J.A. Dumesic, R.C. Van Lehn, and G.W. Huber, Recycling of multilayer plastic packaging materials by solvent-targeted recovery and precipitation, *Science Advances*, Vol. 6, p. eaba7599, 2020.
40. Y.-B. Zhao, X.-D. Lv, and H.-G. Ni, Solvent-based separation and recycling of waste plastics: A review, *Chemosphere*, Vol. 209, p. 707, 2018.
41. M. Schlummer, T. Fell, A. Mäurer, and G. Altnau, The role of chemistry in plastics recycling: A comparison of physical and chemical plastics recycling, *Kunststoffe International*, Vol. 5, p. 34, 2020.

42. I. Georgiopoulou, G.D. Pappa, S.N. Vouyiouka, and K. Magoulas, Recycling of post-consumer multilayer Tetra Pak® packaging with the selective dissolution-precipitation process, *Resources, Conservation and Recycling*, Vol. 165, p. 105268, 2021.
43. D.S. Achilias, A. Giannoulis, and G.Z. Papageorgiou, Recycling of polymers from plastic packaging materials using the dissolution-reprecipitation technique, *Polym Bull*, Vol. 63, 2009.
44. E.B. Nauman and J.C. Lynch, Polymer recycling by selective dissolution, US Patent 5 278 282, assigned to Rensselaer Polytechnic Institute, January 11, 1994.
45. X.-M. Xie and X. Sui, Compatibilization of polymer blends by the addition of random copolymers in A.R. Ajitha and S. Thomas, (Eds.), *Compatibilization of Polymer Blends*, chapter 5, pp. 145–177. Elsevier, 2020.
46. F. Mostafapoor, A. Khosravi, A. Fereidoon, R. Khalili, S.H. Jafari, H. Vahabi, K. Formela, and M.R. Saeb, Interface analysis of compatibilized polymer blends in A.R. Ajitha and S. Thomas, (Eds.), *Compatibilization of Polymer Blends*, chapter 12, pp. 349–371. Elsevier, 2020.
47. K. Ragaert, L. Delva, and K. Van Geem, Mechanical and chemical recycling of solid plastic waste, *Waste Management*, Vol. 69, p. 24, 2017.
48. T. Thiounn and R.C. Smith, Advances and approaches for chemical recycling of plastic waste, *Journal of Polymer Science*, Vol. 58, p. 1347, 2020.
49. J.M. Garcia and M.L. Robertson, The future of plastics recycling, *Science*, Vol. 358, p. 870, 2017.
50. M. Solis and S. Silveira, Technologies for chemical recycling of household plastics – A technical review and TRL assessment, *Waste Management*, Vol. 105, p. 128, 2020.
51. M. Forrest and D. Schneider, Closed-loop control system for recycling products, US Patent 20 050 027 555, assigned to Michael W. Forrest, David Schneider, February 03, 2005.
52. V.S. Cecon, P.F. Da Silva, G.W. Curtzwiler, and K.L. Vorst, The challenges in recycling post-consumer polyolefins for food contact applications: A review, *Resources, Conservation and Recycling*, Vol. 167, p. 105422, 2021.
53. Q.-Z. Su, P. Vera, C. Nerín, Q.-B. Lin, and H.-N. Zhong, Safety concerns of recycling postconsumer polyolefins for food contact uses: Regarding (semi-)volatile migrants untargetedly screened, *Resources, Conservation and Recycling*, Vol. 167, p. 105365, 2021.
54. C.T. de Mello Soares, M. Ek, E. Östmark, M. Gällstedt, and S. Karlsson, Recycling of multi-material multilayer plastic packaging: Current trends and future scenarios, *Resources, Conservation and Recycling*, Vol. 176, p. 105905, 2022.

55. P. Ghisellini, C. Cialani, and S. Ulgiati, A review on circular economy: The expected transition to a balanced interplay of environmental and economic systems, *Journal of Cleaner Production*, Vol. 114, p. 11, 2016.
56. T.I. Butler and B.A. Morris, PE-based multilayer film structures in J.R. Wagner, (Ed.), *Multilayer Flexible Packaging*, Plastics Design Library, chapter 17, pp. 281–310. William Andrew Publishing, 2nd edition, 2016.
57. J.R. Wagner, (Ed.), *Multilayer Flexible Packaging*, Plastics Design Library, William Andrew, 2nd edition, 2016.
58. K. Suzuki, S. Suzuki, S. Suzuki, and W.D. Bishop, Jr., Recyclability enhancement of food containers, US Patent Application 20 130 118 946, assigned to Kiyan Suzuki, Shabnam Suzuki, Sana Suzuki, William David Bishop, Jr., May 16, 2013.
59. L.K. Ncube, A.U. Ude, E.N. Ogunmuyiwa, R. Zulkifli, and I.N. Beas, An overview of plastic waste generation and management in food packaging industries, *Recycling*, Vol. 6, p. 12, 2021.
60. K.S. Marsh, Polymer and plastic packaging in *Reference Module in Food Science*. Elsevier, 2016.
61. C. Cimpan, A. Maul, M. Jansen, T. Pretz, and H. Wenzel, Central sorting and recovery of MSW recyclable materials: A review of technological state-of-the-art, cases, practice and implications for materials recycling, *Journal of Environmental Management*, Vol. 156, p. 181, 2015.
62. M.H. Filyaw, Recycled plastics in food contact in J.S. Baughan, (Ed.), *Global Legislation for Food Contact Materials*, Woodhead Publishing Series in Food Science, Technology and Nutrition, chapter 8, pp. 193–202. Elsevier, 2nd edition, 2022.
63. V. Komolprasert and A. Bailey, Recycled plastics for food applications: Improving safety and quality in E. Chiellini, (Ed.), *Environmentally Compatible Food Packaging*, Woodhead Publishing Series in Food Science, Technology and Nutrition, chapter 13, pp. 326–350. Woodhead Publishing, 2008.

Index

Acronyms

AFM
Atomic force microscopy, 211
CMC
Carboxymethyl cellulose, 218
CMCS
Carboxymethyl chitosan, 218
CNF
Cellulose nanofiber, 213
CNT
Carbon nanotube, 197
CS
Chitosan, 214
DAG
Diacylglycerol, 319
DART
Direct analysis in real time, 359
EDTA
Ethylenediamine tetraacetic acid, 46
EVOH
Ethylene vinyl alcohol, 134
FCM
Food contact material, 69, 84
FTIR
Fourier transform infrared, 211, 372
GC
Gas chromatography, 9
GNP
Graphene nanoplate, 197
HDPE
High-density poly(ethylene), 11, 134, 304, 360
HPLC
High-performance liquid chromatography, 290

HPMC
Hydroxypropyl methyl cellulose, 214
IAS
Intentionally added substances, 69
LDPE
Low-density poly(ethylene), 84, 277, 304, 360
LLDPE
Linear low-density poly(ethylene), 200
MMPP
Multi-material multilayer plastic packaging, 367
MMT
Montmorillonite, 191
MS
Mass spectroscopy, 9
MVTR
Moisture vapor transmission rate, 135
MW
Microwave, 278
NIAS
Non-intentionally added substances, 69
PA
Poly(amide), 126, 359
PBA
Poly(1,4-butylene adipate), 227
PC
Poly(carbonate), 369
PCL
Poly(caprolactone), 198
PE
Poly(ethylene), 109, 126, 222, 329, 372
PEN
Poly(ethylene naphthalate), 76
PET
Poly(ethylene terephthalate), 3, 76, 131, 200, 360
PHB
Poly(3-hydroxybutyrate), 229
PLA
Poly(lactic acid), 138, 218
PMMA
Poly(methyl methacrylate), 197
POM
Poly(oxymethylene), 369
PP
Poly(propylene), 2, 110, 131, 274, 360

PPE
Poly(phenylene ether), 192, 369
PS
Poly(styrene), 7, 144, 192, 277, 372
PTFE
Poly(tetrafluoroethylene), 10
PU
Poly(urethane), 71
PVC
Poly(vinyl chloride), 144, 192, 277, 360
SEM
Scanning electron microscopy, 211
SPI
Society of the Plastics Industry, 385
SSP
Solid-state polycondensation, 58
TDI
Toluene diisocyanate, 72
TTC
Threshold of Toxicological Concern, 84
WLF
Williams-Landel-Ferry, 79
WVTR
Water vapor transmission rate, 134

Chemicals

Boldface numbers refer to Figures

Acerola, **163**
Acetic acid, 71, 213, 292, 296, 304, 331, 366, 378
Acetoin, **194**
Acetylsalicylic acid, **160**
Aflatoxin B1, **284**
Aflatoxin B2, **284**
Aflatoxin G1, **284**
Aflatoxin G2, **284**
Aflatoxin, **24**
1-Allylhydantoin, **230**
o-Aminoazotoluene, **295**
4-Aminobiphenyl, **295**
2-Amino-1,6-dimethylfuro[3,2-e]imidazo[4,5-b]pyridine, **340**
2-Amino-3,4-dimethyl-3H-imidazo-4,5-f-quinolone, **342**
3-Amino-1,4-dimethyl-5H-pyrido[4,3-b]indole, **342**
2-Aminodipyrido[1,2-a:3,2′-d]imidazole, **342**
2-Amino-3-methylimidazo-4,5-f-quinolone, **342**
2-Amino-3-methylimidazo(4,5-f)quinoxaline, **343**
2-Amino-3-methyl-9H-pyrido-2,3-indole, **342**
3-Amino-1-methyl-5H-pyrido[4,3-b]indole, **342**
2-Amino-3,4,8-trimethyl-3H-imidazo(4,5-f)quinoxaline, **343**
2-Amino-3,7,8-trimethyl-3H-imidazo(4,5-f)quinoxaline, **343**
2-Amlno-1-methyl-6-phenylimidazo[4,5-b]pyrldine, **339**
Ammonium stearate, 50
Anisidine, **239**
o-Anisidine, **294**
Anthocyanin, **206**
Apigenin, **315**
6-O-A-L-Rhamnopyranoside, **317**
Ascorbic acid, **211**
Asparagine, **36**
Aspartic acid, **321**
Aurantindin, **99**
Azodicarbonamide, **261**
Barium sulfate, 52
Benomyl, **55**
Benz[a]anthracene, **22**
Benzenamine, **379**
1,2-Benzenedicarboxylic acid, **363**
1,4-Benzenedicarboxylic acid, **363**, **379**
Benzene disulfonyl hydrazide, **261**

Benzidine, **294**
Benzo[b]fluoranthene, **22**
Benzo[j]fluoranthene, **22**
Benzo[k]fluoranthene, **22**
Benzoic acid, 46
Benzoin, **194**
Benzophenone, **89**
d10-Benzophenone, **9**
Benzo[a]pyrene, **22**
Betanin, **206**
Bicyclo[2.2.1]hept-5-ene-2,3-dicarboxylic acid, **136**
Bicyclo[2.2.2]oct-5-ene-2,3-dicarboxylic acid, **136**
3,9-Bis[2,4-bis(1,1-dimethylethyl)phenoxy]-
2,4,8,10-tetraoxa-3,9-diphosphaspiro[5.5]undecane, **196**
2,5-Bis(5-*tert*-butyl-2-benzoazolyl)thiophene, **272**
N,N'-Bis(3-(3,5-di-*tert*-butyl-4-hydroxyphenyl)propionyl)-hydrazide, **273**
Bis(2,6-di-*tert*-butyl-4-methylphenyl)pentaerithritol diphosphite, **270**
N,N-Bis(2-hydroxyethyl)lauramide, **273**
Bis(2-methoxyethyl) phthalate, **259**
Bisphenol A, **8**
Boric acid, 256
Borneol, **324**
Butylated hydroxytoluene, **87**, 113
Butyl benzyl phthalate, **363**
Butyl palmitate, 86
Butyl stearate, **94**
Cadaverine, **13**
Caffeine, **211**
Calcium phosphate, 54
Camphor, **326**
m-Camphorene, **90**
Carrageenan, **102**
Carvacrol, **109**, **163**
Catechin, **231**
Chitosan, **163**, **347**
4-Chloroaniline, **294**
2-Chlorobiphenyl, **18**
3-Chlorobiphenyl, **18**
4-Chlorobiphenyl, **18**
5-Chloro-2-(2,4-dichlorphenoxy)-phenol, 30
6-Chloro-1,2-dihydroquinoxaline, **15**
1-Chlorohexadecane, **95**
Chrysene, **22**
1,8-Cineol, 52

1,8-Cineole, **324**
Cinnamaldehyde, **207**
Citral, **53**
Citric acid, 46, 140, **163**, **321**
Copper oxide, 52
Curcumin, **102**, 145
Cyanidin-3-glucoside, **308**
Cyanidin-3-sambubioside, **308**
p-Cymene, **327**
Dansyl chloride, **290**
Decabromodiphenylethane, 256
Decabromodiphenyl ether, **33**, **35**
Dehydroabietic aldehyde, 85
Delphinidin, **98**
5-Demethyl nobiletin, **314**
Deoxynivalenol, **28**
Diacetoxyscirpenol, **28**
1,2-Diacylglycerol, **320**
4,4′-Diaminodiphenylmethane, **260**, **294**
7H-Dibenz[c,g]carbazole, **23**
Dibenz[a,h]acridine, **22**
Dibenz[a,h]anthracene, **22**
Dibenz[a,j]acridine, **22**
Dibenzo[a,e]pyrene, **23**
Dibenzo[a,h]pyrene, **23**
Dibenzo[a,i]pyrene, **23**
Dibenzo[a,l]pyrene, 21
4,4′-Dibromodiphenyl ether, **34**, **35**
Dibutylamine, **73**
2,6-Di-*tert*-butyl-1,4-benzoquinone, **89**
3,5-Di-*tert*-butyl-4-hydroxybenzaldehyde, **89**
3,5-Di-*tert*-butyl-4-hydroxybenzoic acid, hexadecyl ester, **266**
2,6-Di-*tert*-butyl-4-methylphenol, **275**
7,9-Di-*tert*-butyl-1-oxaspiro(4,5)deca-6,9-diene-2,8-dione, **89**
2,4-Di-*tert*-butylphenol, **87**, **379**
2,6-Di-*tert*-butylphenol, **87**
Di-*n*-butyl phthalate, **363**
Dibutyl phthalate, 85
2,4-Dichloro benzenamine, **379**
3,3′-Dichlorobenzidine, **295**
Dichlorobenzidine, **72**
2,3-Dichlorobiphenyl, **18**
2,4-Dichlorobiphenyl, **18**
3,5-Dichlorobiphenyl, **19**

4,4′-Dichlorobiphenyl, **19**
2,2′-Dichloro-4,4′-methylene dianiline, **260**
Dicyclohexyl phthalate, **258**
Di(2-ethylhexyl) adipate, 256, **365**
Di(2-ethylhexyl) phthalate, **365**
Diethylhexyl sodium sulfosuccinate, **371**
Diethyl phthalate, **363**, **379**
Diheptyl adipate, **258**
Dihydromyricetin, **173**
1,9-Dihydro-2H-pyrido-2,3-indol-2-imine, **342**
Dihydroquercetin, **315**
7-Dihydroxy-3′-methoxy-4′,5′-methylenedioxyisoflavone, **317**
Diisobutyl phthalate, 85, 256
Diisoheptyl phthalate, **258**
2,6-Diisopropylnaphthalene, **90**
3,3′-Dimethoxybenzidine, **294**
Dimethylamine, **291**
3,3′-Dimethylbenzidine, **294**
2,3-Dimethyl-1,2-dihydroquinoxaline, **15**
3,8-Dimethyl-3H-imidazo(4,5-f)quinoxalin-2-amine, **343**
N,N-Dimethylpalmitylamine, **95**
Di-octyl adipate, 256
Diosmin, **313**, **328**
2,4-Di-*tert*-pentylphenol, **87**
1,2-Diphenoxyethane, **87**
Diphenyl sulfone, **95**
Docosanol, 85
1-Docosene, 85
1-Dodecene, **380**
Dodecyl acrylate, **380**
Eicosane, **88**
Enterocin, **164**
Epirosmanol, **326**
Eriocitrin, **311**, **328**
Erucamide, **280**
β-Estradiol, **8**
4-Ethoxy ethylbenzoate, **273**
Ethyl acrylate, 52
Ethylamine, **291**
Ethyl chloroformate, **73**, 292
1-Ethyldecylbenzene, **90**
Ethylenediaminetetraacetic acid, **47**
2-Ethylhexyl (Z)-3-cyano-2,3-diphenylprop-2-enoate, **276**
2,2′-Ethylidene bis(4,6-di-*tert*butyl-phenyl)fluorophosphonite, **272**

Ethyl palmitate, **94**
Eugenol, **206**
Europinidin, **99**
Fluoxetine, **159**
Flvylium salts, 96
Fumonisin, **24**, 26
Furosemide, **159**
Gallic acid, **164**
Garcinol, **308**
Gemfibrozil, **159**
Generaniol, **53**
Genkwanin, **326**
Glucomannan, **102**
6-O-β-D-Glucopyranoside, **317**
Glycerol monolaurate, **164**
Halloysite, **169**
Harmane, **343**
2,2′,3,4,4′,5′,6-Heptabromodiphenyl ether, **33**
2,2′,3,4,4′,5,6-Heptabromodiphenyl ether, **31, 32, 34**
Heptacosane, 86
Heptyl adipate, 256
Heptyl octyl adipate, 256, **258**
Hesperidin, **311, 328**
Hesperitin, **315**
Hexabromocyclohexane, 256
2,2′,3,4,4′,5′-Hexabromodiphenyl ether, **31, 32**
2,2′,4,4′,5,6′-Hexabromodiphenyl ether, **31, 32, 34**
Hexadecane, **88**
Hexadecyl acetate, 86
Hexadecyl palmitate, **94**
Hexadecyl trifluoro acetate, **89**
Hexadecyltrimethoxysilane, **215**
Hexamethylenetetramine, **51**
Histamine, **291**
2-Hydroxyacetophenone, **194**
Hydroxyanisole, **163**
Hydroxyapatite, **169**
p-Hydroxybenzoic acid, **51**
2-(2-Hydroxy-3,5-bis(1,1-dimethylbenzyl)phenyl)benzotriazole, **271**
Hydroxycitric acid, **307**
2-Hydroxy-4-*n*-octylbenzophenone, **266**
2-Hydroxy-1-phenyl-2-*p*-tolyl-ethanone, **194**
1-Hydroxy-2-propanone, **194**
Hydroxypropyl-β-cyclodextrin, **173**

3-(1H-Indol-3-yl)-1,2-dihydroquinoxaline, **15**
Irganox 1076, **87**
Isopentylamine, **291**
7-Isopropyl-1,4-dimethyl tetradecahydrophenanthrene, **91**
7-Isopropyl-1-methyl-1,2,3,4-tetrahydrophenanthrene, 85
Isopropyl myristate, **94**
Kaempferol, **314**
Lactic acid, 101, 222, 286
Limonene, **163**
Linalool, **326**
Linoleic acid, **329**
Lupane, **325**
Lutein, **324**
Luteolin, **327**
Luteolinidin, **99**
Lysine, **321**
Lysozyme, **51**, 345
Malic acid, **321**
Malvidin, **98**
Mentha spicata oil, **163**
Menthol, **109**
Metformin, **160**
Methyl abietate, **95**
Methylamine, **291**
Methyl chavicol, **53**
Methyl cinnamate, **53**
Methyl dehydroabietate, **95**
Methyl elaidate, **92**
4,4′-Methylene-bis(2,6-di-*tert*-butylphenol, **275**
Methyl erucate, **93**
Methyl eugenol, **53**
3-Methylheneicosane, 86
5-Methylheneicosane, 86
3-Methyl heptadecane, 86
Methyl heptadecanoate, **92**
Methyl icosanoate, **93**
Methyl isopimarate, **93**
Methyl jasmonate, **327**
Methyl laurate, **92**
Methyl linolelaidate, **92**
Methyl methacrylate, 52
2-Methyl-5-nitroaniline, **295**
5-Methylnonacosane, 86
3-Methylnonadecane, 86

2-Methyloctacosane, 86
Methyl palmitate, **92**
3-Methylpentadecane, 86
Methyl stearate, **92**
11-Methyltricosane, 86, **88**
Metilox, 85, **87**
Monobenzyl phthalate, **363**
Monobutyl phthalate, **363**
Mono(2-ethylhexyl) phthalate, **363**
Myrcene, **325**
Myricitrin, **315**
Naringin, **311**
Narirutin, **312**
Nicotine, **160**
Nisin, **165**
2,2′,2″-Nitrilo(triethyl-tris(3,3′,5,5′-tetra-*tert*-butyl-1,1′-biphenyl-2,2′-diyl)phosphite), **266**
3-(3-Nitrophenyl)-1,2-dihydroquinoxaline, **15**
Nobiletin, **313**
2,2′,3,3′,4,4′,5,5′,6-Nonabromodiphenyl ether, **33, 35**
2,2′,3,3′,4,4′,5,6,6′-Nonabromodiphenyl ether, **33, 35**
Nonadecanol, 85
9-Nonadecene, 85
3-Nonenol, 85
Norharmane, **343**
Ochratoxin, **24**, 27
2,2′,3,4,4′,5,5′,6-Octabromodiphenyl ether, **33–35**
2,3,3′,4,4′,5,5′,6-Octabromodiphenyl ether, **33, 35**
Octacosane, 86
Octadecane, 86
Octadecanol, 85
Octadecyl 3-(3,5-di-*tert*-butyl-4-hydroxyphenyl)propionate, **269**
Octadecyl trifluoro acetate, 86
Octocrylene, **379**
Oleamide, **280**
Oleanane, **325**
Oleyl palmitamide, **281**
10,10′-Oxybisphenoarsine, **261**
4,4′-Oxydianiline, **294**
Paracetamol, **160**
Pectin, **237**
Pelargonidin, **98**
2,2′,4,4′,5-Pentabromodiphenyl ether, **31, 32**
2,2′,4,4′,6-Pentabromodiphenyl ether, **31, 32**

2,3′,4,4′,6-Pentabromodiphenyl ether, **31, 32**
3,3′,4,4′,5-Pentabromodiphenyl ether, **31, 32**
Pentacosane, 86
Pentaerythritol tetrakis(3-(3,5-di-*tert*-butyl-4-hydroxyphenyl)propionate), **267, 268**
Peonidin, **98**
Perfluorobutane sulfonate, **283**
Perfluorodecane sulfonate, **283**
Perfluorohexanoic acid, **282**
Perfluorooctane sulfonate, 10, **283**
Perfluorooctanoic acid, 10, **283**
Perfluorooctyl acrylate, **230**
Perfluorotetradecanoic acid, **282**
Perfluorotridecanoic acid, **282**
Petunidin, **98**
2-Phenyl benzo pyrylium salts, 96
Phenyl didecyl phosphite, **195**
3-Phenyl-1,2-dihydroquinoxaline, **15**
3-Phenyl-1,2-dihydroquinoxaline-2-carboxylic acid, **15**
β-Phenylethylamine, **291**
Phloretin, **315**
Phloridzin, **315**
Phosphorous acid, tris(2,4-di-*tert*-butylphenyl)ester, **267, 268, 272**
β-Pinene, **326**
Potassium sorbate, 46, **347**
Propionic acid, 46
Propranolol, **159**
Propyl paraben, **347**
Pullulan, **237**
Putrescine, **13**, 290
o-Quinone, **327**
p-Quinone, **327**
Rhoifolin, **312**
Rofficerone, **325**
Rosemarinic acid, **113**
Rosinidin, **99**
Rosmadial, **326**
Rosmanol, **326**
Rosmariquinone, **327**
Rutin, **314, 324**
Saponin, **371**
Sebacic acid, bis(2,2,6,6-tetramethyl-4-piperidyl)ester, **271**
Sodium benzoate, 46, **347**
Sodium chlorite, 140

Sodium ligninsulfonate, **370**
Sorbic acid, 46, **51**
Spermidine, **13**
Spermine, **13**, 290
Sphinganine, **27**
Stearamide, **280**
Stearyl amide, **280**
Tangeritin, **313**
Tannic acid, **370**
γ-Terpinene, **326**
Tetrabromobisphenol A, 256
2,2′,4,5′-Tetrabromodiphenyl ether, **31, 32**
2,3′,4′,6-Tetrabromodiphenyl ether, **31, 32**
2,3′,4,4′-Tetrabromodiphenyl ether, **31, 32**
2,4,4′,6-Tetrabromodiphenyl ether, **31, 32**
3,3′,4,4′-Tetrabromodiphenyl ether, **31, 32**
2,2′,5,5′-Tetrachlorobiphenyl, **19**
3,3′,4,4′-Tetrachlorobiphenyl, **19**
Tetracosanol, 85
Tetradecane, **88**
12-*o*-Tetradecanoyphorbol-13-acetate, **41**
1-Tetradecene, **380**
7-Tetradecen-1-yl acetate, **89**
Tetraiacontane, 86
2,2′,5,5′-Tetramethyl-1,1′-biphenyl, **90**
2,4,7,9-Tetramethyl-5-decyne-4,7-diol, **288**
2,6,10,15-Tetramethylheptadecane, **88**
Theobromine, **41**
Thiobarbituric acid, **114**, 174
4,4′-Thiodianiline, **295**
Thymol, **163**
Titanium dioxide, 52, 167
Tocopherol, **271**
Toluene-2,4-diamine, **73**
Toluene-2,6-diamine, **73**
2,4-Toluene diisocyanate, **73**
2,6-Toluene diisocyanate, **73**
2,4,4′-Tribromodiphenyl ether, **34,** 35
Tri-*n*-butyl acetyl citrate, 86
Tributyl acetyl citrate, **276**
2,2′,5-Trichlorobiphenyl, **19**
2,4′,5-Trichlorobiphenyl, **19**
2,4,4′-Trichlorobiphenyl, **19**
2,4,5-Trichlorobiphenyl, **19**

Triclosan, **261**
9-Tricosene, 85
Triethyleneglycol-bis(3-(3-*tert*-butyl-4-hydroxy-5-methylphenyl)propionate), **270**
2,4,5-Trimethylaniline, **294**
1,3,5-Trimethyl-2,4,6-tris(3,5-di-*tert*-butyl-4-hydroxybenzyl)benzene, **269**
Trioctyl phosphite, **195**
1,3,5-Triphenylcyclohexane, **91**
1,3,5-Tris[(2S and 2R)-2,3-epoxypropyl]-1,3,5-triazine-2,4,6-(1H,3H,5H)-trione, **260**
Tris(2-chlorisopropyl)phosphate, 256
Tris(2-chloroethyl)phosphate, 256, **259**
Tris(2,4-di-*tert*-butylphenyl) phosphite, **275**
Tris(2-ethylhexyl) trimellitate, **276**
L-Tryptophan, **39**
Tyramine, **291**
Undecylcyclohexane, **91**
Ursolic acid, **325**
Vanillin, **347**
Verbenone, **325**
Vitexin, **312**
Xanthine, **41**
6-O-β-D-Xylopyranosyl-(1-6)-β-D-glucopyranoside, **317**
α-Zearalenol, **28**
β-Zearalenol, **28**
Zearalenone, **24**, **28**
Zein, **206**
Zinc silicate, 52

General Index

Fusarium species, 27
Garcinia indica, 307
Accelerated aging, 78
Activation energy, 5, 262, 263
Adjuvants, 56, 75
Aerobic bacteria, 113
Aerobic biodegradation, 78
Aerosol, 46, 68, 74
Aflatoxin-contaminated chicken meat, 26
Allergens, 39
Aluminum-alloy, 106
Amadori rearrangement, 37
Anaerobic bacterial growth, 56
Anthocyanidins, 97
Antibacterial agents, 145, 344
Antimicrobial absorption, 145
Antimicrobial packaging material, 52, 161, 162, 164, 167
Antioxidants, 113, 274, 278, 310, 322, 329
Aquatic environment, 197, 232, 287
Arrhenius law, 5, 263
Atmospheric pressure, 72, 109, 378
Bacterial endospores, 96
Bacteriocins, 46, 51, 164
Bacteriostatic substances, 46
Balkan endemic nephropathy, 27
Barley, 29, 212, 285
Barrier lidding material, 107
Beverages, 3, 111, 156, 236, 287, 362
Bioactive packaging materials, 51
Biodegradable foams, 212
Biodegradable polymers, 2, 97, 169, 224
Biogenic amines, 13, 289
Biopolymers, 97, 142, 156, 191, 198
Biosensors, 2, 40
Browned food, 37
Carbohydrate-rich foods, 36
Cardiovascular disease, 96, 307
Cartonboard packaging, 9
Chelating agents, 46, 47
Chemosensors, 13
Color indicator, 96, 101
Color stability, 97
Color-changing
compounds, 99
films, 96
pH-responsive functions, 96
Colorants, 68, 145, 292, 304
Colorimetric pH indicator, 100, 217
Compostable plastics, 78
Contaminants, 2, 14, 17, 25, 39, 58, 75, 83, 110, 233, 284, 362
Coulometric sensor, 109
Cranberries, 43
Cryogenic grinding, 50
Decontamination, 58, 286
Diabetes, 96, 307, 316
Diuresis, 41
Double-layer indicator, 102
Endotoxins, 25
Enterobacteriaceae, 96
Estrogen receptors, 8, 28
Extractable substances, 55
Fick's law, 262
Field-assisted extraction, 6
Fluorescence detection, 290
Foamed plastics, 149, 151
Food fermentation, 14
Food freshness, 13, 218
Food irradiation, 112
Food simulants, 3, 30, 106, 262, 274, 277, 288, 378
Food spoilage, 13, 142, 218
Food-packaging materials, 9
Foodborne diseases, 1, 12
Foodborne pathogens, 39

Fossil fuels, 21, 49
Fungal spores, 96
Fungicides, 51, 54
Gastrointestinal toxicity, 287
Genotoxic substances, 71, 364
Gram-positive bacteria, 46, 192, 220
Herbicides, 39
Heyns rearrangement, 37
Hypersensitivity pneumonitis, 74
Ice cream, 279
Indicator, 101, 103
Indicator films, 97
Infant feeding bottles, 3
Lactoferrin, 47
Maillard reaction, 37
Microbes, 13, 96, 113, 145, 318
Microbial contamination, 144, 377
Microbial fermentation, 13
Microbial growth, 46
Microbial inactivation, 141, 285
Microbiological contamination, 42, 45
Microplastics, 16, 30, 197, 198, 232–234
Migration, 3
Migration limit, 8
Migration of benzophenone, 10
Moisture regulators, 45
Mutton meat, 113
Nanoplastics, 198, 231, 234, 287
Natural antioxidants, 322
Nephropathy, 27
Nitrogenous polymers, 37
Non-alcoholic beer, 5, 264
Non-recyclable plastic packaging, 48
Nutrients, 221, 306
Ochratoxins, 27
Optical brighteners, 68
Organochlorine pesticides, 21
Organoleptic properties, 112, 164
Organophosphorus flame retardants, 16
Oxygen permeability, 109, 114, 214, 223
Pathogenic microorganisms, 14
Perishability, 42
Phenylketonuria, 27
Phthalate migration, 6
Plastic kitchenware, 296
Plasticizers, 79, 84, 202, 222, 274, 277, 304, 361, 366
Poisonous carcinogens, 25
Porous lidded packages, 107
Product-package systems, 107
Propolis, 101
Raspberries, 43
Seafood, 100, 233
Strawberries, 43
Strecker degradation, 37
Styrene migration, 7
Surface modification, 40, 231
Surfactants, 54, 333, 369
Taste absorbers, 45
Triclosan migration, 30
UV absorbers, 274
Williams-Landel-Ferry equation, 79
WinPerm software, 109

Also of Interest

Check out these other books by the author published by Scrivener Publishing

Food Safety, Plastics and Sustainability: Materials, Chemicals, Recycling and the Circular Economy
By Johannes Karl Fink
Published 2023. ISBN 978-1-394-17456-0

Porous Plastics
By Johannes Karl Fink
Published 2022. ISBN 978-1-119- 89638-8

Contact Lenses
Materials, Chemicals, Methods and Applications
By Johannes Karl Fink
Published 2022. ISBN 978-1-119- 85735-8

Polymers and Additives in Extreme Environments
Application, Properties, and Fabrication
By Johannes Karl Fink
Published 2021. ISBN 978-1-119- 85137-0

Plastics Process Analysis, Instrumentation, and Control
By Johannes Karl Fink
Published 2021. ISBN 978-1-119-79573-5

Flame Retardants
By Johannes Karl Fink
Published 2020. ISBN 978-1-119-75061-1

The Chemistry of Environmental Engineering
By Johannes Karl Fink
Published 2020. ISBN 978-1-119-70774-5

The Chemistry of Bio-Based Polymers, second edition
By Johannes Karl Fink
Published 2019. ISBN 978-1-119-68131-1

Liquid Silicone Rubber
By Johannes Karl Fink
Published 2019. ISBN 978-1-119- 63133-0

Functional Synthetic Polymers
By Johannes Karl Fink
Published 2019. ISBN 978-1-119-59202-0

3D Industrial Printing with Polymers
By Johannes Karl Fink
Published 2018. ISBN 978-1-119-55526-1

Polymer Waste Management
By Johannes Karl Fink
Published 2018. ISBN 978-1-119-53608-6

Materials, Chemicals and Methods for Dental Applications
By Johannes Karl Fink
Published 2018. ISBN 978-1-119-51031-4

Fuel Cells, Solar Panels, and Storage Devices
Materials and Methods
By Johannes Karl Fink
Published 2018. ISBN 978-1-119-48010-5

Chemicals and Methods for Conservation and Restoration
Paintings, Textiles, Fossils, Wood, Stones, Metals, and Glass
By Johannes Karl Fink
Published 2017. ISBN 978-1-119-41824-5

Additives for High Performance Applications
Chemistry and Applications
By Johannes Karl Fink
Published 2017. ISBN 978-1-119-36361-3

Metallized and Magnetic Polymers
By Johannes Karl Fink
Published 2016. ISBN: 978-1-119-24232-1

Marine, Waterborne, and Water-Resistant Polymers
Chemistry and Applications
By Johannes Karl Fink
Published 2015. ISBN 978-1-119-018486-7

The Chemistry of Printing Inks and Their Electronics and Medical Applications
By Johannes Karl Fink
Published 2015. ISBN 978-1-119-04130-6

The Chemistry of Bio-Based Polymers
By Johannes Karl Fink
Published 2014. ISBN 978-1-118-83725-2

Polymeric Sensors and Actuators
By Johannes Karl Fink
Published 2012. ISBN 978-1-118-41408-8

Handbook of Engineering and Specialty Thermoplastics
Part 1: Polyolefins and Styrenics
By Johannes Karl Fink
Published 2010. ISBN 978-0-470-62483-5

Handbook of Engineering and Specialty Thermoplastics
Part 2: Water Soluble Polymers
By Johannes Karl Fink
Published 2011. ISBN 978-1-118-06275-3

A Concise Introduction to Additives for Thermoplastic Polymers
By Johannes Karl Fink
Published 2010. ISBN 978-0-470-60955-2

Printed and bound by CPI Group (UK) Ltd, Croydon, CR0 4YY

17/09/2026

14964279-0003